WHEN YOU FREEZE VEGETABLES
HOW MUCH WILL IT MAKE?

Fresh	Frozen Pints	Fresh	Frozen Pints
ASPARAGUS		**EGGPLANT**	
1 crate (12 2-lb. bunches)	15 to 22	1 lb.	1
1 to 1½ lbs.	1	**GREENS**	
BEANS, LIMA, IN PODS		BEET GREENS	
1 bu. (32 lbs.)	12 to 16	15 lbs.	10 to 15
2 to 2½ lbs.	1	1 to 1½ lbs.	1
BEANS, SNAP, GREEN AND WAX		CHARD, COLLARDS, MUSTARD GREENS	
1 bu. (30 lbs.)	30 to 45	1 bu. (12 lbs.)	8 to 12
⅔ to 1 lb.	1	1 to 1½ lbs.	1
BEETS, WITHOUT TOPS		KALE	
1 bu. (52 lbs.)	35 to 42	1 bu. (18 lbs.)	12 to 18
1¼ to 1½ lbs.	1	1 to 1½ lbs.	1
BROCCOLI		**PEAS**	
1 crate (25 lbs.)	24	1 bu. (30 lbs.)	12 to 15
1 lb.	1	2 to 2½ lbs.	1
BRUSSELS SPROUTS		**PEPPERS, GREEN**	
4 qt. boxes	6	⅔ lb. (3 peppers)	1
1 lb.	1	**PUMPKIN, WINTER SQUASH**	
CARROTS, WITHOUT TOPS		3 lbs.	2
1 bu. (50 lbs.)	32 to 40	**SPINACH**	
1¼ to 1½ lbs.	1	1 bu. (18 lbs.)	12 to 18
CAULIFLOWER		1 to 1½ lbs.	1
2 medium heads	3	**SQUASH, SUMMER**	
1⅓ lbs.	1	1 bu. (40 lbs.)	32 to 40
CORN, SWEET, IN HUSKS		1 to 1¼ lbs.	1
1 bu. (35 lbs.)	14 to 17	**SWEET POTATOES**	
2 to 2½ lbs.	1	⅔ lb.	1

Farm Journal's
FREEZING & CANNING
COOKBOOK

OTHER COOKBOOKS BY FARM JOURNAL

Farm Journal's Country Cookbook
Farm Journal's Complete Pie Cookbook
Let's Start to Cook
America's Best Vegetable Recipes
Homemade Bread
Homemade Candy
Homemade Cookies
Homemade Ice Cream and Cake
Busy Woman's Cookbook
The Thrifty Cook
Country Fair Cookbook
Farm Journal's Homemade Snacks
Farm Journal's Best-ever Recipes
Farm Journal's Great Dishes From the Oven

Farm Journal's
FREEZING & CANNING
COOKBOOK

NEW REVISED EDITION

Prized Recipes from the Farms of America

EDITED BY

Nell B. Nichols
FIELD FOOD EDITOR

and
Kathryn Larson

ASSOCIATE EDITOR, BOOK DIVISION
FARM JOURNAL, INC.

DOUBLEDAY & COMPANY, INC.
GARDEN CITY, NEW YORK

Library of Congress Cataloging in Publication Data
Main entry under title:

Farm journal's Freezing & canning cookbook.

Edition for 1973 published under title:
Freezing & canning cookbook.
Includes index.
1. Cookery, American.   2. Canning and pre-
serving 3. Food, Frozen.   I. Nichols, Nell
Beaubien.   II. Larson, Kathryn.   III. Farm
journal (Philadelphia, 1956–   )   IV. Title:
Freezing & canning cookbook.
TX715.F865   1978   641.4′2
ISBN: 0-385-13444-4
Library of Congress Catalog Card Number 77–81787

Contents

COLOR ILLUSTRATIONS

All photographs created by Farm Journal Food and Art Staffs.

What You Will Find
in This Cookbook

This is the second revised edition of the widely used FREEZING & CANNING COOKBOOK originally produced 15 years ago by the Food Editors of FARM JOURNAL.

Since 1963, when the book first appeared, gardening and the preservation of home-grown foods have become a way of life for more and more Americans—not just farm families. Earlier generations practiced the "home arts" out of necessity—baking yeast breads, canning, quilting, curing meats—and children learned how to do all these things by helping their parents.

Now, with many young homemakers freezing and canning for the first time, our aim in this revision is two-fold: to emphasize the basic methods that yield superior food with excellent keeping qualities; and to include the most up-to-date recommendations resulting from continuing research into better ways to protect foods in storage.

This book is filled with ideas and directions and recipes for putting up homemade "convenience foods"—everything from your own home-grown fruits and vegetables to frozen main dishes, breads and desserts, and home-canned jams, jellies, pickles and relishes. Many of these foods, if bought in the supermarket, would qualify as luxury or gourmet items. The practice of doing-it-yourself will save money—and ongoing inflation makes thrift an important incentive.

Pleasure is another reason for the return to gardening and preserving. Many people who work all day in offices or factories view gardening as recreation—a good way to unwind. With their children, they spend the most peaceful hours of the day—early morning or late afternoon—planting, weeding, tying up tomato vines, marveling that the beans and peas and zucchini seem to grow overnight.

Families who don't tend a garden patch are taking Saturday excursions to pick-your-own farms. Then it's back to the kitchen, to the pickling spices and the jelly kettle, the canning jars and freezer cartons, to take care of the harvest.

Farmer's wife, weekend gardener, roadside stand shopper—we all share the yearning to capture the tastes and smells of a country kitchen: fresh raspberry jam, sparkling grape jelly, tomato ketchup, spicy mincemeat, sweetly sour sliced cucumber pickles so good children used them for sandwich fillings and named them "bread-and-butters."

* * *

THE MAJOR CHANGES in this book are in the canning section. It's more than an up-date: it is an effort to explain more clearly—for those new to canning—why you must follow certain procedures to insure food that is not only good to eat, but is also safe to eat. Practices once considered all right—such as open kettle canning—are no longer recommended and Chapter 9, "Preserving Food Safely," will explain why this is so.

For this information, we turned to a book FARM JOURNAL published in 1976: *Home Canning . . . The Last Word,* by Louise W. Hamilton, Professor of Foods and Nutrition, Gerald D. Kuhn, Professor of Food Science, and Karen A. Rugh, Communications Specialist, all with the Cooperative Extension Service at The Pennsylvania State University.

Their widely-praised book, published in paperback,* not only covers information about basic canning procedures, it also explains the *whys* of canning successes and failures. We are grateful for their permission to repeat information from *Home Canning . . . The Last Word* in this revised edition of our FREEZING & CANNING COOKBOOK.

Of all the foods preserved in canning jars, pickles and relishes are at once the most desirable and the most frustrating for home canners. If you've been less than pleased with your homemade pickles, and especially if you've had some spoilage problems, do review the information and try the recipes in Chapters 13 and 14.

The pickle recipes are all from the earlier edition, but with some important improvements. We have, for example, eliminated an old-fashioned additive, alum, which Grandma used to give pickles icicle crispness. There are other, better ways to insure crisp pickles, which you will read about in Chapter 13.

This is no mere pencil change in the book. Every pickle and relish recipe in Chapters 13 and 14 has been retested. We asked Maxine B. Kuhn, Dr. Kuhn's wife, and an expert canner who has also done laboratory research work on home canning at Penn State, to use produce from the Kuhn garden and remake every recipe.

All summer long we got notes on her progress: "The dills are superb. . . . The long-brine sweet-sours are great. . . . We are extremely pleased with the recipes—some are the best we have ever eaten. All of the packs are beautifully colored—they look like blue ribbon winners. . . .

* Dolphin Books, Doubleday & Company, Inc.

Had the Pepper Slaw for dinner last night—fantastic! We ate a whole pint!"

But she was also critical of the way some of recipes were written—not clear enough, she felt, for people who would be pickling for the first time. Not only has she clarified directions, but she has also modified some of the recipes to improve appearance or flavor. And she has developed some procedures for curing pickles that will improve your chances for success if you heed them. Her advice: "Pretend you are a pediatrician and treat your cukes like babies."

Making pickles, relishes, jams and jellies is important in country cooking. Most of our recipes came, originally, from farm homemakers. Once, when we invited FARM JOURNAL readers to share their best recipes for pickles and preserves, we received more than 30,000 recipes.

At the same time, our letter writers paid tribute to home freezers, telling us how they dovetail their freezing and canning operations for maximum results. Read their freezing slick tricks in the first eight chapters and try their favorite recipes—see how much goodness you can pack into your own zero storage.

Along with the recipe testing which we do in our FARM JOURNAL Test Kitchens, we rely also on freezing and canning research being done by the U. S. Department of Agriculture and the state Land Grant universities; this work is specifically acknowledged in this book. We thank Dr. Kuhn for reviewing Chapters 9–15 on home canning. Recognition must also go to Shirley T. Munson, Associate Professor, Department of Horticultural Science at the University of Minnesota, who consulted with us on the freezing chapters.

This cookbook, filled with long-time country favorite recipes—many of them consistently blue ribbon winners at fairs, is a testimonial to farm women. The extra-good recipes from farm homemakers, from home economists who work with country women and from the FARM JOURNAL food staff make this cookbook an exciting one. Enjoy your freezing and canning!

NELL B. NICHOLS
KATHRYN LARSON

PART ONE

CHAPTER 1

Freezing Foods at Home

Freezer Management · Steps to Successful Freezing · Packaging
Materials · Correct Packaging · Refreezing Foods · Storage Tem-
perature · Freezer Capacity · Freezer and Microwave Oven ·
Defrosting Freezers · Slick Freezing Tricks

Country cooking never was so good as it is today. That's what people who
sit down to farm meals say. Ask the cooks why and they agree that fresh-
tasting frozen foods have much to do with it. They also like the dollars-
and-cents savings—freezers take farm products when they're most plentiful and at their best and keep them fresh for months.

The four farm-to-freezer favorites of FARM JOURNAL readers are fruits,
vegetables, meats and poultry. To such a stockpile, farm women add supermarket ice cream and bread—or their own home-baked breads and
desserts, their own casseroles and cartons of stews and spaghetti sauce.
Farm kitchens are miniature markets; shopping trips to town can be less
frequent.

Frozen foods add variety to meals, too, because just about everything is
available around the calendar. That makes it possible to combine foods
that ripen at different times—a cooking adventure our grandmothers could
not enjoy. "You get some wonderful flavor blends," one farm woman says.
"It's like discovering a brand-new food no one ever tasted before."

To make the best use of time, smart homemakers cook, bake and shop
for their freezers on less busy days. Dishes to heat or thaw-and-serve
make no-watch meals when there's little time to stay in the kitchen. And
they help women to feed unexpected people without fuss and worry, to entertain graciously and to take the right food to church bazaars, community
suppers and home economics extension club teas.

Thumb through the pages on freezing that follow for up-to-date directions. Don't miss the clever slick tricks that farm women have discovered
to add "that extra little touch" to a meal.

Manage Your Freezer—Make It Work

What's in the freezer? That's your first thought when your husband comes to the kitchen to tell you he has a steer in prime condition for slaughtering . . . when you see some good supermarket bargains in frozen foods . . . when a letter arrives with news of visitors on the way . . . when a bumper strawberry crop piles up. You can answer that question quickly if you're a good freezer operator. You'll look at your records.

The first rule for superior freezer management is to keep foods going in and coming out and to keep track of the flow. The second rule is to make the freezer do part of your work.

If your load is especially heavy at mealtime, cooking and freezing ahead will ease the situation. If you occasionally are away from home at dinner time, you can leave frozen foods thawing in the refrigerator, with directions for heating in the oven or microwave oven. Your family will eat fine meals whether you're there or not.

Steps to Successful Freezing

Select foods of top quality: The freezer does not *improve* the quality of food. Its function is to *preserve* quality and nutritive values and prevent spoilage.

Choose vegetables and fruits suitable for freezing and the best varieties for freezing. Growing conditions and the varieties available vary greatly across the country. Check with your County Extension Home Economist or write to your State Agricultural Experiment Station to find out which are best.

Freeze fruits and vegetables when they are at their best for table use. Fruits should be ripe, but firm, and if possible, freeze those that are tree-, vine- or bush-ripened. Freeze garden-fresh, young, tender vegetables in their prime. Some of them, like asparagus, corn, peas, snap beans, lima beans and soybeans, change rap-

idly in the garden. A delay of a day or two in harvesting them may give you a tough frozen product.

Do not let fruits and vegetables wait in the kitchen any longer than necessary. Freeze them as soon as possible to retain their flavors. Many fruits and vegetables lose from one third to two thirds of their vitamin C if allowed to wait a day after picking.

Chill meats, poultry and fish immediately after killing. If you want to age beef or lamb, hang it a shorter time than when you plan to use the meat without freezing. Overaging shortens freezer storage time.

Observe cleanliness to avoid contamination of foods.

Heed the directions for the different foods that follow in this book. Do not slight any of the steps outlined. Even if you know a woman who has frozen peas or corn without blanching, don't try it. Blanching destroys enzymes that continue to change the taste and texture of foods when frozen.

Use the Right Packaging Materials

Success in freezing foods depends a great deal upon proper packaging, wrapping and sealing of the foods. In the dry 0° F. temperature of the freezer, solidly frozen foods can lose moisture; they must be protected by wrapping materials and/or containers that are moistureproof, vaporproof and airtight.

Moreover, the wrap must be *tight*, with no pockets of air trapped between food and wrapping material. Air draws moisture from the food resulting in *freezer burn*—those grayish-white areas on the surface of meats and poultry when wrap is not tight or when it becomes damaged or torn. Also, many foods are sensitive to oxygen. Meat, fish or poultry not properly wrapped may become rancid.

Wrapping materials include heavy-duty aluminum foil, heavy-weight plastic film or wrap, plastic bags and freezer paper (heavy paper bonded to moisture-vaporproof plastic or foil). Read the manufacturer's package instructions to make sure the wrap you buy is suitable for freezing.

Do not use waxed paper, bread wrapping paper, paper bags and regular cellophane; they are not moisture-vaporproof.

Butcher paper or plastic film used to wrap meat, poultry and fish for supermarket display is not adequate protection against freezer burn and rancidity. Rewrap this food if you plan to freeze it.

Most of the freezer containers on the market today give satisfactory results. Rigid plastic containers with wide tops are favorites with farm women. Do not use lightly waxed cartons, like those in which you buy ice cream, cottage cheese or milk. They are not moisture-vaporproof. Instead use waxed cartons made especially for freezing foods. Clean tin cans with tight-fitting lids, wide-mouth glass canning jars and plastic containers are excellent for liquids and semi-liquids. They are moisture-vaporproof, easy to fill and clean. If cared for properly, they are long-lasting. Consider the possibility of reusing freezer containers when comparing the costs of different kinds.

Freeze-and-cook bags, which must be closed with a special heat-sealing appliance, are also available. These bags can endure temperatures from 0° F. or below to 240° F. without change. This makes it possible to freeze and reheat or cook food in the same pouch. When ready to use, simply drop the bag, still sealed, into a pan of boiling water. After the food is fully cooked (usually 15 to 20 minutes), slit open the bag and pour out the contents.

Packaging, Sealing and Labeling

Avoid air pockets by making tight packages. There are two different ways you can wrap meat and other food packages:

DRUGSTORE WRAP: Use enough paper to cover meat 1½ times. Place meat in the center of the paper. Bring the two longest sides together over the meat and fold these edges over about 1″ making a lock seam. Fold again as many times as necessary to bring the paper flat and tight against the meat. To avoid waste, wrapping material need be only long enough to make two folds. Turn the package over and fold the ends into triangles. Then fold ends over against package, pulling tight, and fasten with freezer tape or cord.

Aluminum foil needs no tape or cord. Do not draw the foil too tightly in making the lock seam. Allow a little surplus to mold tightly around the meat, pressing the foil against the meat while the two ends are open. Close the foil ends by pressing them, starting next to the meat. Then fold ends over and make a lock seam with both of them, pressing the seam tightly against the meat to exclude air.

BUTCHER WRAP: Use enough wrap to cover meat twice. Place meat diagonally on wrap. Press corner nearest you snugly over meat. Roll meat and paper toward opposite corner. At halfway point, fold side corners in to center, drawing tightly. Roll to remaining corner. Seal with freezer tape.

Pack food in containers as solidly as possible, leaving the necessary head space for expansion. Seal containers to exclude the air. Some of them should be sealed with freezer tape to make them airtight and to prevent leakage. The tuck-in tops of waxed cartons should also be sealed with freezer tape. Cover glass jars with tight fitting lids.

After filling food into plastic bags, use your hands to press out as much air as possible. Twist open end, fold it over and fasten with a twist-tie or string. Use heavy plastic bags.

Label packages clearly and carefully; use a wax pencil or crayon for writing on cans. Sharpened waxed crayons give easily read labels on waxed surfaces or freezer tape. Give name of product, date when frozen and any information that will help you identify it. Some women like to indicate number of servings on labels.

Spread out packages of food to freeze, allowing at least 1″ between them for the circulation of the cold air in the freezer, which removes the heat more quickly. Freeze only the number of packages that can be spread against the sides of chest freezers and on the refrigerator shelves of upright models. From 2 to 3 lbs. food to 1 cubic ft. storage space may be frozen at a time in most freezers—30 to 45 lbs. in a 15-ft. freezer. It should freeze in 10 to 12 hours or less.

You need not change the cold control when freezing a few pounds of food, but with larger quantities, it is a good idea to set it at −10° F. 24

hours in advance (if your freezer has a cold control). The speed of freezing will not increase if you set the temperature control lower when you place unfrozen food in freezer. Leave packages in freezing position for 24 hours before stacking them close together. If you need to freeze more than 5% of your freezer's capacity at one time, check in your community or nearby city for commercial establishments that do quick freezing. Have food frozen there and bring it home. Stack the packages when the food in them is frozen. *Keep the freezer temperature at 0° F. or lower.* You can use a freezer thermometer to check the temperature of your freezer. *Recommended storage times in this cookbook are based on maintaining freezer temperature at 0° F. or lower. They are not for the maximum time you can keep food in the freezer, but for the period in which top quality is maintained. No frozen food is improved by long storage.*

Refreezing Foods

When the Freezer Stops: Every freezer owner needs to know what to do when the power fails or a mechanical breakdown occurs. The first step is to keep the freezer door closed. Little thawing takes place the first 12 to 20 hours if the freezer is fairly full of food that has been stored at 0° F. (Food may thaw faster if freezer is an upright standing in a warm place.)

If the freezer will be off longer than a day, move the food, if possible, to some place where low storage temperature is available. When power goes on again, a fully loaded freezer may be unable to refreeze such a large quantity of food before spoilage starts.

Dry ice may be used to check thawing. A 50-lb. cake of dry ice, if placed in the freezer soon after it stops operating, prevents thawing for 2 to 3 days. Wear gloves when you handle dry ice to avoid freezer burn. Place pieces of it on cardboard or a board, not directly on food packages.

Refreezing Foods: You may safely refreeze frozen foods that have thawed if they still contain ice crystals. Do not refreeze prepared foods —TV dinners, pot pies, pizza, stews, casseroles, unbaked pies, etc.—if they have thawed to a point where there are few or no ice crystals. Do not refreeze thawed ground meat, poultry, fish or seafood. Certain foods—red meats, for example—may be refrozen if still *very cold*, even if there are no ice crystals. But they must be below 40° F., and they should be held no longer than 1 or 2 days at that temperature.

If frozen foods have gone through slow temperature changes and reached a temperature higher than 40° F., do not eat or refreeze them. Thawed ice cream should always be discarded. If the odor or color of any food is poor or questionable, do not use it.

There is nothing harmful about the refreezing process. The danger is that —because thawed food spoils at a faster rate than fresh food—thawing may result in spoilage before you can refreeze the food. Partial thawing and refreezing reduce the eating quality of foods, particularly fruits, vegetables and prepared foods. The eating quality of red meats is affected less than that of many other foods. Use refrozen foods as soon as possible.

To summarize, here are guidelines on refreezing specific foods:

Red Meats: If their temperature is below 40° F., they are still in good condition. You can detect the beginning of spoilage by the odor and color. Thawed meats do lose juices.

Poultry and Shellfish: It is unwise to refreeze them because they spoil so easily and there is no way to detect whether early deterioration has started.

Fruits: When thawed, they may be refrozen. Even though fruits ferment, they are not a risk, but some of their flavor is lost. And they may shrink and become mushy.

Fruit Juices: Some deterioration occurs when you refreeze thawed concentrated juices. Sometimes it is difficult to reconstitute juices that have been refrozen. There is little change in other frozen juices.

Ice Cream: Do not use if melted or partially melted. Refreezing will result in possible spoilage and unacceptable texture changes.

Vegetables: Thawed vegetables will spoil more quickly than fruits. It is best not to refreeze them. Also, they often become tough when refrozen.

Cooked Foods: Do not refreeze dishes that contain meats and poultry, such as stews, casseroles and pies.

You usually lose quality when foods are frozen. The texture especially is likely to be less desirable. Baked pies, for example, will be soggy.

Nutritive Value of Frozen Foods

Meat, fish, poultry and eggs are equal in food value to the fresh products. The retention of the vitamins and other nutrients in fruits and vegetables depends on how they are handled before freezing, on the storage temperature in the freezer and on how you cook them. This is one good reason for following the directions given in this book. Foods that taste delicious almost always are those that have kept their food values.

Storage Temperature Is Important

The freezer should maintain a temperature of at least 0° F. Some freezers operate −5 to −10° F., which is even better. The deterioration in quality when temperatures are high is due to enzyme activity, not to the development of bacteria. Enzyme activity speeds up chemical changes which produce unpleasant flavors, loss of vitamin C and unattractive color. Some people believe that foods will keep indefinitely if they are "frozen hard." This is not true. Frozen foods should not be stored (except for a very few days) in an open-end ice cube compartment of a household refrigerator. The temperature is not low enough for long storage and it is not adequately controlled. A loss of quality may take place, even though the freezer is cold enough, if the frozen foods are stored too long or were not handled properly before freezing, or if they were not packaged correctly. Follow the directions in this cookbook.

You can check the operating temperature of your freezer by placing an accurate freezer thermometer on top of the food packages.

Quantity Freezers Will Hold

A good estimate is 35 lbs. per cubic foot of usable space. The amount varies with different foods—from 30 to 42 lbs. meat, 20 to 50 lbs. fruit, 15 to 25 lbs. vegetables. Bulky foods, like chicken, take more space.

SIZE OF FREEZER (CUBIC FEET)	APPROXIMATE POUNDS OF FOOD
9	315
11	385
15	525
22	770

Note: During a year 2½ to 3 times this amount of frozen food can be stored in freezers, of course, because of turnover through use.

The Freezer and the Microwave Oven

If you own a microwave oven, you'll find it a special convenience for defrosting frozen foods when you forget to take them out of the freezer ahead of time. Some models have an automatic defrost cycle; follow manufacturer's directions for timing. Without automatic controls, you can defrost foods by turning the appliance alternately on and off. The standing time allows heat on the outside of the food to penetrate and defrost the center. Without standing time, food would cook on the outside while the center is still defrosting. Timing depends on the amount and kind of food being defrosted; follow recommendations given in your microwave cookbook.

The microwave oven is not recommended for blanching vegetables to be frozen. Because of the wide variety of microwave appliances available (both brands and models), there are as yet no reliable guidelines available.

Defrost Your Freezer

Many freezers now defrost automatically. If you have a nonautomatic chest-type model, scrape the frost whenever accumulation is ⅓" thick. To do this, turn off the electricity and move as much food as possible to one part of the chest or take it out and wrap it in a blanket or newspapers. Spread towels on the bottom to absorb the moisture. Scrape off the frost with a wooden or plastic scraper, not metal.

A chest-type freezer should be completely defrosted and cleaned at least once a year. This is a good time to take inventory: to move out food you've stored too long.

It may be necessary to defrost upright models more often. And, if food is spilled and a scraper will not remove it, it will be necessary to defrost nonautomatic models, either chest-type or upright.

To defrost, turn off the electricity. Remove the food and wrap it in blankets. Place towels in bottom of freezer to absorb moisture. Place pans of hot water in freezer; close door or lid.

After thawing, wash inside of freezer with a soda solution, 3 tblsp. baking soda to 1 qt. water, or use water with synthetic detergent added. Rinse, wipe dry and turn on the electricity. When the moisture in the freezer is frozen, return the food.

Slick Freezing Tricks
GOOD IDEAS FROM FARM JOURNAL READERS

We take our hats off to farm women, the greatest of all cooks and experimenters. And we want to acknowledge that they send our Food Department many original recipes and ideas. Take a look at the collection of their slick freezing tricks here and in the following chapters. See if you don't think they're different and fascinating. We let farm women themselves tell you about their ingenious tricks in their own words.

FROM THE WOODS

WILD PERSIMMONS: After the frost touches the little persimmons on the trees in the woods or fence rows, I gather the ripe fruit, pick over, wash and drain it. Then I put it through a food mill or colander and freeze it to make persimmon puddings the year round.

HOOSIER PERSIMMON ROLL

Shape rolls 3 to 4" in diameter

2 c. frozen persimmon purée
2 c. sugar
1 c. brown sugar, firmly packed
1 lb. crushed graham crackers
1 c. chopped pecans
1 lb. miniature marshmallows

· Thaw persimmon purée. Combine ingredients in order listed and mix together. Spoon about one third of mixture on a sheet of waxed paper; shape in a roll. Wrap in aluminum foil or plastic wrap and freeze. Repeat until all mixture is used.

• To serve, cut the slightly thawed roll in slices for individual servings. Top with whipped cream or vanilla ice cream. Makes about 24 servings.

GOOSEBERRIES AND MULBERRIES: I combine these berries, half and half, and freeze them unsweetened, in quart jars. The two kinds of berries complement each other in pies.

MULBERRY DESSERT: I freeze mulberries, after washing and drying, in moistureproof containers with tight lids. About 45 minutes or an hour before I wish to serve them, I stir them into sweetened whipped cream.

SASSAFRAS ROOTS: Dig the roots in the spring, wash and remove the bark. Dry and store in tightly covered glass jars in your freezer. You can make the pretty pink tea iced or piping hot, all year. I also like to add a bit of the root to the sieved fruit when cooking apple butter. Remove it before storing the spicy butter.

TIME AND MONEY SAVERS

SCRAPS OF FOIL: Save foil scraps and wrap them around the sharp bones of meats and poultry when you are wrapping them for the freezer. The foil makes a cushion that protects the freezer paper, plastic bag or other wrap.

NYLON MESH BAGS: Keep small packages together in the freezer by putting them in a large nylon mesh bag. You can find them in a jiffy.

SPECIAL DISHES: If you fix special dishes for children, oldsters or invalids, prepare them in quantity and freeze menu-size portions in freezer. It saves work and dishwashing.

TV SURPRISE DINNERS: Empty trays for TV dinners, washed and dried, make ideal containers for dinner or supper leftovers—meat, chicken, vegetables, etc. I fill a tray, cover tightly with foil and freeze. After I have accumulated enough trays for everyone in my family, we have a "Surprise Dinner." I could label them, but our family prefers and enjoys the element of surprise as everyone eats what he "draws." Sometimes we swap dinners.

BROWN SUGAR: Pour into freezer containers or glass jars; seal and store in freezer. The easiest way to avoid lumps and annoyance on baking day.

MARSHMALLOWS: Keep them packaged in the freezer. They do not dry out and when you cut them to add to salads and desserts, you or your kitchen shears don't get stuck up.

FARM FAVORITES

ICE-CREAM SANDWICHES: My children like pretty pink-dappled peppermint ice cream frozen between layers of crushed cocoa graham crackers. On special occasions, I pass chocolate sauce to pour over this dessert.

HOMEMADE NOODLES: I have plenty of eggs so I make noodles. They freeze successfully and I find it convenient to keep a supply in the freezer. To make them, beat 3 egg yolks and 1 whole egg until very light. Beat in 3 tblsp. cold water and 1 tsp. salt. Then stir and work in with the hands 2 c. flour to make a stiff dough.

Divide dough into 3 parts. Roll each one out on lightly floured, cloth-covered board as thinly as possible—paper thin. Place each piece of dough between towels until it is partially dry like a chamois skin. Roll up like a jelly roll and cut with a sharp knife to make strips the width you like—⅛" for fine noodles and ½" for broad ones. Shake out the strips and let them dry before packaging, sealing and freezing. Makes 6 cups.

HOSTESS TRICKS

PLATES: Chill salad and dessert plates in freezer when weather is warm before serving food on them.

CANDLES: Buy them ahead to use on the table for anniversary celebrations and other special occasions. Store them in the freezer. They keep their shape, burn with a bright flame and are not so likely to drip on the table cover.

COFFEE AND TEA: Freeze leftover coffee or tea in ice cube trays. Place cubes in plastic bags. Use them for quick servings. Or use them in iced tea or coffee to keep it undiluted.

FRESH GINGER: I wash, dry, wrap, label and store fresh ginger roots in the freezer. All I need do when a Chinese-style recipe calls for ginger is to grate the frozen root. Many supermarkets sell fresh ginger.

PIE SHORTCAKES: Roll pastry ⅛" thick and cut in circles with 3" cutter. Place on an ungreased baking sheet. Prick rounds with fork. Bake in a very hot oven (450° F.) 8 to 10 minutes. Cool. Package layers of pastry in rigid freezer containers, cover and seal. Freeze. When you need a dessert in a hurry, stack 3 pastry rounds on a plate with pie filling between and on top. You can use canned or frozen berries, heated and thickened with cornstarch with a little butter and sugar added. Or use canned pie filling. If you have mounds of whipped cream in your freezer, put one on top of each shortcake. They're easier-than-pie and just as delicious and appealing.

CHEESE WAFERS: A week or two before a luncheon party, I freeze dough for Cheese Wafers to serve with my main dish salad—guests love them.

The dough is a snap to fix. I grate 4 oz. sharp process cheese into a bowl and sift in 1½ c. flour, ¾ tsp. salt and a dash of ground red pepper. Next, I cream ½ c. butter until fluffy and combine it with the flour-cheese mixture, blending thoroughly. The mixture looks crumbly and dry, but I quickly knead it into a smooth ball, refrigerate 1 hour, shape rolls 1" in diameter, wrap and freeze.

To bake, I thaw the rolls about an

hour in the refrigerator. Then I slice them about ¼″ thick, line them up on a baking sheet and bake in a hot oven (400° F.) about 10 minutes. They're wonderful hot or cold. My recipe makes about 6 dozen wafers.

FOR YOUR CAT

CATNIP: If you have a cat, gather, wash and dry catnip leaves. Pack them in layers in plastic containers with waxed paper between. Your cat will be happy all winter.

CHAPTER 2

Fruits and Vegetables

Freezing Fruits and Berries · Protecting Color · Fruit Purées and Relishes · Fruit Desserts · Slick Freezing Tricks with Fruits · Freezing Vegetables · Blanching · Cooking Frozen Vegetables · Heat-and-eat Potato Treats · Vegetable Sauce Cubes · Seasoning Ideas and Slick Freezing Tricks with Vegetables

Country women mentally start freezing vegetables in winter with seed catalogs in their hands. And when trees blossom pink and white, that's when they ponder how much fruit to freeze. For they don't have long to wait. When the rosy-red rhubarb stalks and tender asparagus spears push their heads above the earth, harvesting and freezing time begins.

Freezing is the quickest, easiest—and coolest—way to put up fruits and vegetables. And it keeps more of their fresh taste and bright color than other methods of preservation—especially for farm people, because they can pick fruits and vegetables at the exact flavor peak, and in the morning before the sun has a chance to heat them. Hurried into the freezer, time stands still until the frozen produce is used.

Desirable as freezing is, garden and orchard products must pass an entrance examination for admittance to many freezers. Space is at a premium. Every family must decide what to can and what to freeze—to consider that some foods, like strawberries, are best frozen and others, like tomatoes and pears, are best canned. Another question: Is it economical to freeze root vegetables if you have a cool cellar for storing them?

Busy cooks fast-fix frozen fruits and vegetables for meals because they are partly prepared—peaches are sliced, peas are shelled, for instance.

"Compare your meals with your grandmother's," one country cook suggests. "Look at recipes in old cookbooks. Then you'll know why freezers bring blessings to homemakers, along with astonishingly good meals to farm tables." Follow our directions for top-quality frozen products.

Freezing Fruits and Berries

Choose Ripe Fruits: They should be riper for freezing than canning—"eating ripe." Tree- and vine-ripened fruits have superior flavor and more vitamins than fruits picked green, then ripened. Discard immature fruits.

Select Right Varieties: Every region has fruits that retain the maximum amount of color, texture and flavor when frozen. Varieties vary with the place where they grow. Ask your County Extension Home Economist or write your State Agricultural Experiment Station to find out which are best for you.

Prepare Quickly: Sort, wash in cold water and fix fruit as for table use.

Pack to Protect Quality: Directions for packing individual fruits give specific amounts of sugar or sugar syrup. These sugar or sugar syrup packs help retain color, texture and flavor of the fruit. Amounts may seem high, but they were established by experiments to obtain optimum quality when freezing fruits. If you want to use a lesser amount of sugar, you can; or you can pack fruit unsweetened.

SUGAR SYRUP PACK: Dissolve the sugar in water. *Do not heat.* Add the sugar to the water and let it stand, stirring occasionally, until sugar dissolves. You can make it a day or two ahead and store it in the refrigerator. *Always add it cold.* If you mix 1 part light corn syrup with 2 parts sugar syrup, you often get frozen fruit with better texture, flavor and color.

Fruits packed in sugar syrup retain the most vitamins.

SUGAR PACK: Sprinkle the sugar on fruit. Stir carefully with a rubber spatula until the sugary juices coat every piece of fruit. Many homemakers especially like to use this method for fruits to freeze for pies. This method also is excellent for sliced strawberries.

UNSWEETENED PACK: Either pack the fruit dry or pack it in water with color protector added. Use 1 tsp. powdered ascorbic acid to 1 qt. water, or follow directions on label of commercial color protector. Berries packed without syrup or sugar collapse and lose flavor more quickly when thawed, however. Use partly thawed. Whole blueberries and cranberries are fine in dry pack; so is rhubarb.

Keep Colors Bright: Some fruits, like peaches, apricots, sweet cherries and figs, darken unless you add a color protector. You can use powdered ascorbic acid available in many drugstores, following directions in this book for individual fruits. Or you can buy packaged products that protect fruit color in supermarkets. These contain ascorbic acid and another fruit acid. If you use the supermarket product, follow directions printed on the label for correct amounts. Add the ascorbic acid mixture or ascorbic acid powder to the syrup; mix thoroughly, but use care not to beat air into the syrup. Add the color protector or ascorbic acid mixture to the syrup *when you are ready to pack the fruit.* You lose some of these products' effectiveness if you let the syrup stand after you add them. In the sugar pack, add it to the sugar and if it is used in the unsweetened pack, to water. Be sure to mix the color protector thoroughly with sugar or water.

Pack the Right Way: For the sugar syrup pack, fill containers one third full of cold syrup and add fruit. Syrup should cover the fruit; add more if needed. To keep fruit that darkens readily submerged, place a generous piece of crumpled waxed paper under the lid. For the sugar pack, place the fruit in a bowl, sprinkle the sugar on and mix gently with a rubber spatula. Ripe berries are so delicate and soft that it is a good idea to add sugar to a small quantity at a time, about 1½ qts. Or place berries in a large shallow pan, sprinkle on sugar and mix gently until each is coated with sugar and juice.

Leave Head Space: When packing fruit purées or fruits in liquid in wide-topped rigid freezer containers or wide-mouth jars, leave ½" head space

for pints and 1" for quarts. Do not use regular-mouth jars with shoulders; in freezing, such jars may break at the shoulder.

When packing fruit or vegetables without liquid (dry pack) or fruit in sugar pack, leave ½" head space for pints and quarts.

When packing fruit juices, soups and sauces in wide-mouth jars, leave 1" for pints and 1½" for quarts.

Hurry to Freezer: Cover, seal if necessary and label every container with the name of the fruit and the date of packaging. Spread packages in freezer *at once* and freeze at 0° F. or lower. Be sure fruit is completely frozen before stacking to save space.

Storage Time: 9 to 12 months.

How to Thaw: Let frozen fruits thaw in their original, *sealed containers* at room temperature. Quality and nutritive values are retained by fairly rapid thawing. If packages are watertight, you can submerge them in cool water. Serve fruit cold, preferably while it contains a few glistening ice crystals. Or cook like fresh fruit.

Directions for Freezing Fruits
(IN ALPHABETICAL ORDER)

APPLE SLICES: Wash firm-fleshed apples suitable for making pies or sauce. Remove bruises and decayed spots, peel, quarter and core. Cut each quarter into three slices. To prevent discoloration while fixing the apples, slice them into a salt water solution, ½ c. salt to 1 gal. water. When

all the fruit is sliced, drain, rinse in cold water and drain. To freeze apple slices in sugar syrup, use 3 c. sugar to 1 qt. water and add ½ tsp. ascorbic acid powder to syrup before adding it to fruit. If you freeze apples for pies, omit syrup. For each quart of apple slices, blend 1 tsp. ascorbic acid powder with ½ to 1 c. sugar according to desired taste. Coat fruit evenly with the mixture. If you use commercial color protector from the supermarket, follow package directions. Pack, label, date and freeze.

APPLESAUCE: Canned applesauce may have better flavor than the frozen sauce. To freeze the sauce, cook apples in the usual way, then put them through a food mill if you wish. Sweeten to taste. Cool, package, label, date and freeze.

APRICOTS: Use directions for freezing peaches (see Index), selecting fully ripe fruit of uniform color. Unless you peel apricots before freezing them, the canned fruit is better for dessert use. (The skin tends to toughen in freezing.) Frozen unpeeled apricots are satisfactory for making pies, however. Cut the washed fruit in halves, discard pits and steam 4 minutes. Crush and pack with sugar, 1 lb. sugar to 4 lbs. fruit, or about ½ c. sugar to 1 qt. fruit. You will not need to add a color protector if you steam the fruit. If you do not steam the apricots, mix in ¾ to 1 tsp. powdered ascorbic acid for every 10 c. sugar. Mix powdered acid and sugar.

AVOCADOS: Select avocados that are soft-ripe with rinds free of dark blemishes. Peel, cut in halves, remove pits and mash pulp. For a superior product, add ⅛ tsp. powdered ascorbic acid to every quart of purée. Unsweetened avocados are best for salads, sandwiches and dips. Package, label, date and freeze.

BERRIES: Boysenberries, Dewberries, Loganberries and Youngberries: Follow directions for blackberries.

BLACKBERRIES: Wash firm, glossy, ripe berries of rich flavor in very cold water. Lift from water into colander; hold in hands to drain. This avoids crushing berries. For dessert, use uncooked. Pack berries in sugar syrup made with 3 c. sugar to 1 qt. water. Pack without sugar for pies and jams. Or add ¾ c. sugar to 1 qt. berries. Label, date and freeze.

BLUEBERRIES: Wash stemmed, ripe berries in very cold water, lift from water into colander and drain. For dessert use (uncooked) and superior quality, pack in sugar syrup made with 3 c. sugar to 1 qt. water. For pies and other cooked dishes, pack whole berries without sweetening. Or crush or sieve berries and mix with sugar, 1 to 1⅛ c., depending on sweetness of berries, to 1 qt. crushed blueberries; pack. Label, date and freeze.

CANTALOUPE: See Melons.

CHERRIES, SWEET: Use fully ripened, dark-red cherries. Chill them in very cold water to prevent bleeding when pitted. Lift from water, stem and pit. Pack in sugar syrup made with 2 c. sugar to 1 qt. water. Add ½ tsp. powdered ascorbic acid. For the most delicious frozen cherries, add 1 tsp. citric acid or 4 tsp. lemon juice to the syrup in addition to the ascorbic acid. (Frozen sweet cherries are tasty in salads.) Package, label, date and freeze.

CHERRIES, TART: Wash, stem and pit bright-red cherries. For pies and other cooked dishes, pack cherries with sugar, ¾ c. to 1 qt. cherries. Mix until sugar dissolves. To improve color retention, add 1 tsp. powdered ascorbic acid to every cup of sugar. Or crush cherries coarsely and increase sugar—mix in 1 to 1½ c. sugar for every quart of fruit until sugar dissolves. Package, label, date and freeze.

CITRUS FRUITS: Fresh oranges and grapefruit are available the year round in most regions, but you may wish to include them, for the flavor they contribute, to mixed fruits for freezing (see Mixed Fruits). If you have a surplus of grapefruit and oranges, you may wish to freeze them. Select tree-ripened, firm fruit that is heavy for its size. Do not use fruit with soft spots. Wash and peel with a knife. Take out fruit sections, removing all membrane and seeds. Slice oranges if you prefer. If grapefruit has many seeds, cut it in halves across and scoop out half-sections with a knife or spoon. Pack the fruit sections or sliced oranges in containers with a sugar syrup made with 3 c. sugar to 1 qt. water, adding ½ tsp. powdered ascorbic acid. Use excess juices of fruit for part of the water in making the syrup. Label, date and freeze.

COCONUT: Shred fresh coconut meat or put it through a food chopper, or cut in pieces. Pack into containers and cover with milk from coconut. Package, label, date and freeze. Partially thaw and use like fresh or commercially packaged coconut.

CRANBERRIES: Select firm, glossy berries that are plump. Stem, sort and wash in very cold water; drain. Pack without sugar. Or make cranberry sauce; add 2 c. water to 1 qt. berries and cook until skins pop. Put through food mill if you wish. Add sugar to taste—about 2 c. for 1 qt. purée. Package, label, date and freeze.

CURRANTS: Stem large-fruited varieties of currants, wash in very cold water, lift from water and drain. Mix each quart of fruit with sugar to taste, about ¾ c. Stir until most of the sugar dissolves. If you crush the currants slightly, the syrup will penetrate the berries. Or pack currants without adding sugar. Label, date and freeze.

DATES: Select ripe dates with good flavor and tender texture. Wash, slit and remove pits. Leave whole or put through a food mill to make a purée. Package, label, date and freeze.

ELDERBERRIES: Follow directions for blueberries.

FIGS: Use tree-ripened, soft-ripe fruit, making certain centers of figs are not soured. Sort, wash and cut off stems. Peel if desired. Slice or leave whole. Pack in sugar syrup made with 2½ c. sugar and 1 qt. water with ¾ tsp. powdered ascorbic acid added. Or omit ascorbic acid and add ½ c. lemon juice to every quart of syrup. You can pack figs without adding syrup, either dry or by covering with cold water with ¾ tsp. powdered ascorbic acid added to each quart of water. If you prefer to freeze crushed figs, mix 1 qt. fruit with ⅔ c. sugar with ¼ tsp. powdered ascorbic acid added. Package, label, date and freeze.

FRUIT JUICES: Directions for extracting juice from fruits and berries are given in Chapter 10. Chill extracted juice quickly; pour into standard glass canning jars or plastic containers that can be tightly sealed. Leave head space as described at the beginning of this chapter. Most fruit juices will keep about 8 months; citrus juices 3 to 4 months. Label, date and freeze.

GOOSEBERRIES: Remove stems and blossom ends. Wash, lift from water and drain. Pack without sugar or syrup; label, date and freeze.

GRAPES, BUNCH: Sort, stem and wash seedless green grapes (Thompson) or Tokay grapes. Pack Thompson seedless grapes whole or halved; cut Tokays in halves or quarters and discard seeds. Pack in sugar syrup made with 3 c. sugar to 1 qt. water. Label, date and freeze.

GRAPES, MUSCADINE: Sort, stem, wash and pack in sugar syrup made with 2 to 4 c. sugar for 1 qt. water. These grapes include the native American variety, the Scuppernong. Label, date and freeze.

HUCKLEBERRIES: Follow directions for blueberries, selecting berries with tender skins.

MELONS: Choose firm, ripe, fine-textured melons. Immature melons are inferior in quality when frozen. Wash, cut in halves, discard seeds and rind.

Cut melon flesh into ½ to ¾" cubes or balls and pack in sugar syrup made with 2 c. sugar to 1 qt. water. Whole seedless grapes may be added. Package, label, date and freeze.

Serve melon when partially thawed. Use same directions for Cantaloupes, Crenshaw, Honeydew, Muskmelons and Persian melons—also for watermelons, although they are not "good freezers."

MIXED FRUITS: Select ripe fruits that contrast pleasantly in taste, color and texture. Wash, sort and prepare them as for table use. Cut fruits into attractive sizes and shapes for use in salads and cocktails.

Fruits that are good mixers are apricots, cherries, grapefruit, oranges, pineapples and Thompson seedless grapes. Fruits that are *not* especially good mixers are strawberries, raspberries, cantaloupe and apples.

A good way to fix grapefruit and oranges for freezing with other fruits is to arrange the sections, free of membrane, in layers with sugar to sweeten. Cover and set in the refrigerator until the fruit is just about covered with juices. Combine with the desired fruits and freeze in a syrup made with 2 to 3 c. sugar (depending on natural sweetness of fruits) to 1 qt. water with ½ tsp. powdered ascorbic acid added. Package, label, date and freeze.

Thaw fruit combinations in original, sealed containers. For molded salads, completely thaw; for cocktails and mixed salads, partially thaw so fruits retain a few ice crystals.

NECTARINES: Follow directions for peaches, using fully ripe, well-colored, firm fruit. Overripe nectarines may have an unpleasant flavor when frozen.

NUTS: Fresh, unprocessed nuts will keep in the freezer 9 to 12 months. If they are salted, the storage time is cut in half. Hold fresh nuts in damp place overnight to prevent brittleness of kernels in cracking. Pack in glass jars or metal containers, label, date and freeze.

PEACHES: Select fruit slightly riper than for canning, with no green spots. If you buy shipped-in peaches for freezing, let them ripen in a temperature of 65 to 70° F. for best results. Sort, wash and peel fruit. A good way to peel peaches (and apricots) for freezing is to dip 3 or 4 of them at a time in boiling water for 15 to 20 seconds or until the skins loosen. Chill quickly in ice or very cold water and peel and cut in halves or slices, working fast.

Slice the peaches directly into the freezer container filled one third full of cold syrup. Make the syrup with 3 c. sugar to 1 qt. water with ½ tsp. powdered ascorbic acid added just before syrup is used. Completely cover fruit with the cool syrup. For best color and flavor, keep the fruit submerged in the syrup by placing a generous piece of crumpled waxed paper under the lid. Cover, label, date and freeze.

If you cannot package the cut peaches immediately, you can hold them a short time without darkening by submerging them in 1 gal. water containing 1¼ tsp. powdered ascorbic acid. (This is a good way to hold sliced peaches for table use too.)

For a water pack, if you cannot use sugar, substitute cold water for the syrup, adding 1 tsp. powdered ascorbic acid to each quart water. (See also Peach Purée Delicious, this chapter.)

PEARS: Better results are obtained by canning pears than by freezing them.

PERSIMMONS, NATIVE AND CULTIVATED: Select soft-ripe fruit. Sort and wash. Press the native variety through a food mill or sieve, add ⅛ tsp. powdered ascorbic acid to 1 qt. purée and pack without sugar. Select orange-colored, soft-ripe cultivated persimmons and peel. Prepare like native persimmons, but add 1 c. sugar to 1 qt. purée if you wish. Package, label, date and freeze.

PINEAPPLE: Peel and core firm, ripe pineapple with fragrant aroma, indicating ripeness. Slice, dice or cut in wedges or sticks. Pack in sugar syrup made with 3 c. sugar to 1 qt. water. If pineapple juices are available, use them for all or part of the water in making the syrup. For the sugar pack, mix 1 c. sugar with 8 or 9 c. cut-up pineapple. Package, label, date and freeze.

Frozen pineapple must be cooked before you add it to a gelatin mold. Otherwise an enzyme in the fruit will prevent jellying.

PLUMS (including fresh prune plums): They are better canned for dessert than frozen. If you want to freeze plums, choose fruit of deep color, wash, remove pits and cut in halves or quarters or leave whole. Pack in syrup made with 3 c. sugar to

1 qt. water. Cover, label, date and freeze.

If no sugar is added, plums (packed dry) make good jam and pies.

RASPBERRIES: Wash firm, fully ripe red or black raspberries in very cold water, but do not let them soak. Lift berries from water; drain small amounts through fingers, shaking hands. Pack them in sugar syrup or sugar. For sugar syrup, use 3 c. sugar to 1 qt. water; for sugar pack, use 1 lb. sugar to 4 to 5 lbs. berries— 1 c. sugar to 7 to 8 c. berries. The exact amount of sugar needed depends on the sweetness of the berries. Package, label, date and freeze.

Freeze red and black raspberries for jam without adding syrup or sugar. For a purée, add ¾ to 1 c. sugar for 1 qt. crushed or sieved berries. Mix until sugar dissolves.

RHUBARB: Select crisp, tender red stalks early in spring. Remove leaves, woody ends and blemishes. Wash, cut in 1″ lengths and cover with sugar syrup, 3½ c. sugar to 1 qt. water. Or pack in sugar using 1 c. sugar for each 7 to 8 c. cut rhubarb. For a few months of storage, you can freeze rhubarb cut in pieces without sugar. Package, label, date and freeze.

Add 1 c. sugar to 4 c. rhubarb and you will not need to add sugar when making pie.

STRAWBERRIES: Select ripe berries of bright color, rich flavor and pleasing aroma. Wash and hull in cold water. Drain in colander. For sugar pack, measure whole berries; use 1 c. sugar to 7 to 8 c. hulled whole strawberries, or 1 lb. sugar to 4 to 5 lbs. berries, depending on how sweet they are. Slice each berry lengthwise in thirds or halves, depending on size. Use a rubber spatula to mix berries and sugar. Small whole berries are best packed in sugar syrup made with 3 to 4 c. sugar to 1 qt. water, but their flavor is not so good as the sliced. Package, label, date and freeze.

Berries chopped with a stainless steel chopper may be packed in the syrup or sugar, as preferred. Unsweetened strawberries are inferior in flavor when frozen.

Enjoy Summer Fruits in Winter

When you take a jar of Grape Purée out of the freezer, think about the good neighbor down the road—the one who does you so many favors. What could be a more acceptable gift for her at Christmas time than Frozen Grape Purée, a remembrance of summer? Why not take along a couple of good recipes for using the deep purple mixture—Grape Aspic or Fluffy Grape Pie, for instance?

FROZEN GRAPE PURÉE

Wash and steam Concord grapes. In large kettle, heat grapes 8 to 10 minutes at low temperature (not over 145° F.) to loosen skins. *Do not boil.* Put through food mill or wide-mesh strainer. Discard skins and seeds. Pour purée into jars, leaving ½ to 1″ head space; seal, label, date and freeze. Recommended storage time: 6 to 8 months.

Note: Purée may develop gritty texture after long freezer storage. This disappears when heated slightly.

GRAPE ASPIC

Serve it with baked ham or turkey for that important winter buffet

2 tblsp. unflavored gelatin
½ c. orange juice
¼ c. lemon juice
½ c. sugar
½ tsp. salt
2 c. Grape Purée
1 c. water
2 whole sticks cinnamon
4 whole cloves

• Sprinkle gelatin over mixture of orange and lemon juices to soften.
• Mix remaining ingredients; heat 15 minutes; do not boil. Then bring just to a boil; add gelatin, stir to dissolve.
• Pour into individual ring molds; refrigerate until set. Do not freeze aspic.
• To serve, unmold on lettuce; fill centers with seeded Tokay grape halves, mandarin oranges, drained, and grapefruit sections. Pass salad dressing separately. Makes 6 servings.

FLUFFY GRAPE PIE

Top-favorite winter pie in Erie County, New York, where people grow and know how to use Concords

Baked 9″ pie shell
1 c. Grape Purée, thawed
      enough to measure
¼ c. water
1 (3 oz.) pkg. lemon flavor
      gelatin
¾ c. sugar
1½ c. heavy cream, whipped

• Bring Grape Purée and water just to a boil; stir in gelatin until dissolved. Add sugar; mix well; chill until mixture mounds when dropped from spoon, stirring occasionally.
• Beat until fluffy; fold in whipped cream. Pour into pie shell. Refrigerate at least 2 hours or overnight. Do not freeze grape pie. Serve with thin layer of whipped cream.

GRAPE TABLE SYRUP

Serve with pancakes, biscuits or over ice cream or vanilla pudding

1¼ c. Grape Purée
1½ c. sugar
¼ c. corn syrup
1 tblsp. lemon juice

• Combine ingredients in heavy pan; bring to a rolling boil; boil 1 minute, counting time after mixture comes to a boil that cannot be stirred down.
• Remove from heat; skim off foam. Pour into containers. Cool; cover and store in refrigerator. Makes 1 pint.

PEACH PURÉE DELICIOUS

If you have overripe, soft peaches, you can salvage them in a delightful purée. It makes a delicious topping for ice cream and other desserts. The following directions were developed by the Colorado Experiment Station.

Divide peaches into two lots according to degree of bruising. Use ripest, softest fruit. Dip peaches in boiling water for 30 seconds, cool in running water and remove skins. Place in bowl of salted water (about 2½ tblsp. to a gallon) while you trim and pit peaches.

Mash drained fruit with a mesh-type potato masher. Four to five average size peaches make 3 c. purée—to this quantity add ¼ tsp. powdered ascorbic acid to prevent discoloration. Add 1 pkg. powdered pectin; stir to dissolve; let stand 15 minutes, stirring. Add 3 c. sugar and stir to dissolve completely.

Remove peel, pits and bruised spots from less bruised fruit in same way. But instead of mashing, cut each peach in 12 slices and then run the knife around the center of the fruit to cut slices in halves.

Fill pint-size glass or other freezer containers with equal parts peach purée and slices mixed together lightly. Cover, label, date and freeze.

PEACHES IN ORANGE JUICE

A refreshing breakfast fruit. Seedless grapes may be frozen with this

1 (6 oz.) can frozen orange
    juice concentrate
3 cans water
¼ c. sugar
¼ tsp. powdered ascorbic acid
10 medium peaches, peeled and pit-
    ted (about 2¾ lbs.)

• Mix orange juice concentrate as directed on can with 3 cans water. Add sugar and ascorbic acid or other color protector, following directions on package. Stir to dissolve sugar.
• Fill freezer containers one third full of the orange juice mixture. Slice prepared, ripe peaches directly into containers. Leave head space. Add enough orange juice to cover. Place crumpled pieces of freezer paper under the lids to hold the peaches

submerged in the orange juice. Pack, label, date and freeze. Makes 7 half pints.
• To serve, partially thaw in original container. Serve while fruit still contains a few ice crystals.

GINGER PEACHES

• Melt ½ c. butter or regular margarine in skillet. Add ⅓ c. slivered crystallized ginger or ½ tsp. ground ginger, 2 tblsp. grated lemon peel and 1 tsp. lemon juice, then 1 qt. frozen peaches, thawed, or 1 qt. canned peach halves and juice. The lemon-ginger taste is special. Heat through. Best served warm over ice cream. Makes 6 servings.

FROZEN CINNAMON APPLES

Wonderful served with roast pork

1 c. sugar
1 c. water
½ c. red cinnamon candies
8 to 12 medium apples

• Combine sugar, water and candies in saucepan. Cook over medium heat until candies are dissolved, stirring frequently. Simmer 5 minutes.
• Peel apples, cut in halves and remove cores. Cook them, a few at a time, in the syrup until they are just tender. Do not overcook and do not crowd apples in syrup.
• Cool, package, label, date and freeze. Makes 16 to 20 servings for relish, 8 to 12 servings when used for salad.
Recommended storage time: 9 months.
• To use, partially thaw apples and

place one on each serving plate or place them on a relish tray. Or fill centers with cream cheese mixed with chopped nuts, top with spoonfuls of salad dressing and serve on lettuce for a colorful salad.

CRANBERRY CHUTNEY

Keeps up to 1 month in refrigerator; freeze for longer storage

4 c. cranberries
2 small oranges, seeded
1 lemon, seeded
2 apples, cored
1 c. raisins
1 c. brown sugar, firmly packed
½ tsp. ground ginger
2 tblsp. candied ginger
2 tblsp. grated onion
6 tblsp. chopped green
    pepper

• Grind cranberries, oranges, lemon, apples and raisins. Add remaining ingredients; mix well.
• Pack in freezer containers. Seal and freeze. Makes about 3 pints.

GOLDEN GLOW MELON FREEZE

Lovely to look at, luscious to eat— serve it when you want to splurge

1 cantaloupe (3 c. cantaloupe
    purée)
¾ c. sugar
1 tblsp. lemon juice
⅛ tsp. salt
1 envelope unflavored gelatin
¼ c. water

• Use ripe, sound cantaloupe; cut in half. Remove seeds and peel. Slice and press through a food mill or sieve to make purée.
• Mix cantaloupe purée, sugar, lemon juice and salt. Soften gelatin in cool water 5 minutes; then dissolve by heating over boiling water. Add cantaloupe mixture slowly to gelatin, stirring while adding to mix thoroughly.
• *For electric ice cream freezer,* pour into freezer container, surround with 6 to 8 parts ice to 1 part rock salt. Keep freezer turning about 20 minutes until mixture becomes firm and clings to dasher. Remove dasher. Serve immediately. Makes about 1 quart.
• *For refrigerator trays,* pour mixture into trays and freeze until firm. Remove trays and stir rapidly to incorporate air and make it smoother. Return trays to freezer a few hours to harden. Serve immediately or pack in freezer containers, label, date and place in freezer.
Recommended storage time: 3 weeks.

Slick Freezing Tricks with Fruits
GOOD IDEAS FROM FARM JOURNAL READERS

GLAMOROUS GARNISHES: Pick the prettiest, most nearly perfect berries. Wash carefully and drain. Leave the hulls on strawberries. Handle lightly to prevent crushing. Place them in rows, ½" apart, on baking sheets. Set baking sheets in freezer. When berries are frozen firmly, package in freezer containers. Put back in freezer. At serving time, remove only what you need.

APPLES, WHOLE: I'm so busy in summer that I often freeze whole apples for cooking. We like the tart taste of the early fruit later on in pies, sauce and other desserts. This is how I do it—I wash firm apples without blemishes, place in plastic freezer bags and put them in freezer. To fix fruit for cooking, I run cold water over each frozen apple until it thaws enough so I can peel, core and slice it like fresh fruit. You have to work fast because apples that thaw before peeling darken quickly.

CRANBERRIES: Instead of freezing cranberries in the transparent bags in which you buy them, look them over, wash and drain thoroughly. Then pack them in plastic freezer bags or containers and freeze. It is easier to clean the berries before they are frozen. And they are ready to use.

GRAPES AND MELON IN LEMONADE: Select firm, ripe seedless green grapes; leave them whole. Remove stems or leave in small clusters. Choose ripe, meaty muskmelons or cantaloupes that are not stringy. Cut melon in halves, remove seeds and cut flesh in balls with ball cutter or ½ teaspoon measuring spoon, or dice it. Place a mixture of grapes and melon balls in freezer containers. Mix 1 (6 oz.) can frozen lemonade concentrate with 3 cans water as directed on can. Mix thoroughly. Pour over grapes and melon to cover. Leave head space. You need ¾ c. lemonade for each pint of grapes and melon balls. To

serve, partially thaw in original container. Serve while fruit and melon contain a few ice crystals.

RIPE BANANAS: To salvage ripe fruit, I put it through my food mill and add lemon juice or powdered ascorbic acid to prevent darkening, and freeze. Then when I want to make a banana nut bread or a banana cake, the fruit is ready to add to batter.

LEMONS: When I make lemonade for the men in the field, I save the peel to flavor salad dressings, cakes, cookies and other baked foods. I cut off as much of the bitter white part as possible, grind the yellow part fine, package and freeze it.

LEMON JUICE: I buy lemons in large quantities, especially when they are bargains. I extract the juice and freeze it in ice cube trays. The cubes, stored in plastic freezer bags in the freezer, enable me always to have lemon juice on hand.

APPLE/RHUBARB SAUCE: Equal parts of apples and rhubarb cooked together make a delightful sauce. Of course, these two foods are not in season at the same time, but with frozen rhubarb in the freezer, we enjoy this sauce from autumn on through winter.

RHUBARB SAUCE: I like to cook fresh rhubarb, sweeten it to taste and freeze it in plastic or glass containers with tight lids. It's ready to use when you take it out. Some women cut the fresh rhubarb stalks and freeze them, but they have to cook the rhubarb before serving it. That takes time.

REPACKING FROZEN FRUITS: Some farm women who live in non-fruit-growing regions, find it economical to buy large packs of frozen fruits, like cherries, peaches and strawberries from locker plants or wholesale grocers that supply restaurants and other institutions. It is best to repack these fruits in smaller, handy containers.

Let the fruits stand at room temperature until they thaw just enough so you can separate the individual frozen pieces of fruit without breaking them. Pack in smaller containers and return to the freezer at once.

Freezing Vegetables

Plant Varieties for Freezing: Many varieties of all vegetables freeze successfully, but some are not good freezers. Ask your County Extension Home Economist or write to your State Agricultural Experiment Station to find out which ones in your area freeze well.

Hurry from Garden to Freezer: It is the best way to keep top quality. If the weather is hot, gather vegetables early in the morning before they have absorbed much heat from the sun. If you must hold them a short time before you can pack them for freezing, cool them rapidly and keep them cold. The fastest way is to plunge them in cold water; you can hold them in cold water, or drain and refrigerate. If you use the refrigerator to cool vegetables, spread them out so they cool quickly. Once you shell or cut vegetables, continue immediately with blanching and freezing steps; otherwise, vegetables lose quality, texture, flavor and nutrients. Other causes of poor quality are inadequate blanching, or stacking packages too quickly. Spread them out in the freezer; stack them only after they're thoroughly frozen.

Blanching Is Essential: Before you freeze them, vegetables must be heated briefly in boiling water or steam—a process called blanching or scalding—in order to protect their color, flavor and texture in storage.

Blanching helps preserve the fresh-picked quality of vegetables by inactivating the enzymes. Enzymes are the substances that help vegetables grow and mature; if their action were to continue, the food would lose quality and would eventually spoil.

Green vegetables will have a brighter color when blanched and the heat will shrink vegetables slightly so they handle more easily and conserve freezer space. Blanching also removes objectionable odors and bitter flavors from some vegetables.

If you freeze vegetables without blanching, within a month enzyme action will begin to cause undesirable changes—toughening and loss of color and flavor.

Properly blanched and packaged vegetables may be stored in the freezer from 9 to 12 months.

HOW TO BLANCH VEGETABLES

1. Place water in a large aluminum, stainless steel or enamelware kettle with cover. Use at least 1 gal. water to 1 pt. vegetable, except for the leafy greens. For them use 2 gals. water. Bring the water to a full rolling boil.

2. Put the prepared vegetable in a wire basket (a covered one is excellent) or loosely in a cheesecloth bag and *submerge in vigorously boiling water*. Start counting time immediately. The time is so important in blanching that we recommend use of a timer. Keep the heat high, kettle covered and the water boiling. Each vegetable requires a different blanching time. The internal temperature of each vegetable must be brought up to about 180° F. Water may not return to a boil if blanching time is very short. A practical test for sufficient blanching is to cut a few pieces of the product to see if it has been heated to the center. Beans with green and yellow pods will bend without breaking if blanched long enough. But avoid overblanching; it results in soft textures, destroys vitamins and nutrients and produces inferior flavors. Underblanching doesn't inactivate enzymes. Use the same water several times if you wish, but add more from time to time to keep it at the proper level throughout the blanching process.

3. *Increase blanching time if you live in high elevations.* At 2,000 to 4,000 feet, add ½ minute; 4,000 to 6,000 feet, add 1 minute; over 6,000 feet, add 2½ minutes to the time given.

Chill Quickly: Plunge basket or cloth bag of blanched vegetable into cold running water or ice water immediately (add ice as needed to keep water cold). Rapid chilling checks the cooking, saves nutrients and helps make for top quality. A practical test for sufficient chilling is to bite into a few pieces. If they feel cold to the tongue, drain the vegetable in a colander or wire basket and pack. It takes about as long to chill as to blanch vegetables. To work faster, pour hot vegetables from blanching basket into a second basket in the cold water; refill blanching basket immediately.

Use Steam Blanching for Broccoli: Some people believe they get a better product if they blanch broccoli with steam rather than in hot water, as the flowerets tend to disintegrate less with the steam method. However, steaming is not recommended for most vegetables and fruits because it is difficult to blanch thoroughly with the equipment available in most home kitchens. If the vegetables are not well surrounded with live steam, they will not be properly heated and the enzymes will not be destroyed. Pumpkin, winter squash and sweet potatoes sometimes are blanched in steam, as the flavor doesn't leach out into the steam as much as into boiling water, but most homemakers prefer to bake them before freezing.

To blanch with steam, place from 1 to 3″ water in the bottom of a large kettle. Bring to a full rolling boil. Place a single layer of the prepared vegetable in a wire basket or in a cheesecloth bag. Suspend over the rapidly boiling water, preferably on a rack. Keep the cover on the kettle during the steaming and start counting the time when the lid is placed on the kettle. Steaming takes longer than blanching in water.

Leave Head Space: Allow ½″ for pints and quarts in rigid freezer containers. Leave ½″ for dry pack and 1″ for liquid pack (as with syrup) in standard glass canning jars and pack firmly, but not tightly. Vegetables that pack loosely, like broccoli and asparagus, need no head space.

Directions for Freezing Vegetables
(IN ALPHABETICAL ORDER)

ARTICHOKES, GLOBE: Select small artichokes; pull leaves from them and cut top from buds. Trim stem to within 1″ of base. Wash thoroughly.

Blanch artichokes in boiling solution of 1 tblsp. citric acid crystals (or 1 tblsp. lemon juice) and 1 qt. water for 8 to 10 minutes, counting time when solution returns to boiling again. Cool 15 minutes under cold running water or in ice water. Drain and package.

Artichoke hearts may be frozen separately. Blanch them in the boiling solution 2 to 3 minutes. Cool like whole artichokes. Drain, package, label, date and freeze.

ASPARAGUS: Select young, bright-colored, brittle stalks that snap when broken and have compact tips. Harvest early in the morning if weather is warm. Break off woody stems. Wash in cold water. Remove a few bracts (scales) to see if sand and soil has washed out; cleaning asparagus may take many washings. Sort into medium and large stalks. Pack in lengths to fit package or cut in 2″ lengths, discarding woody ends.

For top quality, process as soon as possible after harvesting. Asparagus becomes tough or woody and loses vitamins after picking. The tips are rich in vitamin C and full of flavor.

Blanch medium stalks 3 minutes, large stalks (½ to ¾″ in diameter) 4 minutes. Chill in cold running or ice water. Drain, package, label, date and freeze.

BEANS, GREEN: Home economists in our FARM JOURNAL Test Kitchens experimented with various ways to freeze green beans, and their results prove the following two methods are best. It is especially important to choose varieties recommended for freezing in your area—check with your County Extension Office. Pick young, tender beans that snap when broken and that have small, tender seeds. Wash in cold running water and process them right after picking. (If you can't process them immediately, keep beans in ice cold water until processing time.) Snip off tips and sort for size. Cut or break beans in 1″ pieces, or leave them whole, if you wish.

Standard 3-Minute Blanch: Have ready a kettle with at least 1 gal. rapidly boiling water. Place 1 pt. beans in wire basket and immerse in boiling water. Cover and count time for 3 minutes. Remove basket from boiling water and immerse it in ice water to chill beans. Test coolness by biting into the vegetable. When it is cool to the tongue, it is ready to pack. Drain beans right away and package them. Label, date and freeze immediately.

Cooked, Frozen in Seasoned Bouillon: Combine 2 qts. water, 8 beef bouillon cubes, 1 tsp. onion powder and 1 tsp. salt in 8-qt. kettle; heat to boiling. Add 4 qts. cut green beans; cover and bring to a boil again. Boil 12 minutes. Remove cover and place kettle in ice water bath. Cool quickly, stirring occasionally. Fill containers with beans and liquid; *be sure liquid covers beans.* Seal and freeze. Makes 8 pints. To serve, put frozen block of beans and liquid into heavy saucepan. Cover; heat slowly until defrosted and liquid begins to boil (about 20 minutes).

BEANS, ITALIAN: Select any good garden variety. Wash in cold water and cut or break in 1 to 1½" pieces. Blanch in water 3½ minutes. Chill in cold running or ice water. Drain, package, label, date and freeze.

BEANS, LIMA: Harvest well-filled pods containing green, tender young beans. Do not freeze white beans; they are overmature. Wash and remove beans from pods, using scissors to snip tough pods. Do not wash after shelling. Discard blemished beans. Process them at once because shelled beans lose flavor quickly if allowed to stand.

Blanch small and medium beans 3 minutes, larger beans 4 minutes. Chill in cold running or ice water. Drain, package, label, date, freeze.

BEANS, SOY: Harvest well-developed pods containing green beans. Wash in cold running water.

Blanch in pods 5 minutes. Chill in cold running or ice water; shell, discarding blemished beans; package, label, date and freeze. You need not blanch them after shelling.

BEANS, YELLOW: Follow directions for green beans.

BEETS: Select smooth, tender beets of small to medium size. Remove green tops, leaving about 2" of stem; leave roots on; wash thoroughly. Discard blemished beets.

Blanch by cooking in water until tender, 15 to 25 minutes. Chill in cold running or ice water, slip off skins, roots and stems. Slice or dice larger beets; package, label, date and freeze.

BROCCOLI: Use tender, firm stalks with compact heads. Discard off-color heads and those that have started to blossom. Remove tough leaves and woody stalks. Cut through stalks lengthwise. When cut, heads should be about 1" in diameter. (The cutting makes uniform blanching possible and gives attractive servings when cooked.) Soak stalks, heads down, for ½ hour in salt water (¼ c. salt to 1 qt. cold water). This drives out the small insects. Rinse in cold water.

Blanch 4 minutes or steam 5 minutes. (For steaming see directions that precede.) Steaming usually is preferred. Chill in cold running or ice water; drain. Place heads and stalks alternately in the container to make compact package. Label, date and freeze.

BRUSSELS SPROUTS: Select firm, compact heads of good color. Wash, trim and discard discolored heads. Soak in salt water (¼ c. salt to 1 qt. water) ½ hour to drive out small insects. Rinse in cold water and drain.

Blanch medium heads 4 minutes; larger heads, 5 minutes. Chill in cold running or ice water; drain, package, label, date and freeze.

CARROTS: Harvest smooth, tender carrots before roots are woody. Plan plantings so you can harvest them in cool weather. The small, immature roots often harvested in hot weather contain less carotene and they rarely are of good quality when frozen. Remove tops, wash and scrape. Dice or slice ¼" thick.

Blanch 3½ minutes. Chill in cold running or ice water; drain, package, label, date and freeze.

CAULIFLOWER: Pick well-formed, compact white heads with fresh leaves. Trim, discard leaves and wash. Slit heads into pieces about 1" in diameter. Soak about ½ hour in salt water (¼ c. salt to 1 qt. water) to drive out small insects. Rinse in cold water and drain. Work fast to prevent discoloration. Blanch 4 minutes. Chill in cold running or ice water; drain, package, label, date and freeze.

CELERY: If freezer space is limited, it is questionable whether it is wise to freeze celery. But the green varieties may be frozen for use in cooked dishes. Trim, discarding tough and blemished stalks. Wash and cut in 1" pieces or finely dice.

Blanch 4 minutes. Chill in cold running or ice water; drain, package, label, date and freeze.

CORN, SWEET: Harvest early in morning if weather is warm. Freeze corn when at its "eating best." It remains at this stage of optimum maturity a short time, usually only about 48 hours. A practical test for maturity: press the thumbnail into a kernel. If the milk spurts out freely, the corn is at or near the stage of desired maturity. Immature corn kernels will be watery; when overmature, doughy.

Process as quickly as possible after harvesting. A delay of more than a few hours often results in loss of quality unless ears are refrigerated.

Husk ears, remove the silk and trim ends. Blanch, using a large kettle that holds 12 to 15 qts. boiling water. Keep kettle covered during blanching. For whole kernel corn, blanch the ear 4½ minutes. Cool in cold running water or ice water, drain and cut kernels from cobs at about ¾ depth of kernels. Package, label, date and freeze.

For corn on the cob, blanch small ears (up to 1½" in diameter) 8 minutes, and large ears (diameter more than 1½") 11 minutes. (Measure diameter of corn at large end, after trimming.) Chill quickly in cold running or ice water; drain and freeze on a tray or baking sheet. Then wrap in moistureproof packaging material and seal, label, date and freeze.

EGGPLANT: Harvest eggplant before it is too mature and when the seeds are tender. For top quality, select firm, heavy eggplant of uniform dark color. Peel and slice in ¼ to ⅓" slices, or dice. Drop pieces into cold water containing ¼ c. salt to 1 gal. water to prevent color loss.

Blanch eggplant in boiling salted water (¼ c. to 1 gal. water) 4½ minutes. Chill in cold running or ice water. Drain and package in layers separated by sheets of freezer paper. Label, date and freeze.

Many homemakers prefer to freeze cooked eggplant.

GREENS (spinach, beet, chard, collard, kale, mustard, turnip): Pick young, tender leaves early in morning if weather is warm. Remove large, tough stems and discard blemished leaves. Wash in cold running water.

Blanch 2 minutes, except collards and stem portions of Swiss chard. Steam these 3 to 4 minutes. Very tender young spinach is best blanched 1½ minutes. Chill in cold running or ice water; drain and package with the water that clings to leaves. Label, date and freeze.

A good way to chill greens after blanching is to swish them in ice water.

HERBS, GARDEN: Wrap a few sprigs of leaves in foil or seal in plastic bags and place in a glass jar or carton. Freeze.

Do not blanch leaves—just wash them thoroughly and drain.

KOHLRABI: Pick when young and tender. Cut off tops, wash, peel and cut in ½" cubes.

Blanch 2½ minutes. Chill in cold running or ice water. Drain, package, label, date and freeze.

MIXED VEGETABLES: Prepare each vegetable and blanch separately according to directions. Freeze the vegetables separately if desired, thaw just enough to separate the pieces, combine as desired; package, label, date and refreeze.

MUSHROOMS: Process young, firm mushrooms as soon as possible after picking. They bruise and deteriorate rapidly. Wash and remove base of stem. Freeze small mushrooms whole. Cut large ones in four or more pieces. To prevent browning, add 3 tsp. lemon juice or ½ tsp. powdered ascorbic acid to every quart of water used in blanching.

Blanch medium or small whole mushrooms 4 minutes; cut pieces, 3 minutes. Chill in cold running or ice water, drain and package. If the mushrooms are very mild in flavor, steam them instead of blanching. Before steaming, put them in water containing 1 tsp. lemon juice to 1 pt. water for 5 minutes. Steam whole mushrooms not larger than 1" across for 5 minutes, small or quartered, 3½ minutes and sliced, 3 minutes. (See general directions for steaming earlier in this chapter.) Chill at once in cold running or ice water; drain, package, label, date and freeze.

Or cut washed mushrooms in slices ¼" thick and sauté in butter for 2 minutes. Cool quickly and pack. Pour excess butter over packed mushrooms. Package in meal-size amounts. Label, date and freeze.

OKRA: Select young, tender pods 2 to 4″ long. Remove stems and wash.

Blanch 3 to 4 minutes. The large-podded varieties grown in the West need to be blanched 5 minutes. Chill in cold running or ice water; drain, package, label, date and freeze.

ONIONS: Onions keep well in a cool, dry place. Usually, they are not home frozen, but you can freeze chopped onions successfully. They will hold from 3 to 6 months, but after that they tend to lose flavor. Select mature sweet Spanish or any good garden onion. Peel, wash and cut in quarters. Chop and blanch in water 1½ minutes. Chill in cold running or ice water. Drain, package, label, date and freeze.

PARSNIPS: Select smooth, firm roots of top quality that are not woody. In northern areas, parsnips may be harvested in spring or late fall. Remove tops, wash and peel. Slice, dice or cut in lengthwise strips.

Blanch 3 minutes. Chill in cold running or ice water; drain, package, label, date and freeze.

PEAS, CHINESE OR EDIBLE POD: Select any good variety with bright green, flat tender pods. Wash well and remove stems, blossom ends and strings; leave whole. Blanch in water 2½ to 3 minutes. Chill in cold running or ice water. Drain, package, label, date and freeze.

PEAS, ENGLISH OR GREEN: Pick bright green, crisp pods containing tender peas that are sweet, but not overmature. Peas are at their best for a short time, about 24 hours. If peas are hard to shell, blanch pods in boiling water for 1 minute. Blanch and shell a few at a time; you need not wash them after shelling. Discard small, undeveloped peas. Or before scalding, you may separate overmature peas from tender ones by floating the peas in cold salt water—½ c. salt to 1 gal. water with 55° F. temperature. After 10 seconds, remove the floaters, which are the tender peas; the overmature will sink. *Avoid delay between shelling and freezing to prevent toughening of skins.* Some homemakers like to mix 2 to 3 tsp. sugar in 3 to 3¼ c. peas (1 lb.) after blanching and chilling.

Blanch shelled peas 1½ to 2 minutes. Chill in cold running or ice water; drain, package, label, date and freeze.

PEAS, FIELD (like Black-eyed): Follow directions for green peas, except blanch them 2 minutes.

PEPPERS, GREEN AND PIMIENTOS: Pick crisp, well-developed peppers of good color, green or red. Wash, cut out stem; remove seeds from green peppers. Cut in halves, dice or slice. Pimiento peppers may be peeled by roasting them in a hot oven (400° F.) 3 to 4 minutes or until peel is charred. Cool peeled pimientos and pack dry. Label, date and freeze.

Blanch green pepper halves 3 minutes; sliced and diced, 2 minutes. Chill in cold running or ice water; drain, package, label, date and freeze.

Peppers lose crispness in freezing but are satisfactory for cooked dishes. Diced peppers are one vegetable that may be packed and frozen without blanching.

PEPPERS, HOT: Wash, drain, package without blanching, label, date and freeze.

POTATOES, FRENCH FRIES: Peel and cut potatoes in thin strips, about ⅜″ wide and ⅜″ thick. Fry in hot fat (370° F. on deep-fat-frying thermometer) until potatoes are a *light* brown. Drain on paper towels spread in a baking pan. Cool and package. Label, date and freeze immediately. To serve, spread potatoes on a baking sheet and place in moderate oven (350° F.) 8 to 10 minutes or until brown. Or spread in broiler pan and broil 8 to 10 minutes. Salt to taste.

POTATOES, SWEET: Pick firm, smooth roots of bright color. Wash and bake in moderate oven (350° F.) until soft. Cool, peel and slice into ½″ slices. To keep the bright color, dip slice in solution made by dissolving ¼ c. lemon juice in 1 pt. cold water. For candied sweet potatoes, drain slices and roll in granulated or brown sugar. The color is less bright when they are rolled in brown sugar.

Or freeze sweet potatoes in purée form. Steam or bake them, cool, remove pulp from skins and mash or put through food mill. Add 2 tblsp. lemon juice for every 10 c. purée to help preserve color.

Or make pie filling by your favorite recipe (omit cloves) and freeze.

POTATOES, WHITE OR IRISH: Select any good potato. Wash, peel, cut out deep eyes and any green spots (sun or light burn). Cut in ¼ to ½″ cubes. Blanch 5 minutes. Chill in cold running or ice water. Drain, package, label, date and freeze.

Also see recipes in this chapter for other frozen potato treats.

PUMPKIN: Select good pie pumpkins. Pick at maximum maturity, indicated by good color and a stem that breaks easily from vine. Wash, cut or break into uniform pieces, remove seeds and bake in moderate oven (350° F.) or steam until tender. Cool, scoop pulp from skins and mash or put through food mill. Package, label, date and freeze.

Or make pie filling by your favorite recipe (omit cloves) and freeze.

RUTABAGAS: Freeze tender, young rutabagas. Wash, remove tops, peel and slice, or dice in ¼″ cubes.

Blanch 3 minutes, chill in cold running or ice water; drain, package, label, date and freeze.

SPINACH: See Greens.

SQUASH, SUMMER (Crookneck, Zucchini and Straightneck): Pick when small, 5 to 7″ long, and while rind is tender and seeds are small. Wash, but do not peel. Cut in pieces not more than 1½″ thick.

Blanch 3 minutes for ¼″ slices, 6 minutes for 1½″ slices. Chill in cold running or ice water; drain, package, label, date and freeze.

SQUASH, WINTER: Pick fully ripe or mature squash, with shells hard enough so that you cannot push your thumbnail through them. "Dry" types are recommended. Wash and cut or break into fairly uniform pieces and remove seeds. Bake in moderate oven (350° F.) or steam until tender. Cool, scoop out pulp and mash or put through food mill. Package, label, date and freeze.

Or make pie filling by your favorite recipe (omit cloves) and freeze. Some homemakers like to mix squash and pumpkin purées.

TOMATOES: Whole tomatoes may be wrapped and frozen. Use them for cooking within 2 months. Skins may be tough, but you can remove them. For best results, stew the tomatoes before freezing, but do not add bread or crackers until you heat the frozen tomatoes for serving. Stew the tomatoes and cool them by placing saucepan in larger pan containing ice water. Package, label, date and freeze.

You can freeze uncooked tomato pulp and store it for a few months without great flavor loss.

TOMATOES, HUSK (Ground Cherries): Husk, blanch 2 minutes, chill and pack in sugar syrup. Use 3 c. sugar to 1 qt. water. Package, label, date and freeze.

TURNIPS: Harvest young tender turnips. Remove tops, wash, peel, slice, or dice in ½″ cubes. Blanch 2½ minutes; chill in cold running or ice water; drain, package, label, date and freeze.

VEGETABLE PURÉES: Blanch the prepared vegetables as directed, cool, drain and put through food mill or purée in blender. Package, label, date and freeze.

How to Cook Frozen Vegetables

Cook all vegetables, except corn on the cob, without thawing. (Or thaw vegetables frozen in a solid block just enough so that you can separate the pieces with a fork.) Drop vegetables, except purées, into boiling water over high heat. You will hasten the cooking if you will break the vegetable block apart with a fork as the hot water thaws it. When the water returns to a boil, reduce the heat, cover and simmer until the vegetable is just tender. The time depends on the vegetable, the variety, maturity, the size of the pieces, the length of the blanching period in preparation and the degree

of thawing before cooking. Most vegetables will cook in from 4 to 15 minutes. Time is slightly longer at high altitudes.

Use as little water as possible in cooking frozen vegetables. Usually from ¼ to ½ c. is adequate for 1 pint. Add salt and other desired seasonings to the water. Plan the cooking time so that the vegetable may be served piping hot when cooked to avoid a loss of food nutrients, attractiveness and flavor.

Partially or completely thaw frozen corn on the cob before cooking. Cook it like other frozen vegetables only a very short time, 3 to 6 minutes, because the kernels were practically cooked when blanched. It is especially important to serve at once or corn on the cob will be soggy. Cook second servings separately instead of keeping them warm in hot water.

Usually enough water clings to the leaves of home-frozen greens for cooking, but this is not always true of those commercially frozen. Always follow package directions for cooking commercially frozen vegetables from retail markets. Read package labels. To cook frozen vegetables in a microwave oven, follow manufacturer's directions.

Place puréed or mashed vegetables in top of a double boiler and heat or cook over boiling water; do not add water to the vegetable itself. If it seems dry, add a little melted butter or margarine, or milk. Season to taste. Use imagination.

ORANGE-GLAZED CARROTS

Quick to get in freezer, no work to freeze. Glistening glaze seasons

1½ lbs. carrots, peeled and cut in
    strips
2 tsp. flour
¼ c. brown sugar, firmly packed
½ tsp. salt
1 tblsp. vinegar
1 tblsp. lemon juice
½ c. orange juice
1 tblsp. grated orange peel
2 tblsp. butter

• Blanch carrots in boiling water 5 minutes. Drain.
• Blend together flour, sugar and salt. Add vinegar, juices and orange peel.

Bring to a boil. Add butter; cook 5 minutes.
• Pour over carrots in 1-qt. foil-lined pan; freeze. Remove from pan; wrap, label, date and store in freezer. Makes 6 servings.
Recommended storage time: 6 weeks.
• To serve, turn carrots upside down in same pan; bake, covered, in a moderate oven (350° F.) about 1 hour, until tender. Uncover last 20 minutes.

BROCCOLI SOUFFLÉ

A dish that rises to the occasion—a guest luncheon special

3 tblsp. butter or regular margarine
3 tblsp. flour
1 c. milk
¼ tsp. salt
⅛ tsp. pepper
½ lb. (about 2½ c.) grated process
    cheese
1 (10 oz.) pkg. frozen chopped
    broccoli, partially thawed
½ c. finely chopped onion
3 eggs, separated

• Melt butter in saucepan, stir in flour to make smooth paste; add milk, salt and pepper; cook 5 minutes.
• Add cheese; stir until melted. Fold in broccoli and onion.
• Fold in well beaten egg yolks. Lightly fold in egg whites, beaten until stiff but not dry (you'll still see some white pieces). Pour into foil-lined 2-qt. baking dish with straight sides.
• Cover and freeze quickly. When frozen, remove from dish; store in plastic freezer bag; return to freezer. Recommended storage time: 6 to 8 weeks.

• To serve, remove foil; place in same dish and set in shallow pan filled with ½" hot water. Cover with foil and bake in moderate oven (350° F.) about 45 minutes, then uncover and bake about 1 hour, until puffy and browned. Makes 6 to 8 servings.

Note: Takes less baking time if you freeze this in individual baking cups; makes about 8 cups. To bake, set cups in shallow pan filled with ½" hot water; bake in moderate oven (350° F.) about 45 to 50 minutes.

To bake without freezing, bake in unlined dish set in pan of hot water in moderate oven (350° F.) for 1 hour.

BAKED STUFFED ZUCCHINI

Perfect example of why families of Italian descent like this squash

8   small to medium zucchini
2   medium onions
1   clove garlic
12 sprigs parsley
3   tblsp. olive oil
1   c. Swiss chard or spinach,
       cooked and drained
1   tsp. orégano leaves
1½ tsp. salt
⅛   tsp. pepper
½   c. grated Parmesan cheese
3   eggs, beaten
⅔   c. dry bread crumbs

• Scrub zucchini; cook in boiling water 5 minutes. Drain and cool. Cut in halves lengthwise; scoop out center pulp, leaving ¼" shell all around.
• Chop onions, garlic and parsley in blender or food chopper. Sauté in olive oil.

• Put zucchini pulp and chard through food chopper or in blender; drain off excess liquid. Add to onion mixture and sauté a few minutes. Add seasonings and cheese; mix well. Add eggs and blend; then add crumbs.
• Sprinkle zucchini shells lightly with salt. Fill with pulp mixture. Sprinkle lightly with more bread crumbs. Makes 6 to 8 servings.
• To freeze, place on a tray. When frozen, pack, label, date and freeze. Recommended storage time: 6 to 8 weeks.
• To serve, take from freezer and place in covered baking pan. Bake in moderate oven (350° F.) 45 minutes. Uncover and bake 10 minutes longer.

HAM/VEGETABLE CASSEROLE

Frozen vegetables win praise when teamed with ham and baked this way

2   c. frozen corn or 1 (10 oz.) pkg.
2   c. frozen peas and carrots or
       1 (10 oz.) pkg.
2   c. frozen succotash or 1
       (10 oz.) pkg.
1   tsp. salt
¼   tsp. pepper
3   tblsp. chopped onion
1   can condensed cream of
       mushroom soup
Ham slices, individual servings,
       cut ¼" thick

• Place frozen vegetables in greased 3-qt. casserole. Sprinkle with salt, pepper and onion. Spread soup over the top.
• Cover with foil and bake in a hot oven (400° F.) 40 minutes. Stir veg-

etables, spreading them evenly over the bottom of the casserole.

· Cover with ham. Return to oven and bake, uncovered, 10 minutes. Turn meat and continue baking 10 minutes. Makes 8 servings.

VARIATION

BACON/VEGETABLE CASSEROLE: Substitute 10 to 15 slices frozen Canadian-style bacon or fresh side meat (lean) for the ham.

Heat-and-Eat Frozen Potato Treats

With so many meat-and-potato men in the United States, it's not surprising that home economists in our FARM JOURNAL Test Kitchens received many letters asking for recipes for potato dishes that freeze successfully. Our Heat-and-Eat Potato Treats are their delicious answer to the questions. The trick is to cook and freeze the potatoes in ready-to-use shapes so you can take them from the freezer and place in the oven at once to reheat. It is important not to let the food thaw before heating. Thawing makes potatoes mushy.

STUFFED BAKED POTATOES

· Scrub 6 medium-size baking potatoes; puncture skin with fork to prevent bursting. Bake in hot oven (400° F.) until done, 45 to 60 minutes.

· Cut slice from top of each potato; scoop out inside; mash. Add ½ c. milk or cream, 3 tblsp. butter and salt and pepper to taste.

· Refill potato shells. Cool quickly; place in flat containers and wrap, label, date and freeze. Or freeze on trays; store in plastic freezer bags.

· To serve, bake unthawed in moderate oven (350° F.) 30 minutes; brown last few minutes at 400° F.

OVEN-FRIED POTATO SLICES

· Slice 5 medium, peeled potatoes ¼″ thick; blanch 2 minutes; drain.

· Pour ⅓ c. salad oil in each of two large shallow pans. Add half of potatoes to each; turn to coat slices with oil. Arrange one layer deep.

· Bake in very hot oven (450° F.) until light brown, 20 to 25 minutes, turn once. Drain. Makes 6 servings.

· Cool quickly; freeze on baking sheets. Store in plastic freezer bags; label, date and freeze.

Recommended storage time: 1 month.

· To serve, spread frozen slices on baking sheets. Heat in hot oven (400° F.) 5 to 10 minutes, drain on paper toweling to remove excess fat. Add salt and pepper to taste.

FROZEN HASHED BROWN POTATOES

Cook two skilletfuls of these golden, crisp-crusted potatoes for 8 or 9

To Freeze: Boil baking-type potatoes in their jackets until just tender but still firm (10 to 15 minutes). Drain, cool and peel. Grate potatoes on a coarse grater.

· Line a 10″ skillet with aluminum foil, bringing foil up to cover sides. Mix 1½ tsp. salt with 4 c. grated po-

tatoes. Pack in foil-lined skillet, pressing down firmly. Remove from skillet with foil. Seal, label and freeze. Repeat to stock your freezer.
Recommended storage time: Up to 2 months.

To Cook: Heat ½ c. shortening over medium heat (350° F. in electric skillet) in same skillet in which potatoes were shaped for freezing. Remove foil and add the disk of frozen potatoes (shortening will spatter, so quickly cover the skillet). Cook 5 minutes over medium heat. Uncover and continue cooking 10 to 12 minutes, or until potatoes are browned on the bottom. Cut into 4 wedges; turn each piece separately with spatula or pancake turner. Continue cooking 5 minutes, or until attractively browned. Makes 4 generous servings.

BASIC MASHED POTATOES

Freeze these plain or make interesting variations

4   lbs. boiling potatoes
**1   c. milk (amount varies with
        moisture in potatoes)**
¼   c. butter or regular margarine
1½ tsp. salt

• Peel potatoes. Boil until soft; drain. Press potatoes through a ricer or mash.
• Heat milk, butter and salt together.

Gradually whip in until the potatoes are smooth and fluffy.
• Form into shapes that are ready to use with no thawing. Freeze by directions in variations that follow.
Recommended storage time: Up to 2 months.

VARIATIONS

POTATO PUFFS: Add ½ c. grated Cheddar cheese or 2 egg yolks to Basic Mashed Potatoes. Chill. Form into balls; roll in mixture of ¾ c. corn flake crumbs and 3 tblsp. toasted sesame seeds. Freeze on tray until firm. Package, label and return to freezer. Makes 48 small or 24 medium balls.

To Serve: Place frozen puffs on baking sheet. Brush lightly with melted butter. Bake in hot oven (400° F.) 20 minutes for small puffs and 30 minutes for medium puffs. To serve without freezing, omit brushing with butter and bake puffs only until they brown, 5 to 10 minutes.

SNOWCAPS: Add 2 egg yolks to Basic Mashed Potatoes. Spoon hot potatoes in mounds on baking sheet. Cool; freeze until firm. Remove from baking sheet; place in plastic bags. Seal, label and return to freezer. Use to top meat and vegetable casseroles. Makes 24 snowcaps.

Instant Vegetable Sauce Cubes

We have developed a variety of sauce cubes you can make ahead and freeze in plastic bags. When your vegetable is almost cooked, toss in a handful of frozen cubes and stir them into the liquid. Here are directions for making these flavorful cubes and some ways to use them.

TOMATO SAUCE CUBES

Four of these frozen cubes add to-mato-basil flavor to green beans

¼ c. butter or regular margarine
½ c. flour
1 (6 oz.) can tomato paste
1 tsp. prepared mustard
2 tsp. sugar
2 tsp. onion powder
2 tsp. salt
½ tsp. dried basil
¼ tsp. pepper
¼ tsp. garlic powder
1 tblsp. olive oil
¾ c. water

• Melt butter in saucepan; remove from heat. Add flour and stir until moistened. Add remaining ingredients, except water; stir until smooth. Gradually stir in water.
• Pour sauce into 8½ × 4½ × 2¾" loaf dish. Freeze until consistency of ice cream. Cut into 32 cubes; remove to chilled tray. Freeze until solid. Package, label, date and return to freezer. Recommended storage time: 1 month.
• To use, cook 1-pt. pkg. frozen beans in ½ c. lightly salted water until tender. Add 4 Tomato Sauce Cubes. Stir and continue cooking until sauce is smooth and thick.

CHEESE SAUCE CUBES

Try this for a satin-smooth sauce

8 oz. sharp process cheese
¾ c. water
½ c. butter or regular margarine
½ c. flour
1½ tsp. salt
1 tsp. dry mustard
2 tsp. Worcestershire sauce
½ c. instant nonfat dry milk
2 tblsp. water

• Grate cheese; add to ¾ c. water in saucepan. Heat over low heat, stirring occasionally, until melted and smooth.
• Melt butter in another saucepan; remove from heat. Add flour; stir until smooth. Stir in salt, mustard and Worcestershire sauce. Add melted cheese mixture; stir until smooth. Add milk and 2 tblsp. water; stir until smooth.
• Pour sauce into 8½ × 4½ × 2¾" loaf dish. Freeze until consistency of ice cream. Cut into 32 cubes; remove to chilled tray. Freeze until solid. Package, label, date and return to freezer. Recommended storage time: 1 month.
• To use, cook 1-pt. pkg. frozen lima beans in ¾ c. lightly salted water until tender. Add 8 Cheese Sauce Cubes. Stir and cook until sauce is smooth and thick.
• To prepare sauce separately for broccoli or cauliflower, stir 6 cubes into ¾ c. milk. Cook over medium heat until thick and smooth. Spoon over vegetable. Makes ¾ cup.

LEMON BUTTER SAUCE CUBES

Cook this sauce separately from the vegetable. Spoon over asparagus or broccoli just before serving

½ c. butter or regular margarine
¼ c. flour
1 tblsp. sugar
2 tsp. salt
1 tsp. nutmeg
½ tsp. grated lemon peel
3 tblsp. lemon juice
¾ c. water
Few drops yellow food color

• Melt butter; add flour; stir until moistened. Stir in remaining ingredients. Cook, stirring constantly, until mixture begins to bubble and thicken.
• Pour sauce into 8½ × 4½ × 2¾″ loaf dish. Freeze until consistency of ice cream. Cut into 32 cubes; remove to chilled tray. Freeze until solid. Package, label, date and return to freezer. Recommended storage time: 1 month.
• To use, combine 8 cubes and ¼ c. water in saucepan. Heat and stir until smooth and thick. Keep hot over low heat—don't allow to boil. Use with 1 (10 oz.) pkg. asparagus or broccoli.

CURRY SAUCE CUBES

A hint of curry enhances delicately flavored peas

½ c. butter or regular margarine
2 tsp. salt
¾ tsp. curry powder
½ tsp. onion powder
⅔ c. flour
¾ c. instant nonfat dry milk
¾ c. warm water
2 tblsp. grated Parmesan cheese

• Melt butter; remove from heat. Add salt, curry and onion powders; stir to dissolve. Add flour; stir until moistened. Gradually stir in milk and water. Stir in cheese.
• Pour into 8½ × 4½ × 2¾″ loaf dish. Freeze until consistency of ice cream. Cut into 32 cubes; remove to chilled tray. Freeze until solid. Package, label, date and return to freezer.
Recommended storage time: 1 month.
• To use, cook 1-pt. pkg. frozen peas in ⅓ c. lightly salted water until tender. Add 8 Curry Sauce Cubes. Stir and continue cooking until sauce is smooth and thick. Do not overcook delicate peas. (For less sauce, use ⅓ c. water and 4 cubes.)

Vegetable Seasoning Ideas

Imaginative country cooks collect and use seasoning ideas to give cooked frozen vegetables unusual and intriguing flavors. Here are favorites from farm kitchens. You start with the hot, cooked vegetable.

ASPARAGUS: Instead of buttering it, pour on heated Italian salad dressing.

BROCCOLI: Dress with butter or cream sauce seasoned with a suggestion of curry powder. Or add chopped pimiento-stuffed olives to the sauce.

CAULIFLOWER OR GREEN BEANS: Butter the hot vegetable and scatter crisp, brown bread cubes on top. To fix the bread, cut either rye or white bread slices into tiny squares. Melt a small amount of butter in a skillet and add the bread. Cook, stirring to brown evenly. Season with salt and pepper.

CORN: Butter as usual but sprinkle with chili powder, garlic salt or both.

GREEN BEANS: Sauté ¼ c. cut-up almonds or cashew nuts and 2 tblsp. finely chopped onion in 2 to 3 tblsp. butter. Toss with 1 (10 oz.) pkg. green beans, cooked. Or toss buttered hot beans, with thin strips of cooked ham just before serving.

LIMA BEANS: Add canned tomatoes and a touch of orégano. Heat piping hot. This dish waits patiently if dinner is delayed. Or add 2 tblsp. finely chopped pimiento, 1 tblsp. butter and ½ c. dairy sour cream to 2 c. cooked lima beans. Heat and season wth salt and pepper.

PEAS: Stir finely chopped chives into the cream sauce before combining with the peas. Or add 2 to 3 tsp. mint jelly to buttered peas.

SPINACH: To hot, drained, chopped and buttered spinach, add a little light cream pointed up with a dash of prepared horseradish. For company, garnish with hard-cooked egg slices. Or sprinkle buttered spinach with lemon juice and grated onion.

SWEET POTATOES OR SQUASH: Fold a little grated orange peel into the mashed vegetable. Or season sweet potatoes with butter and maple syrup.

Slick Freezing Tricks with Vegetables
GOOD IDEAS FROM FARM JOURNAL READERS

GREEN TOMATOES: Slice and dip in cornmeal; package and freeze. To serve, pan- or deep-fry without thawing. Excellent with pork.

CUCUMBERS: Peel, seed and put them through food chopper. Pack in ice cube trays, freeze. Wrap each cube and place in a freezer. *Do not thaw;* crush cubes at last minute to toss in salads for summertime flavor.

SWEET POTATOES: When we dig sweet potatoes, some are bruised or cut. I trim, cook and freeze them. It's quite a saving and the salvaged sweets make many tasty dishes.

VEGETABLES IN BULK: After blanching corn, carrots, peas and snap beans I spread them on trays and freeze them. Then I pour the frozen vegetables into plastic freezer bags or other containers, seal and return them to the freezer. That way I can take out as many cupfuls as I need for a meal. When you freeze corn on trays, the freezer has a strong "corn odor." If the food in the freezer is well wrapped, it will not be affected.

BAKED BEANS: I bake 5 pounds of beans at a time, using two huge bean pots and baking them all night at 300° F. By morning, they are deep brown and luscious. When cooled, I put them in the freezer.

THANKSGIVING DINNER: A couple of weeks or so in advance, I freeze cooked squash in a casserole, cranberry/orange relish, and apple, pumpkin or mince pies. They are ready to combine with turkey for dinner.

CELERY: When cleaning celery for table use, save the leaves and little scraps. Put them in a plastic freezer bag, twist to exclude as much air as possible, seal and freeze. You will always have a touch of celery to season poultry stuffing, soups, stews and Chinese dishes. A great convenience!

PARSLEY: When there is plenty of parsley in the garden, I grind it in the food chopper (fine blade), saving the juice that is extracted. I freeze the juice in an ice cube tray, adding a little water to it. When frozen, I wrap each cube in foil and store in the freezer. Handy flavor for adding to soup, white sauce and gravies in winter.

CUTTING CORN FROM COB: When you are cutting corn from the cob to freeze, stand the tip of the ear in the center hole of a tube cake pan. The deep pan catches the corn and it's easy to hold the ear steady while cutting kernels. Speeds up the work!

CHAPTER 3

Meat, Poultry, Fish
and Game

Directions for Freezing Beef, Veal, Lamb, Ground Beef, Pork Sausage, Ham, Bacon, Game Animals, Lard · Recipes for Barbecued Ribs, Barbecued Brisket and Swedish Flank Steaks · Directions for Freezing Chickens, Turkeys and Other Poultry · Game Birds · Freezing Cooked Poultry, Fish and Shellfish

Farm people with very long memories can recall when fresh beef and pork were pretty scarce in summer. A Texas rancher says, "We had so much chicken and salt pork I used to get hungry for winter." But locker plants and home freezers have changed farm menus forever. Today's country woman simply steps to her freezer to make selections from her storehouse of meat, poultry, fish and game. The suburban homemaker, too, uses her freezer to stockpile meat, for convenience and to take advantage of special buys at the market.

Locker plants and other commercial establishments slaughter, chill and quick-freeze much of the beef, veal, pork and lamb farm people take home to their freezers. Other foods have to surrender space to meats in the home's zero-storage plant—these perishable protein foods are money! As the supply runs low, thrifty women fill the space with short-term casseroles and baked goods—some from the supermarket, some made at home from the good recipes in following chapters.

Sportsmen freeze fish after a big catch; venison, when they shoot a deer. When animals are slaughtered, farmers cut roasts, steaks and chops the sizes they want and package cuts in the most useful way—some packs family-size, bigger packs for company. Thoughtful packaging can save cooking time later. Frozen chicken, meaty pieces separated from bony pieces, insure good meals with less work. (Old-timers remember the frantic beginning of Sunday dinner, the way it used to be: First, catch the chicken . . .)

While most meats, poultry and fish are frozen fresh, country cooks like to have a few main dishes cooked or partly cooked, ready for reheating on days when there's almost no time to spend in the kitchen. You'll find recipes for fix-ahead meat dishes in Chapter 7. A few of our favorite ways to fix venison, brisket, short ribs and flank steak are included in this chapter. And if you render lard at home, be sure to notice how to handle it so it will stay fresh and sweet in or out of the freezer.

Freezing Meats

While beef, veal, pork and lamb often are produced on the farm and the meats go into home freezers, most farmers find it advisable to have animals slaughtered, chilled and frozen in locker plants or other commercial establishments. There they can be handled in a sanitary way and inspected by local authorities. And the animals can be slaughtered and frozen throughout the year as needed.

The home freezer is not designed to freeze several hundred pounds of meat at a time. If you want to freeze more than 5% of the freezer's capacity at one time, it is best to have the freezing done at a locker plant and to bring the frozen meats home for storage.

The quality of frozen meat depends on (1) the original selection of good quality meat, (2) the way it is handled when prepared for the freezer, (3) use of the right wrapping materials and (4) always keeping the freezer temperature at zero or below. Recommended storage time: Unground beef, 9 months; ground beef, 4 months; lamb, 9 months; pork, 4 months; veal, 6 months; sausage, unsalted, 2 to 3 months.

Animal Selection: Select healthy, quality animals with adequate finish (fattening). Though freezing for several weeks has a tenderizing effect on meats, remember that *a tough steak will never become a tender one in the freezer.* Don't expect the impossible.

Slaughtering: If you do this yourself, use sanitary methods spelled out in Farmers' Bulletins published by the U. S. Department of Agriculture. There are separate bulletins for beef and pork; ask your County Extension Office.

Chilling:

1. *Beef.* Hang the meat and chill it quickly to minimize the development of off-flavors and odors. Chill the carcass at 32 to 34° F. After chilling, age the hind quarters in temperatures of 36 to 38° F. from 10 to 12 days, forequarters 6 to 8 days. Do not age beef as long as for use unfrozen. Age young beef of standard and utility grades or lower quality 5 to 6 days.

2. *Pork.* Hang the carcass whole to chill with body cavity open and the leaf fat pulled loose but left attached to the hams. Chill 24 hours at 34° F. Failure to chill pork quickly may result in souring at the ham bone, which causes spoilage during the curing. Cut and package pork as soon as it is chilled to the bone, always within 3 days after slaughtering. Meat from hogs fed fish or fish meal develops rancidity more quickly when frozen.

3. *Veal and lamb.* Hang carcasses whole during chilling. Allow 10 to 16 hours for them to reach an internal temperature of 38° F. Veal, like pork, should be cut, wrapped and frozen as soon as it chills to the bone. Lamb and mutton may be aged like beef, but for less time—6 to 7 days.

Cutting: There is no one best way to cut up a carcass—it will depend on how you want to use the meat. If you employ a professional meat-cutter, tell him what you want so he can make cuts the right size for your family meals. Remove excess fat and bone. Fat becomes rancid quicker than lean meat; the bones waste freezer space and make wrapping more difficult. Meat tastes just as good without them and boning will save from 5 to 35% in storage space. Do not make packages larger than necessary. They are awkward to handle and take longer to freeze and thaw than smaller ones.

Wrapping Materials: Poor wraps often cause unpleasant flavors in frozen meats. Use moisture-vaporproof materials to prevent freezer burn or the light gray spots that sometimes form on meats. Be sure the wraps also efficiently exclude oxygen to slow down the development of rancidity. Use a pliable material that will make a tight wrap and crowd out air pockets. The wrapping needs to be odorless, strong when wet, grease-proof and it should not adhere to the meat. Among the best materials for packaging meats are heavy-duty aluminum foil, plastic wrap and freezer paper laminated to moisture-vapor-proof plastic or foil. A single layer of the wrapping material is adequate

if well sealed. Usually a packaging material 18″ wide is satisfactory, although 24″ width is needed for large cuts.

Wrapping: The best method is one that excludes the most air, and usually the "drugstore wrap" is considered best. It also takes less wrapping material than the "butcher wrap." Both are described in Chapter 1 (see Index).

To Stack Meats: Several chops, steaks and ground meat patties often are packaged together for a meal. Place two thicknesses of freezer paper between pieces of meat. This allows quick and easy separation when used. If the paper is waxed, only on one side, put the shiny side next to the meat.

Ground Meats: Pack in rolls or blocks. Season if desired, but do not add salt to meat you'll keep longer than 2 to 3 weeks. Place double pieces of freezer paper, shiny side next to the meat, between small blocks and rolls for easy and quick separation and use.

Recommended storage time: 4 months.

Pork Sausage: Use of the proper packaging materials and spices helps keep sausage in good condition when frozen. You can buy sausage seasonings containing antioxidants at many retail food markets, feed stores and some locker plants. Smoked sausage, everything being equal, keeps longer than fresh sausage. Bologna does not freeze successfully—its texture changes.

Recommended storage time for sausage: With antioxidant added, 3 to 4 months. Unsalted sausage, *properly wrapped,* may keep that long without the antioxidant, but often it loses quality before then.

Ham and Bacon: Freezer life of these cured meats depends on their freshness, the cure and the smoking. Salt has an undesirable effect on frozen meats. Usually hams and bacon, unsliced, keep from 2 to 3 months when frozen. Whole or half hams and bacon sides hold up better than the slices when frozen and they may be sliced as needed with a power saw. Or you can freeze the fresh hams and pork sides and cure and smoke them as needed. Ham and bacon purchased in retail markets, when frozen, may keep a very short time, depending on its freshness.

Game Animals: Handle them much like beef, only age them a shorter time—5 to 6 days—especially if they are lean. Or if they are not in top condition when they reach home, cut, package and freeze them without aging. Bleed, dress and cool the carcass at once after killing. Sprinkle the inside body cavity with pepper to keep flies away. Trim off and discard portions damaged by gun shot. Hang to cool in a breezy place; it often is desirable to spread the ribs apart with

a stick to speed up the cooling. Since the hide helps to prevent contamination of the meat, usually it is best to leave it on. Wrap carcass loosely on the trip home to protect from dust. Recommended storage time for big game: 9 months.

To Thaw Meat: Leave it in the original wrapper. Thaw meat for 5 to 8 hours in the refrigerator.

To Cook Frozen Meats: Meats are equally flavorful and juicy whether cooked before or after thawing. Often it is convenient to start cooking thin cuts, like some steaks, chops and cutlets, while they are frozen or partially frozen. The cooking of roasts and steaks more than 1½″ thick will be more uniform and quicker if they are thawed or partially thawed first. Sometimes unthawed meat becomes too dark and overcooked on the outside and is undercooked inside.

Thaw ground meats if you want to shape them in patties or loaves. Thaw or partially thaw sliced meats, such as liver and cutlets, which you dip in flour before frying or broiling.

Cook thawed meats like fresh. To cook without thawing, allow extra time—10 to 15 minutes per pound for roasts and 13 to 23 minutes to total cooking time for thick steaks.

How to Render Lard

Trim the fat from the pork carcass and render it into lard as quickly as possible after slaughtering. Cook the leaf fat, backfat and fat trimmings.

Caul and ruffle fats from internal organs yield lard of darker color; if you use them, cook them separately.

Remove fat, wash, chill promptly.

Cut fat into small pieces or grind it to hasten the rendering. Place in a large, heavy kettle and *cook slowly,* starting with a small quantity to make stirring easier. When the fat begins to melt, add more pieces, but do not fill kettle or it might boil over. Stir frequently and cook slowly to avoid sticking and scorching.

As cooking begins, the temperature will stay at 212° F. As the water evaporates from the fat, the temperature will rise slowly. Do not let it go higher than 255° F.

During the cooking the brown cracklings will start to float. When the lard is almost rendered, they sink to the bottom of the kettle. *Be careful not to let them stick and scorch.* You can stop the cooking while the cracklings still float, but complete rendering removes more water and results in lard that keeps better. If moisture is eliminated by proper rendering, water souring should not develop during storage.

Let the lard cool slightly and settle before emptying the kettle. Dip the clear lard carefully into 5- or 10-lb. containers. *Store immediately in a cool place, at near freezing or freezing temperature.* Chilling it quickly produces a fine-grain lard.

Strain the lard at the bottom of the kettle through three thicknesses of cheesecloth to remove settlings. (Put cracklings through a press, if you wish.)

Keep lard in metal containers or greaseproof freezer cartons, sealed with a tight cover, and store in a dark cool place or in freezer or locker. Air and light often cause chemical changes that result in rancidity.

Note: You can greatly increase the storage life of lard by adding an antioxidant (available at most locker plants). Another way to prolong the fresh flavor is to stir 1 (3-lb.) can of hydrogenated vegetable shortening into every 50 lbs. of lard while it is cooling.

Favorite Beef Recipes

BARBECUED RIBS

You're lucky if you have this in the freezer when company comes

1 large onion, chopped
2 tblsp. salad oil
½ c. molasses
1 c. ketchup
½ c. water
¼ c. Worcestershire sauce
¼ c. prepared mustard
1 tblsp. salt
3 lbs. short ribs

• Sauté onion in oil; combine remaining ingredients, except ribs, and add to onion. Cook 10 minutes.

• Divide ribs for serving; brush pieces with sauce; put in shallow baking pan.

• Bake in moderate oven (350° F.) 1½ to 2 hours, or until tender. Brush on sauce every half hour and turn.

• Cool, wrap, label, date and freeze. Recommended storage time: 2 to 3 months.

• To serve, thaw and heat in moderate oven (350° F.) about 30 minutes. Makes 4 servings.

BARBECUED BRISKET

Leftovers will make good sandwiches

3 to 4 lbs. beef brisket
1 c. ketchup
1 c. water
¼ c. vinegar
1 tblsp. sugar
1 tblsp. prepared horseradish
1 tblsp. prepared mustard
1 tsp. salt
¼ tsp. pepper
2 onions, finely chopped
2 tblsp. chopped celery

• Place brisket in casserole.
• Mix remaining ingredients; pour over meat. Refrigerate overnight.
• Bake in moderate oven (350° F.) 1 hour per pound, or until tender.
• Cool. Place meat and sauce in freezer container; seal, label, date and freeze.
Recommended storage time: 2 to 3 months.
• To serve, thaw meat and heat, uncovered, in moderate oven (350° F.) 40 to 50 minutes, or until heated throughout. Makes 4 to 5 servings.

SWEDISH FLANK STEAKS

Rolled steak stuffed with ground meat

½ lb. veal, ground twice
½ lb. lean pork,
        ground twice
1 egg, beaten
2 tblsp. chopped parsley
½ small onion, grated
1 tblsp. light cream
2 tblsp. bread crumbs
1 tsp. salt
½ tsp. freshly ground pepper
½ tsp. Worcestershire sauce
2 lbs. flank steak, cut in pieces
        3 to 4" wide and 5" long
½ lb. salt pork, cut into strips
        ⅓" wide, 4" long

• Mix all ingredients except steak and salt pork. Spread each steak with stuffing; cross with one strip of salt pork. Roll the long way with ends of pork extending beyond roll.
• Tie with string; wrap, label, date and freeze.
Recommended storage time: about 6 to 8 weeks (depends on freshness of salt pork).
• To serve, defrost flank steaks and coat with mixture of ½ c. flour, 2 tsp. salt and ¼ tsp. pepper; place in skillet and brown in 2 to 4 tblsp. salad oil; transfer to baking pan.
    Sauté 1 c. sliced mushrooms in skillet, adding more oil if needed. Add 1 (12½ oz.) can chicken broth.
    Mix 3 tblsp. flour and ½ c. milk; blend into broth; cook until thickened.
    Pour over steaks; cover and cook in moderate oven (350° F.) 1 to 1½ hours, or until very tender. Makes 6 servings.

Note: You can use the large flank steak to make one big roll.

Exciting Venison Treats

Country cooks excel in cooking game. Here are examples of what they do with venison from their freezers. When we tested these recipes in our Countryside Kitchens, the men on our taste panel said they had never tasted better venison. Each is superior—try them all.

VENISON MEATBALLS IN SAUCE

Heavenly served over biscuits

 2   c. grated raw potatoes
1½ lbs. ground venison
 ⅔   c. chopped onion
1½ tsp. salt
 ⅛   tsp. pepper
 ¼   c. milk
 1   egg
 ¼   c. butter
 3   c. water
 2   to 3 tblsp. flour
 2   c. dairy sour cream
 1   tsp. dill seeds
 1   (10 oz.) pkg. frozen peas, cooked

• Combine potatoes, venison, onion, salt, pepper, milk and egg; shape into 1½" balls. Brown slowly in butter in large skillet. Add ½ c. water, cover and simmer until done, about 20 minutes. Remove meatballs.
• Stir in flour, then remaining water, simmer to thicken. Reduce heat; stir in cream and dill; add meatballs and peas. Heat but do not boil. Makes 8 servings.

VENISON TERIYAKI

Developed by Oregon State University

 2   lbs. venison sirloin or round (1½" thick)
 1   can condensed beef consommé
 ⅓   c. soy sauce
 ¼   c. chopped onion
 1   clove garlic
 2   tblsp. lemon juice
 2   tblsp. brown sugar
 1   tsp. seasoned salt

• Slice meat diagonally across grain about ¼" thick.
• Mix other ingredients for marinade; pour over meat; refrigerate overnight.
• Drain meat; broil 3 to 4" from heat about 5 minutes first side, baste with marinade; broil 3 minutes other side, basting (do not overcook). Heat marinade to pass with meat. Makes 6 servings.

VENISON/NOODLE SKILLET

Flavor improves with reheating

 1   lb. ground venison
 3   tblsp. shortening
 ½   c. diced onion
 ½   c. diced green pepper
 1   c. diced celery
 1   (1 lb.) can red kidney beans
 2   c. broad noodles, uncooked
 1   qt. tomatoes
 1   (4 oz.) can button mushrooms
 2   tsp. seasoned salt
 1   tsp. chili powder
 1   tsp. salt
 ⅛   tsp. pepper

• In large skillet brown meat in shortening; sauté onion, green pepper and celery until transparent.
• Add remaining ingredients; mix well. Cover tightly; bring to a boil. Reduce heat; simmer 20 minutes. Makes 8 servings.

VENISON STEAK CASSEROLE

Old-fashioned stew goodness

2 lbs. venison round steak
    (1" thick)
6 tblsp. flour
1 tsp. salt
½ tsp. pepper
⅛ tsp. orégano leaves
1 clove garlic, crushed
3 tblsp. shortening
6 medium potatoes, sliced
2 medium onions, sliced
2 carrots, sliced or in strips
1 green pepper, cut in squares
3 c. beef bouillon

• Cut steak into serving pieces.
• Combine ¼ c. flour, ½ tsp. salt, ¼ tsp. pepper, orégano and garlic; pound into meat. Brown in shortening; place in one layer on bottom of 3-qt. oval baking dish.
• Layer half of potatoes, onions, carrots and green pepper on top; sprinkle with half remaining salt, pepper and flour. Repeat. Pour bouillon over top.
• Bake, covered, in moderate oven (350° F.) 1 hour; uncovered, ½ hour. Makes 6 servings.

VENISON LIVER KABOBS

Traditional hunters' camp food

• Cut ½" thick slices of venison liver in 1½" squares. Sprinkle with salt and pepper. String on skewer alternating with bacon, pineapple chunks and 1½" squares of green pepper.
• Broil 3 to 5" from heat about 5 minutes per side; do not overcook. Brush often with mixture of ½ c. melted butter and ¼ tsp. garlic salt.

VENISON BACKSTRAP BARBECUE

This cut of meat lies along both sides of the backbone; it is equivalent to the rib eye in beef and may extend, if you wish, into loin. A camp favorite

• Strip out backstrap in one piece (remove bones if animal is large).
• Put on long skewer or lay on rack over broiler pan; wrap strips of bacon loosely around meat, fasten with toothpicks.
• Broil about 5 minutes per side for backstrap 2" thick. Brush often with your favorite barbecue sauce (reduce sugar one fourth), or mixture of ½ c. melted butter and ¼ tsp. garlic salt or 2 tblsp. lemon juice.

BUTTERFLY VENISON STEAKS: Cut raw backstrap into 2" slices; slice each piece in half sideways, cutting almost through; open flat to butterfly. Broil 3 to 4" from heat about 5 minutes per side, brushing often with barbecue sauce.

VENISON HERB ROAST

Crust keeps moisture inside roast

3 to 4 lbs. venison rump, loin or
    rib roast
1 tblsp. salad oil
½ tsp. salt
¼ tsp. pepper
¼ c. flour
2 tsp. marjoram leaves
1 tsp. thyme leaves
2 tsp. rosemary leaves
1 clove garlic, crushed
1 c. apple juice
1 c. water

• Dry meat well; cut several slits in meat about ½″ deep. Rub roast with oil, sprinkle with salt and pepper.

• Combine flour, marjoram, thyme, rosemary and garlic; pat mixture on roast and stuff into slits. Insert meat thermometer.

• Pour apple juice and water into shallow pan. Set roast in liquid (no rack); bake, uncovered, in slow oven (325° F.) about 1 hour until flour mixture adheres to meat. Baste frequently with liquid in roaster. Finish baking about 1 hour (or 25 to 30 minutes per pound total time), basting often. Meat thermometer should read 160° F. for medium, 170° F. for well done. Makes about 6 servings.

VENISON MINCEMEAT

Team with pastry to make famous Christmas pie of the Old West

4   lbs. venison "trim" meat with
    bones
Water
¾   lb. beef suet
3   lbs. apples, peeled and quartered
2   lbs. seedless raisins
1   (15 oz.) box seeded raisins
1   (12 oz.) box currants
1   tblsp. salt
1   tblsp. ground cinnamon

1   tblsp. ground ginger
1   tblsp. ground cloves
1   tblsp. ground nutmeg
1   tsp. ground allspice
1   tsp. ground mace (optional)
2   qts. apple cider, grape or other
    fruit juice
1   lb. brown sugar

• Trim fat from venison. Cover with water; simmer until meat is tender. Refrigerate venison in cooking liquid overnight. Remove all fat from top of liquid. Separate meat from bones and put meat through food chopper using coarse blade. (There should be enough ground venison to make at least 2 qts. ground meat.)

• Grind suet and apples.

• Combine all ingredients in large kettle. Simmer 2 hours to plump fruit and blend flavors. Stir often to prevent sticking.

• Cool, pack in pint containers. Label, date and freeze. Makes about 11 pints.

Recommended storage time: 3 months.

Note: You can also can this mincemeat. After simmering, pack and process like State-of-the-Union Mincemeat (see Index for recipe).

Freezing Chickens and Turkeys

Select the Best: Freeze healthy, well-finished birds that have grown rapidly. Young birds have tender flesh, but older ones, like mature hens, often have more flavor. Among the good freezers are broiler-fryer chickens at 8 to 12 weeks of age, roasters at 3 to 5 months, mature hens (stewing chickens), fryer-roaster turkeys at 12 to 15 weeks and young tom and hen turkeys at 5 to 7 months.

Starve before Killing: If practical, starve chickens and turkeys overnight or at least 6 to 8 hours before killing. When you draw them, there will be less danger of rupturing the digestive tract, contaminating the birds and producing off-flavors. Do not feed birds fish or fish oil for several weeks before killing.

Suspend to Kill: Hang birds by legs from a rope or shackle to avoid bruises. Since the head hangs down, the bleeding will be better. Stretch the head in a line with the neck and make a clean cut across the throat just back of the jaw. Use a sharp knife.

Scald to Remove Feathers Easily: When bleeding ceases, scald the birds by swishing them from 40 to 60 seconds in hot water, 140° F. Scalding removes the outer skin layer and the birds will dry out quickly if allowed to stand. *Keep them wet.* If you plan to cool the birds by air or think you will have delays in getting them into the freezer, scald them in cooler water, 125 to 130° F. (You will find the pin feathers more difficult to pull out if you use the cooler water.) Pull out the wing and tail feathers first. After feathers are removed, singe off the hairs, cut off head and feet and wash birds thoroughly inside and out.

Draw Birds Promptly: Freeze only drawn birds. Draw them at once after plucking feathers, to check development of off-flavors. Cut out the oil sac, which is on top of the tail. Then, cut a circle around vent below the tail, freeing it for removal with the internal organs. Make a crosswise slit large enough for drawing between this circle and the rear of the breastbone. If you want to freeze the bird whole, leave a band of skin between the two cuts.

Remove internal organs and vent. Slit skin lengthwise at the base of the neck. Then slit the skin down and remove the crop and windpipe. Cut the neck off short and save it. Leave neck skin on bird. Wash bird, inside and out, in cold running water. Trim off excess abdominal fat.

Trim and cut blood vessels from the heart. Cut away the gall sac from liver, being careful not to break. Cut through one side of the gizzard to the inner lining. Remove and discard inner lining and contents. Wash giblets.

Chill Thoroughly: Lower the temperature of the washed birds as quickly as possible to around 36° F. by packing them in crushed ice. Use plenty of ice, about 1 lb. to 1 lb. chicken or turkey. Or pack the birds in a slush of ice and water. It is especially important to chill the inside of the bird quickly for spoilage is likely to start there. Thorough chilling helps insure tenderness.

Age Whole Birds: Turkeys and large chickens will be more tender after freezing if you age them by icing them for 12 to 24 hours. After the first 3 or 4 hours of chilling, frequent draining is necessary to prevent birds from absorbing too much water and the flavors from leaching out. Add more ice as it melts.

Before packaging birds, drain them thoroughly for 10 to 15 minutes.

Fold the neck skin under the bird. Truss the wings and legs close to the body with butcher's twine.

Whole Chickens and Turkeys: Do not stuff birds before freezing. It takes time to heat the frozen dressing in frozen birds and food-spoiling and food-poisoning bacteria may develop. (This caution does not apply to commercially frozen chickens and turkeys which are handled by methods unavailable in home kitchens.)

To Freeze: Place the trussed bird in a moisture-vaporproof plastic freezer bag, press out all the air and fasten with a twist-seal. Or submerge the bag containing the chicken or turkey in water, leaving open end out; water presses air out of bag. Fasten with a twist-seal. Or mold heavy-duty foil tightly around the bird to exclude the air. Overwrap with freezer wrap. Freeze immediately at 0° F. or below.
Recommended storage time for whole birds: 9 months.

Package giblets separately to speed freezing, and use them within 2 to 3 months.

To Thaw: If the chicken or turkey is in a watertight package, you can thaw it in cold running water. Place it in a steady stream of water. It takes from 1 to 2 hours to thaw chickens weighing less than 4 lbs.; 2 to 2½ hours for chickens 4 lbs. or over; 4 to 6 hours for 4- to 11-lb. turkeys; 6 to 7 hours for 12- to 24-lb. turkeys.

Since running cold water is not plentiful in some homes, the USDA recommends thawing watertight packages of poultry in cold water. It is important to change the water several times.

It takes longer to thaw poultry in the refrigerator, but it requires little or no watching. You need to allow 12 to 16 hours for chickens under 4 lbs.; 24 to 36 hours for turkeys weighing 4 to 11 lbs.; 48 to 72 hours for 12- to 24-lb. turkeys.

Thaw frozen poultry in its original wrap. Cook it promptly when thawed or keep it refrigerated until cooking time.

To Cook: Roast thawed birds like freshly dressed.

Note: Do not be disturbed if there is a darkening around the bones of frozen, young birds. This is caused by a seepage of the hemoglobin from the bone marrow. It does not affect the taste of poultry and is harmless.

If you want to freeze a large number of chickens, turkeys or other poultry at a time, it may be best to have them frozen at a commercial freezing plant. This eliminates the danger of overloading the home freezer and getting a slow, uneven freeze. It also avoids the possibility of increasing the freezer temperature enough to affect the quality of the other frozen foods in the freezer.

Cut-up Chicken and Turkey: Birds for frying, or cooking in water will take up less space in the freezer if cut up before packaging. After the pieces are washed, dry them thoroughly. Paper toweling is excellent for this.

To Freeze: Sort the pieces, the meaty ones like the legs, breast and wings from the bony ones like the neck, back and wing tips. If you are processing several birds, package the different cuts separately. Place the meaty pieces close together in layers with 2 sheets of freezer paper between or lay each piece in a fold of freezer paper. Pack close together in moisture-vaporproof plastic freezer bags or cartons. Seal, label, date and freeze. Place broiler halves together with 2 pieces of freezer paper between. Use paper waxed on both sides. Wrap with foil, or place in moisture-vaporproof plastic freezer bags. Press out as much air as possible before tying with a twist-seal. Package bony pieces for making broth.

Recommended storage time for chicken pieces: 9 months.

Package giblets separately to speed freezing, and use them within 2 to 3 months.

Some homemakers freeze chicken in water; research indicates the flavor of the chicken is inferior to that packaged in plastic freezer bags and aluminum foil.

To Thaw Chicken Pieces: Leave them in the original wrapping. Thaw in the refrigerator until pliable or until completely thawed; complete thawing will take about 9 hours. Stewing chicken need not be thawed.

To Cook: Cook as you would freshly dressed poultry. Cook stewing hens without thawing if you desire.

Freezing Domestic Ducks and Geese

Handle like chickens and turkeys, but scald in warmer water, 160° F., for 1 to 1½ minutes, with ½ c. vinegar added to every 10 gals. of hot water. Many farm cooks find use of a commercial wax helpful in removing down and pin feathers. Pack and freeze like chickens and turkeys. Recommended storage time: Ducks and geese, 9 months.

To Thaw: It takes from 2 to 2½ hours to thaw 3- to 5-lb. ducks in cold running water, 1 to 1½ days in the refrigerator. Geese weighing 4 to 14 lbs. thaw in 3 to 5 hours in cold running water, 1 to 2 days in refrigerator.

Freezing Game Birds

Take good care of game birds. Remove body heat as quickly as possible after killing. Avoid heaping birds together, for this retains heat. Allow space between birds, rather than piling them in a car trunk, for air circulation and quicker cooling. Deterioration is rapid when birds are stacked. Often it is best to have birds frozen at a locker plant near the hunting grounds before returning home if it's a long trip. Observe state game laws for "tagging" and length of time in storage.

Remove feathers and clean birds the same way as poultry. Do not scald

wild ducks—pluck them dry. Scald pheasants in warm water, 155 to 160° F.

Freezing Cooked Poultry

It is handy to have cooked poultry in the freezer ready to use. Stewing is an excellent method to prepare birds for freezing. When stewing chicken, save the broth. Pour it into glass jars or other rigid, moisture-vaporproof containers and freeze.

To save freezer space, remove bones from stewed chicken. Chicken covered with broth or gravy keeps

longer when frozen than if no moisture is added.

Leftover roasted chicken and turkey, sliced or cut up, freezes successfully if packaged in moistureproof packages.

Recommended storage time: Slices or pieces of cooked poultry covered with broth or gravy, 6 months; without broth or gravy, 1 month. Fried chicken, 4 months.

Freezing Fish and Shellfish

Freeze fish as soon as possible after catching; it spoils quickly. Kill immediately if practical. If fish flop around in the bottom of a boat, they bruise themselves. Let the blood drain from the flesh, remove viscera and gills and pack in ice. Prepare all fish for freezing as for table use; wash in cold water and drain. Properly wrapped fish will not affect the flavor of other food in the freezer. Aluminum foil and plastic wrap are excellent wrappings for fish.

Small Fish: Freeze them whole. One of the best ways is to freeze them in a block of ice. Place the fish in a clean, watertight container, like a 2-lb. coffee can or loaf pan. Cover with water and freeze. If you use a

loaf pan, remove the block of frozen fish; package it in freezer wrap. There is no need to remove the frozen fish from a coffee can. Just adjust the lid. Hurry back to freezer before thawing starts. When ready to use, thaw in cold running water.

Large Fish: They will handle easier if cut into steaks or filets ready for the pan. Place two layers of freezer paper between steaks, filets or pieces of fish and wrap in meal-size packages for convenient use.

Ice Glazing: Whole fish, prepared for table use, may be glazed with ice for freezer storage. Place the fish on a tray and freeze. Remove from freezer, dip fish in near freezing water, and return to freezer. Repeat several times or until a coating of ice about ⅛" thick covers the fish. Wrap in aluminum foil or other materials that exclude the most oxygen (as for meats).

Freeze fish in the coldest part of the freezer, near the bottom of chest type and directly on the refrigerated shelves of upright types. Recommended storage time: 3 to 4 months.

Fish Roe: Wash thoroughly, place in moisture-vaporproof containers, label, date and freeze.

To Thaw and Cook Fish: Partially or completely thaw fish in its original wrapper. You can thaw fish in the refrigerator, there will be less drip. It will take about 20 hours to thaw a 4½-lb. whole fish, 8 hours for 1-lb. fish steaks or filets and 8 hours for a 1-lb. container or package of shellfish. If fish is in a watertight package, the best and quickest way to thaw it is to place it in cold running water. It takes 1¼ hours for a 4½-lb. whole fish, ½ hour for 1-lb. fish steaks or filets and ½ hour for a 1-lb. package of shellfish. Cook like fresh fish while still chilled, but allow additional cooking time and use a lower cooking temperature if the fish is only partially thawed.

Shellfish: It is desirable to freeze them near the source of supply because they deteriorate quickly.

Clams: Keep cold. Shuck at once and save liquid. Wash in salt water (⅓ c. salt to 1 gal. water). Drain. Wash thoroughly, eviscerate and wash again. Pack them in freezer containers and cover with their natural liquor or brine (⅓ c. salt to 1 gal. water). Place waxed paper or cellophane under the lid to keep the pieces submerged in the liquid. This prevents darkening. Label, date and freeze. Recommended storage time: 3 to 4 months.

Crabs: Keep cold. Clean at once and wash in cold water. Eviscerate and wash again. Cook in salted water (½ tsp. salt to 1 qt. water) 15 to 20 minutes. Cool quickly, shell and pack meat in freezer containers. Label, date and freeze. Frozen crabmeat is apt to be tougher. Recommended storage time: 3 to 4 months.

Lobsters: You can freeze them alive, but the meat is difficult to remove from the shell. Or plunge them in boiling salt water for 1 to 2 minutes, just long enough to cook the meat next to the shell. Cool quickly, package and freeze. Or cook the lobsters 20 minutes in salted water as if they were to be used at once, or cook them 10 minutes. Cool quickly. Wrap, label, date and freeze. Recommended storage time: 2 months.

Oysters: Handle like clams. Recommended storage time: 3 to 4 months.

Scallops: Handle like clams, only do not save liquid when shucking them. Label, date and freeze. Recommended storage time: 3 to 4 months.

Shrimp: Freeze cooked or uncooked shrimp. The cooked shrimp is easier to use, but uncooked has the best flavor. Wash. Pack in ice as soon as possible after catching.

To Raw Pack: Wash uncooked shrimp thoroughly, remove heads and sand veins; wash in brine (1 tsp. salt to 1 qt. water); drain. Place in freezer containers or pile together in blocks and pour ice water over them and freeze to obtain a protective glaze. Tightly wrap glazed shrimp in foil or other moisture-vaporproof packaging material. Label, date and freeze. Recommended storage time: 3 to 4 months.

To Freeze Cooked Shrimp: Place washed shrimp in boiling salt water (⅔ c. salt to 1 gal. water) and cook. Drain, cool quickly and, if you desire, remove veins and shells. Rinse in cold water, drain and package with head space in moisture-vaporproof containers. Label, date and freeze.

People who live inland may wish to cook frozen shrimp and refreeze it. Recommended storage time: Cooked peeled shrimp, 2 to 3 months; cooked unpeeled shrimp, 4 to 6 months; Shrimp Creole and Shrimp Cocktail, about 6 weeks.

Slick Freezing Tricks for Meat
GOOD IDEAS FROM FARM JOURNAL READERS

MEAT: Save and freeze small leftover bits of meat until you have enough to make a meat pie.

HAMBURGERS AND MEAT LOAVES: I shape ground beef into patties, place them on baking sheets so they do not touch, and freeze overnight. In the morning, I place them in coffee cans with tight-fitting lids and seal. Then I can take out as many as I need for a meal. Frozen in this way, they do not stick together. I find that several patties, used to make a meat loaf, will thaw much faster than a solid piece of ground beef.

HAM: Chop or cube baked ham left near the bone; freeze in 1½ c. portions. If unexpected company comes, mix ham with 1 can condensed cream of mushroom soup and 2 sliced, hard-cooked eggs; serve over rice, potatoes, or in toast cups.

LARD: If you render your own lard, don't store it in jars. While it is in liquid form, pour it into square layer cake pans and cool it. When cold, cut in blocks. Wrap each block in moistureproof packaging material; freeze. It saves a lot of digging fat out of jars.

PORK TENDERLOIN: I pound and flatten rather thick slices of tenderloin before freezing. Then they may be dipped in flour and put on to cook without thawing. I freeze the slices by placing them on baking sheets without touching. Then I put them in flat packages to take up little freezer space. They do not stick together and I can take out as many as I need.

CHAPTER 4

Dairy Foods and Eggs

Homemade Butter · Freezing Cream · Storing Ice Cream · Recipes
for Coconut/Honey, Quick Vanilla, Vanilla Custard Ice Creams ·
Vanilla Almond Crunch · Velvety Lime Squares · Peach Melba
Parfait · Grape Sherbet · Fresh Strawberry Sherbet · Freezing
Cheese · Cheesecake · Raspberry Swirl · Hot Cheese Dip · Freezing
Eggs · Frozen Egg Dessert · Meringue Shells · Banana-Berry Dessert

Think of the best country meals you've enjoyed. You will find that liberal
use of milk, cream, butter and eggs helped make them memorable. Farm
women's reliance on these foods started when every place had cows and
poultry. Now dairy and poultry are specialized business and many farmers
keep no cows or chickens. Even so, their wives often continue their habit
of using a free hand in cooking with dairy foods and eggs.

When surplus cream, butter and eggs pile up—whether they're produced
on the farm or purchased at a bargain—country women store some in
their freezers. Many country children still have a chance to eat ice cream
from dashers. Home-cranked ice creams get space in freezers for special
occasions—sometimes for Christmas gifts. (Do make Mariella's Ice
Cream and the tropical Coconut/Honey Ice Cream from recipes in this
chapter.) Wrap the ice cream and sherbets you buy to store according to
directions in this chapter to keep their best flavors and textures.

Relatively few country kitchens have churns, but read how some farm
women whip up a pound of butter in their electric mixers—often in 5 minutes. This way they always have the perfect seasoning for garden-fresh
new potatoes, succulent peas and sweet corn.

Fresh eggs, when plentiful, go into freezers in recipe-measured amounts,
ready to use. This chapter shows how good country cooks keep an abun-
dance of dairy foods and eggs on hand even if they live on "cowless,
chickenless" farms. They depend on the freezer.

Homemade Country Butter

Good country cooks know that foods made or seasoned with butter have superior flavor. One-cow families may not have much heavy cream to churn. But if you have a quart of it and an electric mixer, you can make a pound of butter in 5 minutes. On farms with several cows, there is often enough cream to make extra butter for freezing. Properly made and packaged, it will keep 6 months at 0° F.

Either sweet or sour cream may be used. (One way to sour the cream quickly is to stir ¼ c. cultured buttermilk into it.) Each one results in different flavors and each has its champions. Experts on freezing advise that you *pasteurize home-produced cream because butter made with it is less likely to become rancid or develop off-flavors in storage.*

Many farm homemakers write the Food Department of FARM JOURNAL that they do have success with freezing butter made from unpasteurized cream. However, it becomes rancid very quickly.

To PASTEURIZE CREAM: Heat it in a regular home pasteurizer or in a double boiler, stirring frequently, to 160 to 165° F. Chill quickly to 50° F.

MIXER-MADE BUTTER

Churn the butter if you have a churn or pour 1 qt. heavy cream (35% butterfat) into the large bowl of your mixer. Cover the bowl with foil, inserting the beaters through a hole made in the center of the foil. This eliminates spattering and cleanup work.

Beat cream at high speed until flecks of butter begin to form. Then turn to low speed and beat until the butter "comes"—or until buttermilk separates from butter.

Pour off the buttermilk.

Add cold water equal in quantity to the buttermilk. Use water 2 or 3° F. cooler than the buttermilk. Agitate at lowest speed. Pour off water and repeat, using slightly cooler water each time, until water drained off is clear. This is important. *Butter that is not washed thoroughly will not retain its prime quality long when frozen.*

Work out water by pressing butter against the sides of bowl with a wooden spoon if you do not have a butter paddle.

Salt the butter slightly, ½ tsp. to 1 lb. butter, if cream is pasteurized. (You can add more salt—up to a total of 2 tsp.—but the butter will not keep as well.) *Do not salt butter made from unpasteurized cream.*

Press butter into glass jars or plastic containers with tight-fitting lids, or mold it and cover it with a moistureproof wrap. Freeze at once. One quart cream makes about 1 pound butter, although it depends on how heavy the cream is.

Recommended storage time: Up to 9 months.

A good Colorado cook saves time by freezing butter measured in the recipe amounts she most often uses (½ and 1 c.). And a North Dakota homemaker freezes the buttermilk she cannot use immediately. She bakes with it later.

BUYING BUTTER FOR FREEZING

Be sure the butter you buy to freeze is fresh. Sometimes butter offered for sale in retail markets has been in storage several months. You may be able to get freshly churned butter from a creamery. Rewrap it in moisture-vaporproof packaging material, such as aluminum foil. Oxidation is a common cause of frozen butter changing in flavor. You can prevent this by using a good wrap. Seal tightly with freezer tape if you do not use foil. Fresh creamery butter may be stored in freezer for 9 months.

Frozen Cream Has Many Uses

Heavy cream, or that with 30 to 35% butterfat content, freezes satisfactorily. Use pasteurized cream. You will find it convenient to use if packaged in small containers—enough cream to use at one time. Completely fill containers. To discourage fat separation, thaw it in the refrigerator.

You will find that frozen cream is smoother when 1 to 1½ tblsp. sugar is added to every cupful, but cooks say this limits the ways it can be used. Frozen cream, sweetened or with considerable fat separation, will not whip. You can use it in cooking and making ice cream.
Recommended storage time: 4 months.

WHIPPED CREAM: Mounds of whipped cream, sweetened with confectioners sugar, freeze successfully. Good cooks like to keep them on hand to serve on desserts. Drop the whipped cream on a chilled baking sheet and freeze. Package in tightly covered containers or in plastic bags, twisted to exclude air. To use them, place unthawed on top of the dessert. They thaw rapidly and will defrost adequately by the time the dessert reaches the table. Recommended storage time: 1 month.

MILK: Do not freeze milk unless absolutely necessary. Undesirable changes occur in flavor and texture. If you do freeze milk, freeze only pasteurized homogenized milk. Store at 0° F. no longer than 3 weeks. Use for cooking.

Keep Ice Cream on Hand

If you buy ice cream to keep in the freezer, you can store it for 3 weeks in the original container without drying out. If you want to keep it 6 weeks, overwrap it with good packaging material. And when you serve ice cream from the carton, either lay a piece of plastic wrap, cut to fit, on top of the remainder, or fill the empty space with crumpled cellophane to exclude the air. Large ice crystals may form if ice cream experiences high storage temperatures.

Storing Homemade Ice Cream: Remove it from the freezer can, pack in rigid containers with tops as wide as the bottoms for easy removal, leaving ½″ head space. Top with a piece of plastic wrap, cut to fit, and cover tightly. Store it in the bottom or the coldest part of the freezer. Soften it before serving by placing it in the refrigerator for 20 to 30 minutes, the time depending on size of the container. You will find ice cream mellowed this way tastes better than when hard frozen.
Recommended storage time: Up to 6 weeks.

Many ice cream desserts are favorite freezer foods for hostesses. Among them are cream puffs, eclairs and cake rolls filled with ice cream. Serve these desserts directly from the freezer.
Recommended storage time: Up to 1 month.

How to Freeze and Ripen Ice Cream

Whatever recipe for ice cream you use (recipes follow), the mix should be thoroughly chilled before you begin to freeze it. Also be sure your freezer can, dasher and lid are spotlessly clean. Wash them in hot soapy water, rinse and scald with boiling water. Then chill the can.

1. Set chilled can into empty freezer bucket; pour in chilled ice cream mixture. Can should be ⅔ to ¾ full—no more—to allow room for expansion.

2. Adjust dasher and cover, following directions for your freezer.

3. Pack crushed ice and rock salt around the can, using about 1½ c. rock salt for every 8–10 lbs. crushed ice. (You'll need about 3 c. rock salt and 20 lbs. ice to freeze and ripen 1 gal. ice cream. Freezing is faster if you use more salt, but you will not get as smooth ice cream.)

4. Turn dasher slowly until the ice melts enough to form a brine. Add more ice and salt, mixed in the proper proportions, to maintain the ice level. Turn the handle fast and steadily until it turns hard. Then remove the ice until its level is below the lid of the can. Take the lid off; remove the dasher.

5. To ripen the ice cream, plug the opening in the lid. Cover the can with several thicknesses of waxed paper or foil to make a tight fit for the lid. Put the lid on the can.

6. Pack more of the ice and salt mixture, using 4 parts ice to 1 part rock salt, around the can, filling the freezer. Cover the freezer with a blanket, canvas or other heavy cloth, or with newspapers. Let ice cream ripen at least 4 hours, sherbet 1 hour. Or put the can in the home freezer.

COCONUT/HONEY ICE CREAM

For a lush, tropical flavor blend, make with orange blossom honey

1½ c. honey
4   eggs, slightly beaten
3   c. heavy cream
2   tsp. vanilla
½   tsp. lemon extract
½   tsp. salt
3   c. milk
1   (3½ oz.) can flaked coconut
2   (8¼ oz.) cans crushed pineapple

• Add honey to eggs; mix well. Add cream, flavorings, salt and milk; stir until well blended. Chill.
• Pour into freezer can; put dasher and cover in place. Pack chopped ice and rock salt around can; freeze. (See "How to Freeze and Ripen Ice Cream" in this chapter.)
• When partly frozen, add coconut and pineapple; continue freezing until crank turns hard. Remove dasher. Let ice cream ripen, using 4 parts ice to 1 part salt. Or spoon into freezer containers; seal, label, date and store in freezer. Makes 1 gallon.

MARIELLA'S ICE CREAM

A Kansas dairyman's wife freezes this ice cream for Christmas giving

1   qt. milk, scalded
4   eggs, beaten
2½ c. sugar
2½ to 3 c. heavy cream
1   qt. cold milk
2   tblsp. vanilla
¼   tsp. salt
3   drops lemon extract

• Stir hot milk slowly into eggs and sugar; cook slowly over direct heat until thick, stirring; cool.

• Add remaining ingredients to cooled egg mixture; stir until smooth. Pour into freezer can; put dasher and cover in place. Pack chopped ice and rock salt around can; freeze. (See "How to Freeze and Ripen Ice Cream" in this chapter.) To store, spoon lightly into airtight freezer containers (do not pack). Seal, label and date. Store in freezer. Makes about 1 gallon. Recommended storage time: Up to 1 month.

VARIATIONS

CHOCOLATE: Stir 1 (5½ oz.) can chocolate syrup into vanilla mixture.

STRAWBERRY: Mix 3 (10 oz.) pkgs. frozen strawberries, thawed, and ½ c. sugar; stir into vanilla mixture; add red food color for pink ice cream.

Note: You can make 4 batches consecutively. By time the custard for the fourth one is cooked, the first is cool enough to put in freezer can. When that gallon is frozen, the second batch is ready for freezing. This assembly line method makes economical use of ice. A 50-lb. bag of cracked ice is enough to freeze 4 gallons. You save time by using the same bowls and pans for each batch. A good selection is to make 2 gallons vanilla, 1 chocolate and 1 strawberry.

QUICK VANILLA ICE CREAM

You're lucky if you have a chance to clean dasher—stores well in freezer

1   tblsp. unflavored gelatin
½   c. cold water
7   c. light cream

When berries are ripe, freeze coral-red Fresh Strawberry Sherbet for desserts. Pack it with ice cream, parfait-style, or in molds for parties and special dinners. Or serve with salads. Recipe in Chapter 4.

At a big party, country hostesses like to dazzle guests with a choice of cake rolls which they make ahead and freeze. Recipe for Frozen Cake Roll in Chapter 6 includes four creamy fillings and toppings.

You can make Turkey Curry from leftover turkey and freeze it ready to heat; or you can freeze the cooked turkey meat and make curry later. Served over rice with lots of condiments, it's a festive way to recycle the bird. Recipe in Chapter 7.

Put a Lemon Meringue Pie Supreme in your freezer when you've got a busy week ahead. It's a truly glamorous dessert with nippy filling and tall meringue. The whole pie freezes successfully. Recipe in Chapter 6.

1½ c. sugar
1 c. evaporated milk
1 tblsp. vanilla

• Soften gelatin in cold water.
• Scald 2 c. cream; add gelatin and stir to dissolve. Add sugar; stir to dissolve. Combine remaining cream, evaporated milk and vanilla; add slowly to gelatin mixture, stirring.
• Pour into 1-gal. freezer container. Pack chopped ice and rock salt around can; freeze. (See "How to Freeze and Ripen Ice Cream" in this chapter.) Makes about 3 quarts.
• Pack in rigid containers with tight lids. Label and date. Store in freezer. Recommended storage time: 3 to 6 weeks.

VANILLA CUSTARD ICE CREAM

Velvety, rich, delicious—a good keeper

1 qt. milk
2 c. sugar
¼ c. flour
½ tsp. salt
4 eggs, slightly beaten
1 tblsp. vanilla
1½ qts. light cream, or dairy half-and-half

• Scald milk. Mix sugar, flour and salt. Add enough hot milk to sugar-flour mixture to make a thin paste. Stir paste into hot milk. Cook over low heat, stirring constantly, until mixture thickens slightly, about 15 minutes.
• Add hot mixture gradually to beaten eggs and cook over low heat, stirring constantly, until mixture thickens

slightly, about 2 minutes (do not overcook or eggs may curdle).
• Cool quickly in refrigerator. Do not allow mixture to cool at room temperature.
• Add vanilla and light cream to cooled mixture. Pour into 1-gal. freezer can; fill only two thirds full.
• Pack chopped ice and rock salt around can; freeze. (See "How to Freeze and Ripen Ice Cream" in this chapter.) Makes 1 gallon.
• Package in rigid containers with tight lids. Label and date. Store in coldest part of freezer. Recommended storage time: 3 to 6 weeks.

VANILLA ALMOND CRUNCH

Try a splash of chocolate sauce on it for a special sundae

1 (4 oz.) pkg. slivered almonds
¼ c. butter, melted
1 c. crushed rice cereal squares
½ c. light brown sugar
½ c. flaked coconut
⅛ tsp. salt
½ gal. vanilla ice cream, softened

• Toast almonds in the melted butter. Remove half of almonds from butter and set aside.
• Combine crushed cereal, brown sugar, coconut and salt with remaining almonds and butter. Pat mixture gently into 13×9×2" pan. Bake in moderate (375° F.) oven 5 minutes. Cool.
• Spread ice cream over cooled crust. Decorate top with reserved almonds. Freeze until firm. Wrap, label, date and return to freezer. Recommended storage time 1 month.

• Remove dessert from freezer 20 minutes before cutting. Makes 10 servings.

VELVETY LIME SQUARES

Garnish with a perfect pecan half

1 (3 oz.) can flaked coconut
½ c. vanilla wafer crumbs
2 tblsp. butter, melted
2 tblsp. sugar
2 (3 oz.) pkgs. lime gelatin
2 c. boiling water
1 (6 oz.) can frozen limeade concentrate
1 qt. plus 1 pt. vanilla ice cream, softened
⅛ tsp. salt
Few drops green food coloring

• Carefully toast ½ c. coconut in moderate (375° F.) oven until lightly browned, about 5 minutes. Set aside.
• Combine remaining coconut, crumbs, butter and sugar. Lightly press into 7×11×1½″ pan and bake in moderate (375° F.) oven 6 to 7 minutes. Cool.
• Dissolve gelatin in boiling water. Add limeade, ice cream and salt; stir until dissolved. Pour into crust. Top with reserved toasted coconut and garnish with pecans, if you wish. Freeze until firm. Wrap, label, date and return to freezer. Recommended storage time: 1 month.
• Remove dessert from freezer 20 minutes before cutting. Makes 6 to 8 servings.

PEACH MELBA PARFAIT

Midsummer's dessert dream—fresh peaches, raspberries and ice cream

1 (3 oz.) pkg. raspberry flavor gelatin
2 c. boiling water
1 (10 oz.) pkg. frozen raspberries (1¼ c.)
1 (3 oz.) pkg. peach flavor gelatin
½ c. cold water
1 pt. vanilla ice cream
1½ c. diced fresh peaches

• Dissolve raspberry gelatin in 1 c. boiling water in bowl. Add unthawed raspberries. Stir occasionally to separate berries. (Gelatin will begin to set.)
• Pour 1 c. boiling water over peach gelatin in another bowl; stir to dissolve. Add ½ c. cold water. Add ice cream in 8 chunks; stir to dissolve. Refrigerate until thick enough to mound up.
• Spoon slightly thickened raspberry mixture into 8 dessert dishes.
• When ice cream mixture has thickened, fold in peaches and spoon over raspberry mixture. Chill at least 30 minutes. Makes 8 servings.

STRAWBERRY BANANA SHERBET

True strawberries-and-cream dessert

1 c. mashed fresh strawberries
⅓ c. mashed ripe banana
2 tblsp. lemon juice
⅓ c. orange juice
1 c. sugar
1 c. light cream

• Beat all ingredients together until thoroughly blended. Pour into refrigerator tray or loaf pan. Freeze 1 hour or until firm around the edges.

• Remove sherbet from freezing compartment and stir. Cover with foil. Freeze 1½ hours longer, or until firm. Makes 6 servings.
• Pack in rigid container with tight lid; label and date. Store in bottom or coldest part of freezer.
Recommended storage time: 1 month.

COUNTRY GRAPE SHERBET

Sour cream adds distinctive flavor

1 c. dairy sour cream
1 c. milk
1½ c. sugar
1 egg, well beaten
1 c. grape juice (Concord)
¼ c. lemon juice

• Combine ingredients and beat until sugar dissolves.
• Pour into 2 refrigerator trays or loaf pans; freeze until nearly firm.
• Turn mixture into chilled bowl and beat until fluffy and smooth. Work fast and do not let mixture melt.
• Return to trays and freeze until firm. Makes 6 to 8 servings. Pack in airtight containers; label, date and store in freezer.
Recommended storage time: Up to 1 month.

FRESH STRAWBERRY SHERBET

Pretty, coral-red dessert to make when the strawberries are ripe

4 qts. fresh strawberries, sliced
4 c. sugar
2⅔ c. milk

⅔ c. orange juice
⅛ tsp. ground cinnamon

• Mix strawberries and sugar; let stand until juicy (about 1½ hours). Mash or purée in blender. Strain out seeds (optional, but we prefer seedless sherbet).
• Add milk, orange juice and cinnamon. Mix well. Freeze in crank-type freezer. (See "How to Freeze and Ripen Ice Cream" in this chapter.) Or pour mixture into refrigerator trays or loaf pans; freeze about 3 hours; stir 2 or 3 times. Makes about 1 gallon. Pack in airtight containers; label, date and store in freezer.
Recommended storage time: 1 month.

VARIATIONS

STRAWBERRY PARFAIT RING: Spoon strawberry sherbet and soft vanilla ice cream in alternating layers in chilled ring mold. Freeze. When frozen, unmold on serving plate; return to freezer. To serve, fill center with fresh berries for color-bright garnish.

STRAWBERRY/LEMON PARFAIT: Spoon strawberry and lemon sherbets alternately into parfait glasses or tumblers. Return to freezer until time to serve. Garnish with fresh berries.

CHOCOLATE/STRAWBERRY RING: Spoon strawberry sherbet into ring mold. Freeze. When frozen, unmold on plate; return to freezer. To serve, fill with scoops of chocolate ice cream.

Choose the Right Cheese to Store

Much research has been done on freezing cheese. These studies show that certain kinds freeze well, while others become crumbly and mealy. Pasteurized process cheese keeps well in freezer for 1 to 2 months. And regardless of the kind of cheese, small portions freeze best. Package the cheese in ½ lb. or smaller lots in pieces from ½ to 1″ thick. Larger portions freeze too slowly at 0° F. Among the types that freeze with good results are Cheddar, Brick, Port du Salut, Swiss, Provolone, Mozzarella, Liederkranz, Camembert, Parmesan and Romano. Some Limburger, Gouda, Club and Colby cheeses freeze successfully, but others do not. Cream cheese becomes crumbly when frozen but dips and sandwiches made with it as an ingredient freeze well. Cottage cheese does not freeze successfully.

To Package: Wrap the cheese in airtight packages. Heavy-duty aluminum foil and plastic wrap are excellent. Camembert cheese, which comes in small, foil-wrapped packages, need not be rewrapped. Freeze process cheese in original polyethylene package, but overwrap with foil.

If you have grated cheese or scraps of cheese, freeze them in plastic bags, twisting to exclude the air, or in other airtight packages. They're handy for seasoning and topping casseroles.

Recommended storage time: 6 months.

CHEESECAKE

When guests taste, the women will ask, "May I have your recipe?"

1½ c. zwieback crumbs
¼ c. butter or regular margarine
2 (8 oz.) pkgs. cream cheese
1 c. sugar
4 eggs, separated
1½ tsp. grated lemon peel
4 tsp. lemon juice
½ tsp. vanilla
½ c. sifted flour
½ tsp. salt
1 c. milk
Sour Cream/Raspberry Topping
  recipe follows

· Blend together crumbs and butter. Press into bottom and 1″ up the sides of a 9″ spring-form pan. Bake in a slow oven (325° F.) 5 minutes. Cool.
· Mix together cheese, sugar and egg yolks. Beat until light. Blend in lemon peel, juice and vanilla.
· Mix in flour and salt, sifted together. Blend in milk. Fold in stiffly beaten egg whites.
· Pour into crumb-lined pan. Bake in a slow oven (325° F.) 1 hour. Turn off heat and allow cake to cool in oven 1 hour. Remove from oven; cool. Makes 12 servings.
· Freeze, then wrap, label and date. Return to freezer.
Recommended storage time: Up to 1 month.
· To use, thaw in refrigerator. Just before serving, add Sour Cream/Raspberry Topping.

SOUR CREAM/RASPBERRY TOPPING: Spread 1 c. dairy sour cream on top of cheesecake. Decorate edge with 1 c. fresh (or drained, partly frozen) raspberries.

RASPBERRY SWIRL

Cool and tangy finale to a meal

¾ c. graham cracker crumbs
3 tblsp. butter, melted
2 tblsp. sugar
3 eggs, separated
1 (8 oz.) pkg. cream cheese
1 c. sugar
⅛ tsp. salt
1 c. heavy cream
1 (10 oz.) pkg. frozen raspberries, partially thawed

• Combine thoroughly crumbs, melted butter and 2 tblsp. sugar. Lightly press mixture into well-greased 7×11×1½″ pan. Bake in moderate (375° F.) oven about 8 minutes. Cool thoroughly.

• Beat egg yolks until thick. Add cream cheese, sugar and salt; beat until smooth and light.

• Beat egg whites until stiff peaks form. Whip cream until stiff and thoroughly fold with egg whites into cheese mixture.

• In a mixer or blender, crush raspberries to a pulp. Gently swirl half of fruit pulp through cheese filling and spread mixture into crust. Spoon remaining purée over top; swirl with a knife. Freeze until firm. Makes 6 to 8 servings.

• Cover tightly, label, date and return to freezer.

Recommended storage time: 1 month.

• To serve, let stand at room temperature 30 minutes to soften. Cut in squares.

HOT CHEESE DIP

Crackers and nippy cheese party-style

2 (5 oz.) jars sharp cheese spread
1 (8 oz.) can minced clams (not drained)
1 tblsp. chopped green onion
2 tblsp. chopped green pepper
2 tsp. Worcestershire sauce
⅛ tsp. paprika

• Thoroughly mix together all ingredients. Put in freezer container or ovenproof dish. Makes about 2 cups.

• Wrap, label, date and freeze.

Recommended storage time: 2 to 4 weeks.

• To use, thaw enough to remove from container or leave in ovenproof container. Heat in a slow oven (325° F.) about 45 minutes. Serve in chafing dish or casserole over candle warmer with crackers or cubes of French bread on toothpicks.

Store Eggs in Freezer

Grandmother put eggs down in waterglass in springtime when they were abundant. She was thrifty and knew that eggs, kept for weeks when they were scarce, came in handy on baking day. Uncracked eggs, if held below 40° F., have a long storage life. But today's country cooks often freeze top-quality eggs when they have a big supply. It's convenient to have them —they stretch the budget.

You can freeze eggs whole, or whites and yolks separately. The important point is to use *fresh, clean, sound-shelled eggs;* do not use cracked eggs. If eggs must be stored before breaking, refrigerate them in the coldest part of the refrigerator. Here are the steps to follow:

1. Wash fresh eggs in water warm to the hands (120° F.) with detergent added to help loosen dirt. Dry quickly.

2. Break eggs, one at a time, into a cup before adding to the bowl. Do not freeze eggs that have an odor or blood spots.

3. Stir whole eggs with a fork or electric mixer on low speed to mix whites and yolks, but do not beat; avoid adding air. To 1 c. eggs (about 12 eggs) add 1 tblsp. sugar or light corn syrup or ½ tsp. salt, depending on how you want to use them. This prevents gumminess when thawed. To use them, thaw in the refrigerator, or under cold running water. Stir them before using. Do not let them wait longer than 24 hours in the refrigerator before using and do not refreeze. Thawed frozen eggs spoil rapidly.

4. Freeze and package egg whites in recipe amounts—for an angel food cake or meringue shells. Be careful not to include even a particle of the yolk. If you do, remove it carefully with the tip of the spoon. Even a small drop will prevent the whites from beating well.

5. Break egg yolks by stirring with a fork, but do not beat them. Into 1 c. yolks, mix 2 tblsp. light corn syrup or

1 tsp. salt, depending on how you want to use them.

6. Package small amounts or *measured-recipe amounts* in moistureproof containers. Leave ½″ head space in half-pint and pint containers, ¾″ in narrow-top jars; seal tightly. Label every package so you will know the quantity and whether eggs contain salt or sweetening (sugar or corn syrup). You will use the sweetened eggs for desserts like custards, cakes and other dishes containing sugar; those with salt for omelets, scrambling and other dishes made without sweetening.

Recommended storage time: 6 to 8 months.

MEASUREMENTS FOR FROZEN EGGS

1 to 1½ tblsp. frozen yolks equal 1 egg yolk

2 tblsp. whites equal 1 egg white

3 tblsp. yolks and whites equal 1 whole egg

1 c. equals 5 whole eggs, 12 yolks, or 8 whites

FROZEN EGG DESSERT

Good way to use extra eggs (developed at University of Connecticut)

¾   **c. butter or regular margarine**
1½   **c. confectioners sugar**
4   **large eggs**
¾   **c. chopped pecans**
¾   **tsp. vanilla**
3¾   **c. vanilla wafer crumbs**

· Cream butter and sugar; add eggs, one at a time, beating well after each addition. Mix in nuts and vanilla. Add 2¾ c. wafer crumbs; mix.

• Spread remaining crumbs over bottom of buttered 8″ square pan. Spread creamed mixture evenly over crumbs. Makes 16 (2″) squares.

• Wrap, seal, label, date and freeze. Recommended storage time: about 1 month.

• Serve without thawing, topped with whipped cream.

VARIATIONS

CHOCOLATE EGG DESSERT: Substitute 2 (1 oz.) squares unsweetened chocolate, melted, for the vanilla.

LEMON EGG DESSERT: Substitute ¼ c. lemon juice and 2 tsp. grated lemon peel for the vanilla.

BANANA-BERRY DESSERT

Put this heart-shaped meringue in the freezer days before your party

Meringue Shell:
    5   egg whites (room temperature)
    ¼   tsp. salt
    ¼   tsp. cream of tartar
    1   tsp. vanilla
    1¼ c. sugar (sifted)

• Beat egg whites, salt, cream of tartar and vanilla until frothy. Gradually add sugar, beating until stiff peaks form (about 10 minutes).

• Cover a baking sheet with heavy brown paper. Draw a heart shape, 12″ high on the paper. Spread meringue to edges of heart shape; build up sides 2″ to make a shell.

• Bake in very slow oven (250° F.)

1½ hours. Turn off heat; leave meringue in closed oven 3 hours. Remove from paper with spatula.

Cream Filling:
    2   (3 oz.) pkgs. cream cheese
    1   tblsp. lemon juice
    ½   c. sifted confectioners sugar
    ½   c. mashed ripe bananas
    Few drops red food color
    1   (10 oz.) pkg. frozen strawberries, defrosted
    1   c. heavy cream, whipped

• Soften cream cheese; beat in lemon juice, sugar, bananas, food color. Drain defrosted strawberries, reserving juice (refreeze juice, if you wish). Fold berries and whipped cream into cheese mixture. Spoon filling into meringue shell. Wrap carefully and freeze.

Recommended storage time: Up to 1 month.

Glazed Banana Slices:
    2   tblsp. corn starch
    ¼   c. sugar
    6   tblsp. water
    1   tsp. lemon juice
    Reserved strawberry juice
    ¼   tsp. red food color (about)
    2   bananas

• *The day you serve:* Combine corn starch, sugar, water, lemon juice, reserved strawberry juice and food color in small saucepan. Cook over medium heat until thick and clear; cool slightly.

• Remove meringue from freezer; unwrap. Peel bananas; score with a fork; slice. Arrange slices on filling, following heart outline. Spoon glaze over bananas. Refrigerate 1 to 3 hours. Makes 12 servings.

INDIVIDUAL MERINGUE SHELLS: For each shell, drop ⅓ c. meringue mixture on heavy brown paper; use back of spoon to hollow out center and build up sides. Bake 1 hour; turn off heat; leave meringues in closed oven 3 hours or overnight. Makes about 15.

• To freeze, wrap each shell in plastic wrap and seal with freezer tape. Overwrap with heavy-duty foil. Label and date. To thaw, remove foil wrapping and let shells stand at room temperature in plastic wrap about 1 hour. Recommended storage time: up to 1 month.

CHAPTER 5

Homemade Breads

Freezing Yeast Breads · Recipes for Breads to Bake and Freeze ·
Recipes for Homemade Frozen Dough · Coffee Breads · No-knead
Batter Breads · Freezing Quick Breads · Fruited Nut Breads · Pop-
overs · French Toast · Doughnuts

Many mothers believe children are short-changed if they never stand wide-
eyed in the kitchen when rounded, crusty loaves of gold-brown bread
come from the oven. How pleasant to sniff the yeasty aroma—and to
remember it. Food fragrances, research shows, are among the most
cherished memories people have of their childhood homes.

Home freezers deserve much credit for keeping bread baking a flourish-
ing art in country kitchens. True, lots of the good daily bread comes from
bakery counters, but when a farm woman finds time, loaves, rolls and
coffee cakes move from oven to cooling rack and into the freezer.
Reheated, these fresh-again breads, sliced and spread with butter and jelly
or jam, make coffee hours and mealtimes something special.

You will find recipes for unusual, superior breads in this chapter. And
all of them freeze successfully. Some are prize winners—Honey Whole
Wheat Bread, which contains potatoes, for instance. It won blue ribbons
for a California FARM JOURNAL reader at county and state fairs.

Be sure to try the breads baked from frozen dough. While there's little
variety in commercially frozen doughs, it's a different story if you make
your own. In addition to Freezer White Bread, we have recipes for Pumpernickel and Honey/Wheat Germ, Apricot Braid, Cinnamon Swirl and
Orange Blossom Buns—six good frozen doughs.

Don't miss the frozen popovers that become light as balloons and crisp-
crusted when baked; the muffin-like Petite Brioche. Note the distinctive,
quick fruit and nut breads for dainty party sandwiches.

Watch your family and friends help themselves again and again to your
homemade breads. Keep a supply in your freezer!

Freezing Yeast Breads

BAKERY BREAD: If it comes to the kitchen in a good wrap, you can put it directly in the freezer. It will stay in *good condition about 4 months.* In case you want to store it longer, overwrap the loaves and rolls with moistureproof packaging material or place them in plastic freezer bags.

HOMEMADE BREADS: Use your favorite bread, roll and coffee cake recipes or use the wonderful ones that follow in this chapter. Let the shaped dough for loaf bread rise until doubled. Then bake in a hot oven (400° F.) 45 to 50 minutes. It will be less crumbly after frozen than bread baked an hour in a moderate oven (375° F.). Bake all breads to a light golden brown to prevent upper crust from separating from inside of loaf. (Specialty breads and rolls, rich in sugar and eggs, often need to be baked at temperatures lower than 400° F. Use temperatures specified when using recipes in this chapter.) Remove from pans, cool quickly to room temperature, wrap in foil or plastic wrap or place in plastic freezer bags, seal, label, date and freeze.
Recommended storage time: 9 months to 1 year.
Bake rolls and buns when dough doubles, in a hot oven (400° F.) 15 to 20 minutes. Cool and wrap like bread loaves or place in a rigid container, overwrap, label, date and freeze. Many women like to wrap, freeze and warm rolls in foil.
Recommended storage time: 9 months.
Bake coffee cakes when dough is

light, in a moderate oven (375° F.) 20 to 30 minutes, or until a light golden brown. Cool and wrap like bread loaves. Frost before serving.
Recommended storage time: 9 months.

To use, thaw the bread in its wrapper at room temperature if you wish to serve it cold. A 1-lb. loaf will thaw in about 3 hours—or in seconds in the microwave oven. You need not thaw bread slices for toast. Thaw rolls in their wrapper to serve cold. To heat them, place unwrapped rolls in a covered container or wrap in aluminum foil. Heat in a moderate oven (350° F.) about 20 to 25 minutes, or until hot. Do not heat them longer than necessary or they will dry out. Put unthawed coffee cakes in a moderate oven (350° F.) and heat, uncovered, about 20 to 25 minutes. Frost if desired.

To FREEZE BREAD DOUGH: We do not recommend freezing doughs for yeast breads unless you use recipes especially developed for this purpose. In this chapter, recipes for Freezer White Bread, Honey/Wheat Germ Bread, Round Pumpernickel Loaves, Apricot Braid, Cinnamon Swirl Loaves and Orange Blossom Buns are examples of breads that may be successfully frozen before baking.

DOUGHNUTS: Fry doughnuts of yeast-leavened dough, drain on layers of paper toweling; cool, wrap in meal-size amounts or individually in foil.

Recommended storage time: 2 months.

To use, let thaw in foil wrap and heat in a slow oven (300° F.) about 15 minutes. Or unwrap, brush lightly with butter, place on a baking sheet and heat quickly in a hot oven (400° F.).
Recommended storage time: Up to 2 months.

HOMEMADE WHITE BREAD

Winter's best snack—bread warm from oven with butter and jelly

3 pkgs. active dry yeast
½ c. warm water (110 to 115° F.)
½ c. sugar
4 tsp. salt
⅓ c. melted shortening
5 c. water
16 to 18 c. flour

• Sprinkle yeast on warm water; stir to dissolve.
• Combine sugar, salt, shortening and water in 5-qt. bowl. Stir in 8 c. flour. Add yeast and enough of remaining flour (8 to 10 c.) to make stiff dough that cleans the bowl when you stir.
• Knead on lightly floured surface until smooth and satiny (5 to 8 minutes). Place in greased bowl; turn to bring greased side up. Cover; let rise in warm place (80 to 85° F.) until doubled, about 1½ hours. Punch down. Turn out on floured board.
• Divide in half. Set aside one half to rise again. Divide other portion in half; shape each half into smooth ball; let rest 10 minutes. Shape into two loaves; place in greased 9×5×3"

loaf pans. Grease top lightly; cover and let rise until doubled, about 1 hour. Bake in hot oven (400° F.) 40 to 50 minutes. Immediately turn out of pans on rack. Cool thoroughly.
• When portion of dough set for second rising is doubled, punch down and shape loaves, let rise and bake as for first portion. Makes 4 loaves.
• Wrap loaves individually as soon as they cool. Label, date and freeze.
Recommended storage time: 9 months to a year.
• To serve, let thaw in wrapper at room temperature (on rack to allow air circulation) and heat in foil wrapper in moderate oven (375° F.) 20 minutes. Foil may be opened for last 5 minutes to crisp the crust.

VARIATIONS

WHOLE WHEAT BREAD: Substitute whole wheat flour for half the white.

RAISIN COFFEE BREAD: Make 3 loaves and knead ½ c. raisins into fourth loaf just before shaping. Place in greased 8" square pan to rise. Before baking, brush top with melted butter and sprinkle with mixture of cinnamon and sugar.

POTATO BREAD: Reduce amount of water to 3 c. Add 2 c. instant mashed potatoes to water mixture before adding flour and yeast.

MILK BREAD: Substitute 5 c. scalded milk for 5 c. water. Add sugar, salt and shortening. Cool to lukewarm before adding yeast to batter.

CRUSTY FRENCH BREAD

Keep a golden loaf in the freezer to make garlic bread for the barbecue

2 c. warm water (110 to 115° F.)
1 pkg. active dry yeast
1 tblsp. sugar
2 tsp. salt
5½ c. flour (about)
1 egg white, beaten

• Pour water into a large bowl. Sprinkle on yeast and stir until dissolved. Add sugar, salt and 3 c. flour. Stir to mix; then beat until smooth and shiny. Stir in about 2½ c. more flour.

• Sprinkle a little flour on bread board or pastry cloth. Turn dough out on board and knead until satiny and smooth, 5 to 7 minutes. Shape in a smooth ball.

• Place in greased bowl; turn dough so top is greased. Cover with waxed paper, then with a clean cloth. Let rise until doubled, about 1 hour.

• Punch down, divide in half and shape each portion in a ball. Place balls on lightly floured board, cover and let stand 5 minutes.

• Rub a little shortening on the palms of hands. Start shaping the loaves by rolling ball of dough at center and gently working the hands toward ends of loaf. Do this several times to get well-shaped, long, slender loaves.

• Place the two loaves about 4″ apart on lightly greased baking sheet. With a sharp knife or razor blade, cut diagonal gashes about ¾″ deep, 1½″ apart into top of each loaf. Cover and let rise until a little more than doubled, about 1 hour. Bake in hot oven (425° F.) 30 to 35 minutes.

• Remove from oven. Brush with egg white. Return to oven for 2 minutes. Remove; cool on racks. Makes 2 loaves.

• Wrap loaves individually as soon as they cool. Label, date and freeze. Recommended storage time: 9 months to a year.

• To serve, let thaw in wrapper at room temperature (on rack to allow air circulation) 3½ hours or heat in foil in moderate oven (375° F.) 20 minutes.

CRUSTY ROLLS: Bake 1 loaf of French Bread and use other half of dough to make rolls. Cut the dough into pieces the size of an egg. Shape each piece into a smooth ball by folding the edges under. Place them 3″ apart on a lightly greased baking sheet. With scissors, cut a cross ½″ deep in top of each roll. Cover and let rise until doubled. Bake in a hot oven (425° F.) 15 to 20 minutes. Remove from oven, brush rolls with egg white and return to oven for 2 minutes. Remove from baking sheet. Serve hot or cold. Makes 12.

• To freeze, package rolls in foil or place in a freezer box and overwrap. Label, date and freeze. Recommended storage time: 9 months to a year.

• To serve, thaw package at room temperature (on rack to allow air circulation) about 1 hour. To serve hot, place foil-wrapped frozen rolls in hot oven (400° F.) 20 minutes. If wrapper on rolls cannot be heated, place in paper bag and heat in moderate oven (350° F.) until hot, about 15 to 20 minutes.

CHEESE BRAIDS

Unusual seasoning: black pepper.
Slice thin, serve with crisp salads

2 c. warm water (110 to 115° F.)
2 pkgs. active dry yeast
¾ c. instant dry milk
¼ c. sugar
2 tsp. salt
7 to 8 c. flour
¼ c. soft lard or butter
3 eggs, beaten
2 tsp. coarsely ground black
    pepper
2 c. coarsely shredded dry
    Cheddar cheese (½ lb.)
Melted fat
1 egg yolk
2 tblsp. cold water

• Measure ½ c. warm water into large bowl. Add yeast and stir to dissolve. Add remaining warm water, dry milk, sugar, salt and 3 c. flour. Stir until smooth.

• Add lard, eggs, pepper and enough remaining flour to make dough that cleans the bowl when stirred.

• Turn out on floured board; knead until smooth and satiny, about 5 minutes. Flatten with hands into rectangle. Sprinkle with ½ c. cheese; roll up like jelly roll, flatten again. Repeat process until all cheese is worked in. Knead lightly to be sure cheese is evenly distributed. Place in greased bowl; turn dough so top is greased. Cover with damp cloth and let rise in warm place (80 to 85° F.) until doubled, about 1½ hours.

• Punch down. Divide in six equal parts. Cover; let rest 5 minutes. Roll each part into strips as long as baking sheet with tapered ends. Lay 3 strips

on each of two greased baking sheets. Starting in middle, braid loosely to each end; turn ends under and press down lightly to baking sheet. Brush with melted fat. Cover; let rise until light, about 40 minutes (do not let size double). Brush tops with mixture of egg yolk and cold water.

• Bake in moderate oven (350° F.) 20 minutes; reduce heat to 325° F. and bake about 25 minutes more. Brush again with egg yolk glaze about 5 minutes before end of baking time. Makes 2 loaves.

• Wrap or package loaves individually as soon as they cool. Seal, label, date and freeze.

Recommended storage time: 9 months to a year.

• To serve, let thaw in wrapper at room temperature (on rack to allow air circulation) or heat in foil wrapper in moderate oven (375° F.) 20 minutes. Foil may be opened for last 5 minutes to crisp the crust.

HONEY WHOLE WHEAT BREAD

A bread for spreads—try it with butter, jelly, jam, apple butter

5 c. warm water (110 to 115° F.)
2 pkgs. active dry yeast
6 tblsp. lard or other shortening
¼ c. honey
4 c. whole wheat flour
½ c. instant potatoes (not
    reconstituted)
½ c. instant dry milk
1 tblsp. salt
6½ to 8 c. all-purpose flour
Melted fat

• Combine ½ c. water and yeast in bowl. Stir to dissolve yeast.

• Melt lard in 6-qt. saucepan; remove from heat. Add honey and remaining water.

• Mix whole wheat flour, instant potatoes, dry milk and salt. Add to saucepan. Beat until smooth.

• Add yeast mixture and beat until smooth. Then with a wooden spoon mix in enough all-purpose flour to make a dough that cleans the pan. Knead on lightly floured board until smooth and satiny, and small blisters appear, 8 to 10 minutes.

• Place in greased bowl; turn dough so top is greased. Cover; let rise in warm place (80 to 85° F.) until doubled, 1 to 1½ hours. Punch down dough; divide in thirds; cover and let rest 5 minutes. Shape into 3 loaves. Place in greased 9×5×3″ loaf pans; brush with melted fat. Cover; let rise until doubled, about 1 hour.

• Bake in hot oven (400° F.) about 50 minutes. Remove from pans; cool on racks. Makes 3 loaves.

• Wrap or package loaves individually as soon as they cool. Label, date and freeze.

Recommended storage time: 9 months to a year.

• To serve, let thaw in wrapper at room temperature (on rack to allow air circulation) or heat in foil wrapper in moderate oven (375° F.) 20 minutes. Foil may be opened for last 5 minutes to crisp the crust.

Note: You may use 1 c. unseasoned mashed potatoes in place of the instant potatoes. Combine with the honey-water mixture.

ORANGE BREAD

Watch—it browns quickly. Team with date-nut filling for party sandwiches

½   c. warm water (110 to 115° F.)
2   pkgs. active dry yeast
1½ c. warm orange juice (110 to 115° F.)
5   to 6 c. flour
½   c. sugar
2   tsp. salt
¼   c. soft lard or butter
¼   c. grated orange peel
Melted butter

• Measure warm water into large bowl. Add yeast and stir to dissolve. Add orange juice and 2 c. flour; beat thoroughly. Add sugar, salt, lard, orange peel and about 4 c. flour, to make dough that cleans bowl when you stir.

• Turn out on lightly floured board and knead until satiny and elastic, 5 to 8 minutes. Place in greased bowl, turning dough so top is greased. Cover with damp cloth; let rise in warm place (80 to 85° F.) until doubled, about 1½ hours.

• Punch down dough; divide in half. Cover; let rest 5 minutes. Shape into loaves and place in greased 8½ × 4½ × 2¾″ ovenproof glass loaf pans. Brush tops with butter; cover and let rise until doubled, about 1 hour.

• Bake in moderate oven (350° F.) about 45 minutes. (Increase oven temperature to 375° F. if you use metal pans.) Cool on racks. Makes 2 loaves.

• Wrap or package loaves individually as soon as they cool. Label, date and freeze.

Recommended storage time: 9 months to a year.

• To serve, let thaw in wrapper at room temperature (on rack to allow air circulation) or heat in foil wrapper in moderate oven (375° F.) 20 minutes. Foil may be opened for last 5 minutes to crisp the crust.

ORANGE-CINNAMON SWIRL: Roll half of dough into rectangle ¼" thick, 6" wide, 20" long. Brush with 1 tblsp. melted butter. Mix 3 tblsp. sugar and 1½ tsp. ground cinnamon. Sprinkle evenly over dough, reserving 1 tblsp. for top of loaf. Roll like jelly roll. Seal ends. Place in greased 8½ × 4½ × 2¾" ovenproof glass loaf pan. Brush top with melted butter; sprinkle with remaining sugar mixture. Let rise until a little more than doubled, about 1¼ hours. Bake in moderate oven (350° F.) about 45 minutes. (Increase oven temperature to 375° if you use metal pans.)

RAISIN BRAN BREAD

Light tender bread. Dough rises more slowly than white bread

1¾	c. milk, scalded
1	c. all-bran cereal
2	tblsp. shortening
2	tblsp. light molasses
1	tblsp. light brown sugar
1½	tsp. salt
1	pkg. active dry yeast
¼	c. warm water (110 to 115° F.)
1½	c. raisins
¼	c. water
5⅓	c. flour (about)

• Stir into milk the cereal, shortening, molasses, sugar and salt; cool.

• Sprinkle yeast on warm water; stir to dissolve.

• Steam raisins in ¼ c. water over medium heat; drain.

• Add 2 c. flour to bran mixture; beat vigorously by hand or with mixer. Stir in yeast. Add raisins and enough remaining flour until dough forms around spoon.

• Turn out on well-floured surface; cover, let rest 5 minutes. Knead dough lightly until springy, about 5 minutes. Dough may be sticky.

• Place in greased bowl; turn to bring greased surface to top. Cover; let rise until doubled, about 1½ hours. Punch down, let rise until doubled.

• Turn out on floured surface; cut in half; cover, let rest 5 minutes. Shape into 2 loaves. Place in well-greased 9 × 5 × 3" loaf pans. Let rise in warm place (80 to 85° F.) until doubled.

• Bake in hot oven (400° F.) 10 minutes; then in moderate oven (375° F.) about 30 minutes. Cool 5 minutes; remove to racks. Brush with butter. Makes 2 loaves.

• Wrap or package loaves individually as soon as they cool. Label, date and freeze.

Recommended storage time: 9 months to a year.

• To serve, let thaw in wrapper at room temperature (on rack to allow air circulation) or heat in foil wrapper in moderate oven (375° F.) 20 minutes. Foil may be opened for last 5 minutes to crisp the crust.

Yeast Breads from Homemade Frozen Dough

Many questions came to our FARM JOURNAL Test Kitchens through the years about a dough to freeze that would produce excellent bread every time. The following six recipes are the answers of our food editors. They are the result of many tests. Do follow the directions when making these treats. They were developed especially for the frozen dough method.

FREEZER WHITE BREAD

Nice evenly shaped loaves, good results every time—a large recipe

12½ to 13½ c. flour
½   c. sugar
2   tblsp. salt
⅔   c. instant nonfat dry milk solids
4   pkgs. active dry yeast
¼   c. softened butter or regular margarine
4   c. very warm water (120 to 130° F.)

• In a large bowl thoroughly mix 4 c. flour, sugar, salt, dry milk solids and yeast. Add butter.
• Gradually add water and beat 2 minutes at medium speed of electric mixer, scraping bowl occasionally. Add 1½ c. flour. Beat at high speed 2 minutes, scraping bowl occasionally. Stir in enough additional flour to make a stiff dough. Turn out onto lightly floured board; knead until smooth and elastic, about 15 minutes. Cover with a towel; let rest 15 minutes.
• Divide dough into 4 equal parts. Form each piece into a smooth round ball. Flatten to form 6" round. Place on greased baking sheets. Cover with plastic wrap. Freeze until firm. Transfer to plastic freezer bags and store in freezer. Recommended storage time: Up to 1 month.
• Remove dough from freezer, place on ungreased baking sheets. Cover; let stand at room temperature until fully thawed, about 4 hours. For round loaves, let thawed dough rise on ungreased baking sheets until doubled, about 1 hour. Or roll each ball into a 12×8" rectangle. Shape into loaves. Place in greased 8½ ×4½ ×2½ " loaf pans. Let rise in warm place until doubled, about 1½ hours.
• Bake in moderate oven (350° F.) about 35 minutes, or until done. Remove from pans and cool on wire racks. Makes 4 loaves.

HONEY/WHEAT GERM BREAD

Light-textured bread from frozen dough. Recipe makes four loaves

9½ to 10½ c. unsifted flour
4   pkgs. active dry yeast
2   tblsp. salt
½   c. honey
¼   c. soft butter
2   c. milk
2   c. water
2   c. wheat germ

• Combine 4 c. flour, yeast and salt in 6-qt. bowl. Add honey and butter. Heat milk and water in saucepan until very warm (120–130° F.). Add to dry ingredients. Beat with electric mixer at medium speed 2 minutes. Add 2 c. flour. Beat at high speed 2 minutes or until thick and elastic. Stir in wheat germ. Stir in enough remaining flour to make a soft dough. Turn out on floured surface; knead until smooth and elastic, about 10 minutes. Cover with plastic wrap and then a towel. Let rest 20 minutes.

• Punch down. Divide in fourths. Roll each into 12×8″ rectangle. Roll up like jelly roll from 8″ side. Seal lengthwise edge and ends well. Wrap in plastic wrap. Place in ungreased 8½×4½×2½″ loaf pan. Freeze until firm. Remove from pans; wrap in aluminum foil. Store in freezer. Recommended storage time: Up to 1 month.

• To bake, unwrap frozen loaf and place in greased 8½×4½×2½″ loaf pan. Cover with plastic wrap. Thaw at room temperature about 2 hours or overnight in refrigerator. Brush with oil. Cover with towel; let rise in warm place until the corners of pan are filled and dough is about 1″ above center of pan, about 3 hours.

• Bake in moderate oven (350° F.) 40 to 45 minutes or until done. If bread browns too quickly, cover loosely with aluminum foil. Makes 4 loaves.

ROUND PUMPERNICKEL LOAVES

Freeze dough, bake when needed

7 to 8 c. unsifted white flour
3 c. unsifted rye flour
2 tblsp. salt
4 large shredded wheat biscuits, broken up
¾ c. cornmeal
3 pkgs. active dry yeast
3½ c. water
¼ c. molasses
2 (1 oz.) squares unsweetened chocolate
1 tblsp. butter
2 c. mashed potatoes
2 tsp. caraway seeds

• Combine flours. Mix together 2 c. flour mixture, salt, shredded wheat biscuits, cornmeal and yeast in 6-qt. bowl. Combine water, molasses, chocolate and butter in saucepan. Heat until very warm (120–130° F.). Add to dry ingredients. Beat with electric mixer at medium speed 2 minutes. Add potatoes and 1 c. flour mixture. Beat at high speed 2 minutes. Stir in caraway seeds and enough flour to make a soft dough. Turn out on floured surface; cover with bowl. Let rest 15 minutes.

• Knead until smooth and elastic, about 15 minutes. Divide in thirds. Form each into a ball. Place on greased baking sheet. Flatten to form 6″ round. Cover with plastic wrap. Freeze until firm. Wrap in aluminum foil. Return to freezer. Recommended storage time: Up to 1 month.

• To bake, unwrap loaf and place on greased baking sheet. Cover with plastic wrap. Thaw at room temperature 3½ hours. Remove plastic wrap; cover with towel. Let rise in a warm place until doubled, about 2 hours.

• Bake in moderate oven (375° F.) 35 minutes or until done. Makes 3 loaves.

APRICOT BRAID

Keep attractive, flavorful loaves in freezer, ready to bake for guests

5½ to 6½ c. flour
¾ c. sugar
1 tsp. salt
3 pkgs. active dry yeast
½ c. softened butter or regular margarine
1 c. very warm water (120 to 130° F.)
3 eggs (at room temperature)
Apricot Filling (recipe follows)
Crumb Topping (recipe follows)

• In a large bowl thoroughly mix 1¼ c. flour, sugar, salt and yeast. Add butter.
• Gradually add water and beat 2 minutes at medium speed of electric mixer, scraping bowl occasionally. Add eggs and ¼ cup flour. Beat at high speed 2 minutes, scraping bowl occasionally. Stir in enough additional flour to make a soft dough. Turn out onto lightly floured board; knead until smooth and elastic, about 8 to 10 minutes.
• Divide dough into 3 equal pieces. Roll one piece into a 12×7″ rectangle. Transfer to greased baking sheet. Starting at a short side, spread one third of Apricot Filling down center third of rectangle. Cut 1″ wide strips along both sides of filling cutting from filling out to edges of dough. Fold strips at an angle across filling, alternating from side to side. Sprinkle with one third of Crumb Topping. Cover tightly with plastic wrap; place in freezer. Repeat with remaining pieces of dough, filling and topping. When firm, remove from baking sheets and wrap each loaf with plastic wrap, then with aluminum foil. Store in freezer.

Recommended storage time: Up to 1 month.
• Remove dough from freezer; unwrap and place on ungreased baking sheets. Let stand covered loosely with plastic wrap at room temperature until fully thawed, about 2 hours. Let thawed dough rise in warm place, free from draft, until more than doubled in bulk, about 1½ hours.
• Bake in moderate oven (375° F.) 20 to 25 minutes, or until done. Remove from baking sheets and cool on wire racks. Makes 3 coffee cakes.

APRICOT FILLING: Combine 2¼ c. dried apricots (11 oz. pkg.) and 1½ c. water in saucepan. Bring to a boil; cook until liquid is absorbed and apricots are tender, about 20 minutes. Sieve; stir in 1½ c. brown sugar, firmly packed. Cool.

CRUMB TOPPING: Combine ½ c. flour, 3 tblsp. sugar and ¾ tsp. ground cinnamon. Mix in 3 tblsp. softened butter or regular margarine until mixture is crumbly.

ORANGE BLOSSOM BUNS

Just form dough into balls and coat with Orange Sugar—easy

5½ to 6½ c. flour
¾ c. sugar
1 tsp. salt
3 pkgs. active dry yeast
½ c. softened butter or regular margarine
1 c. very warm water (120 to 130° F.)
3 eggs
Melted butter or regular margarine
Orange Sugar (recipe follows)

• In a large bowl thoroughly mix 1¼

c. flour, sugar, salt and yeast. Add ½ c. softened butter.

• Gradually add water and beat 2 minutes at medium speed of electric mixer, scraping bowl occasionally. Add eggs and ¼ c. flour. Beat at high speed 2 minutes, scraping bowl occasionally. Stir in enough additional flour to make a soft dough. Turn out on lightly floured board; knead until smooth and elastic, 8 to 10 minutes.

• Divide dough into 3 equal parts. Divide one part into 8 equal pieces; form each piece into a smooth ball. Dip each ball into melted butter, then coat with Orange Sugar. Place in greased 8" round cake pan. Cover pan tightly with plastic wrap, then with aluminum foil; place in freezer. Repeat with remaining pieces of dough. Store in freezer.

Recommended storage time: Up to 1 month.

• Remove dough from freezer. Let stand covered loosely with plastic wrap at room temperature until fully thawed, about 3 hours. Let thawed dough rise in a warm place, free from draft, until more than doubled in bulk, about 2 hours and 15 minutes.

• Bake in moderate oven (350° F.) 25 to 30 minutes, or until done. Remove from pans and cool on wire racks. Makes 2 dozen.

ORANGE SUGAR: Mix together 1 c. sugar and 2 tblsp. grated orange peel.

CINNAMON SWIRL LOAVES

Shaped like a loaf; tastes like coffee cake. Festive when frosted

7¼ to 8¼ c. unsifted flour
⅔ c. sugar

3   pkgs. active dry yeast
1   tsp. salt
⅔ c. soft butter
1   c. milk
¾   c. water
3   eggs
Melted butter
Sugar
Ground cinnamon

• Combine 2 c. flour, sugar, yeast and salt in 6-qt. bowl. Add butter. Heat milk and water in saucepan until very warm (120–130° F.). Add to dry ingredients. Beat with electric mixer at medium speed 2 minutes. Add eggs and 1 c. more flour. Beat at high speed 2 minutes or until thick and elastic. Stir in enough remaining flour to make a soft dough. Turn out on floured surface; knead until smooth, about 10 minutes.

• Divide in thirds. Roll each to 14×8" rectangle. Brush with melted butter. Sprinkle with 3 tblsp. sugar and ½ tsp. cinnamon. Roll up like jelly roll, from 8" side. Pinch seam to seal. Wrap in plastic wrap. Place in ungreased 8½ × 4½ × 2½" loaf pan. Freeze until firm. Remove; wrap in aluminum foil. Return to freezer.

Recommended storage time: Up to 1 month.

• To bake, unwrap frozen loaf and place in greased 8½ × 4½ × 2½" loaf pan. Cover with plastic wrap. Thaw at room temperature about 2½ hours or overnight in refrigerator. Brush dough lightly with oil. Cover with towel. Let rise in warm place until the corners of the pan are filled and the dough is about 1" above center of the pan, about 2 hours.

• Bake in moderate oven (375° F.) 25 to 30 minutes or until done. If you wish, drizzle your favorite confectioners sugar icing over top of slightly warm loaves. Makes 3 loaves.

SAVORY BUNS

Fix hamburgers with seasoned buns —they'll ask where you found the buns

    2 pkgs. active dry yeast
    ½ c. warm water (110 to 115° F.)
    1½ c. milk
    ½ c. soft shortening
    ½ c. sugar
    1 tsp. salt
    ½ tsp. celery salt
    ½ tsp. spaghetti sauce mix
    2 eggs, beaten
    7 to 7½ c. flour

• Dissolve yeast in warm water in large mixing bowl.
• Scald milk; add shortening, sugar, salt, celery salt and spaghetti sauce mix; cool until lukewarm. Add milk mixture, eggs and half the flour to the yeast. Beat with spoon until smooth.
• Add remaining flour gradually, mixing first with spoon and then by hand. Turn onto a lightly floured surface and knead until smooth and elastic, about 5 minutes.
• Place in greased bowl, turning to grease all sides. Cover with towel and let rise in warm place (80 to 85° F.) until doubled, about 1½ hours. Punch down, let rise 30 minutes.
• Divide dough into thirds. Roll each third into a square about ½" thick. Cut in 9 small squares. Tuck square corners under slightly to shape buns. Place on a greased baking sheet. Pat buns out so they are about ½" thick.

Cover with a towel; let rise until doubled, about 30 minutes.
• Bake in hot oven (400° F.) 12 to 15 minutes. Cool on racks. Makes 27.
• Freeze buns as soon as cool. Place in plastic freezer bags or wrap in foil, label and date. Return to freezer. Recommended storage time: 9 months.
• To serve cold, thaw in package at room temperature 2 to 3 hours. To serve hot, place in covered container or aluminum foil wrap and heat in moderate oven (350° F.) until hot, about 20 to 25 minutes.

PETITE BRIOCHE

You'll like the hint of lemon

    1 c. milk
    ½ c. butter or regular margarine
    1 tsp. salt
    ½ c. sugar
    2 pkgs. active dry yeast
    ¼ c. warm water (110 to 115° F.)
    4 eggs, beaten
    1 tsp. grated lemon peel
    5 c. sifted flour (about)
    Melted butter

• Scald milk; stir in butter, salt and sugar. Cool until lukewarm.
• Sprinkle yeast over warm water; stir to dissolve.
• Add eggs and lemon peel; mix well. Combine with cooled milk mixture. Add flour gradually to make a soft dough.
• Knead lightly until dough is smooth and satiny. Place in greased bowl; cover with cloth and let rise in warm place until doubled, about 2 hours. Punch down, turn out on lightly floured board. Knead lightly.

• Shape two thirds of dough into smooth balls about 2″ in diameter. Shape other third into smaller balls, about 1″ in diameter. Place large balls in greased muffin-pan cups; set one small ball firmly on top of each of larger ones. Brush surface with butter; let rise until doubled in bulk. Bake in hot oven (425° F.) about 10 minutes. Remove from pans at once. Cool on racks. Makes about 3 dozen.

• Wrap rolls, as soon as they are cool, in moistureproof material or place in plastic freezer bags or freezer containers. Label, date and freeze. Recommended storage time: 9 months.

• To serve cold, thaw in package at room temperature approximately 2 to 3 hours. To serve hot, place frozen rolls in covered container or aluminum foil wrap and heat in moderate oven (350° F.) 20 to 25 minutes.

COTTAGE CHEESE ROLLS

Delicate, sweet rolls with distinctive flavor—perfect coffee go-with

2¾ c. sifted flour
¼ c. sugar
1 tsp. salt
1 c. small curd creamed cottage cheese (room temperature, drained)
½ c. soft shortening (half butter)
2 pkgs. active dry yeast
¼ c. warm water (110 to 115° F.)
1 egg

Filling:
⅔ c. finely chopped pecans
⅔ c. brown sugar, firmly packed
3 tblsp. melted butter
½ tsp. vanilla

• Measure flour, sugar, salt and cheese into a large bowl. Cut in shortening, like making pastry.

• Add yeast to warm water; let stand a few minutes. Blend egg into yeast mixture. Thoroughly mix yeast into flour mixture. Dough will be sticky.

• Pat dough out on a lightly floured surface. Turn over a few times to coat with flour. Roll out to a 12″ square.

• Mix together filling ingredients. Spread evenly over dough. Roll up like jelly roll. Seal edges. Cut into 18 slices and place in 2 greased 8″ round layer cake pans. Let rise in a warm place (80 to 85° F.) until doubled, about 45 minutes.

• Bake in moderate oven (375° F.) until browned, 20 to 25 minutes. Cool on rack. Makes 18 rolls.

• Wrap rolls, as soon as they are cool, in moistureproof material or place in plastic freezer bags or freezer containers. Seal, label, date and freeze. Recommended storage time: Up to 6 months.

• To serve cold, thaw in packages at room temperature about 2 to 3 hours. To serve hot, place in covered container or aluminum foil wrap and heat in moderate oven (350° F.) until hot, about 20 to 25 minutes. Sprinkle with confectioners sugar.

Wisconsin Coffee Specials

Travel around the world and you'll not find yeast-leavened coffee breads that excel those baked in many Wisconsin country kitchens. Here are two recipes from the Milwaukee area that explain why the cooks there are famous for their yeast breads.

KOLACHES

Fruit topknots on this coffee bread tempt and reward

¾   c. milk
¼   c. sugar
1   tsp. salt
¼   c. soft shortening (part butter)
3½ to 4 c. flour
1   egg (room temperature)
2   pkgs. active dry yeast
¼   c. warm water (110 to 115° F.)

Filling:
1   c. dried prunes
1   c. dried apricots
½   c. sugar
¼   tsp. ground cinnamon
1   tblsp. lemon juice

• Scald milk. Pour into large bowl with sugar, salt and shortening. Cool until lukewarm. Add 1 c. flour; blend in egg. Beat until smooth.

• Add yeast to warm water in small bowl. Let stand 3 to 5 minutes. Stir to dissolve. Add yeast to bowl containing milk-flour mixture. Beat until smooth. Stir in remaining flour, a little at a time. First use a spoon, then the hands. Squeeze dough through fingers to blend. The dough should be soft, but stiff enough to clean sides of bowl.

• Knead dough on lightly floured surface until smooth and elastic. Place in greased bowl; turn once to grease all sides. Cover and let rise in warm place (80 to 85° F.) until doubled, 1 to 1½ hours, or until dent made deep in dough with finger remains.

• While dough rises, prepare filling. Cover prunes and apricots with water and simmer until tender, 20 to 30 minutes. Cool, drain and chop fruits fine. Combine with remaining ingredients.

• Punch down dough. Let rise 15 more minutes. Shape into 1½" balls. Place smooth side up about 3" apart on lightly greased baking sheet. Cover and let rise 10 minutes.

• Make a dent in each ball by pressing finger to the pan. Leave about ½" edge around outside of circle. Spoon filling into center. Brush sides of rolls with melted butter.

• Let rise 20 to 30 minutes. Dent remains when sides of dough are pressed gently with fingers.

• Bake in hot oven (400° F.) until golden brown, 10 to 15 minutes. Remove from baking sheet and cool on racks. Makes about 3 dozen.

• Place cool rolls on baking sheet and put in freezer. When frozen, wrap in foil, label, date and return to freezer. Recommended storage time: Up to 6 months.

• To serve, thaw in wrapper on rack at room temperature about 2 to 3 hours. Or heat frozen rolls, uncovered, in moderate oven (350° F.) 20 to 25 minutes and sprinkle with confectioners sugar.

HOLIDAY COFFEE CAKE

Good yuletide selection but a fine choice for all special occasions

1   c. buttermilk
⅓   c. sugar
1½ tsp. salt
2   pkgs. active dry yeast
¼   c. warm water (110 to 115° F.)
4¾ to 5¼ c. flour
2   eggs
¼   c. soft shortening

Filling:

½   c. strained honey
⅓   c. finely chopped nuts
¼   c. sugar
1   tblsp. grated orange peel
1   tblsp. orange juice
1   tsp. ground cinnamon
⅓   c. raisins
1   tblsp. butter or regular
     margarine

Frosting:

1 c. sifted confectioners sugar
Warm water or light cream

• Heat buttermilk to lukewarm, stirring constantly. (If overheated, it will separate.) Pour into large bowl; add sugar and salt.
• Dissolve yeast in water.
• Beat half the flour into buttermilk mixture; add yeast mixture and continue beating until well blended.
• Add eggs, shortening and more flour, a little at a time. Blend until dough starts to clean sides of bowl. This dough will be soft.
• Turn out on floured surface; knead until smooth and elastic. Round up

dough and place in lightly greased bowl. Turn once to grease all sides.
• Cover and let rise in warm place 1 hour or until deep dent, made by pressing finger into dough, remains.
• Prepare filling by mixing together all ingredients except butter.
• Punch down dough. Turn out on floured surface and divide into 2 equal parts. Keep one half covered with cloth. Roll out other half to make a 10″ square. Brush with butter and sprinkle with half the filling. Roll up like jelly roll. Pinch edges into the roll. Turn pinched edge to bottom and seal by rolling back and forth.
• Cut dough into 1″ slices. Place layer of slices in greased 10″ tube cake pan so slices barely touch. For second layer, place slices alternately around the pan; the third layer as the first. Prepare remaining dough and add slices to pan as before.
• Cover and let rise in warm place about 30 minutes or until doubled. Bake in moderate oven (350° F.) 45 to 60 minutes. Turn out of pan and cool on rack. (Frost after freezing.)
• Wrap with foil, label, date and freeze.
Recommended storage time: Up to 6 months.
• To serve, thaw in wrapper overnight if you want to serve it for breakfast. To serve warm, place thawed coffee cake in moderate oven (350° F.) 20 to 25 minutes.
• Combine sugar and enough warm water for spreading consistency. Spread on heated coffee cake.

No-knead Batter Breads

Batter breads are timesavers. You skip the steps of kneading and shaping the loaves. The bread will not be as fine-textured as a well-kneaded loaf—but it will have the wonderful home-baked taste. Beating takes the place of kneading in these easy-to-mix doughs —a job best turned over to your electric mixer. After the dough rises, you stir it down; then transfer it to the baking pan to rise again.

Wrap and freeze baked breads as soon as they're cool. Batter breads are at their best when fresh-baked, and the freezer will help preserve that quality. After defrosting them, you may want to heat them briefly just before serving, to revive that "fresh-from-the-oven" fragrance and taste.

APPLE CRUMB KUCHEN

Serve warm from the oven with coffee

2¼ c. sifted flour
½ c. sugar
½ tsp. salt
1 pkg. active dry yeast
¼ c. milk
¼ c. water
⅓ c. butter
2 eggs
4 medium apples, pared, cored and sliced
Crumb Topping (recipe follows)
Vanilla Icing (recipe follows)

• Mix together 1 c. flour, sugar, salt and yeast in large bowl.
• Combine milk, water and butter in saucepan. Heat to 120–130° F. Gradually add to dry ingredients; beat 2 minutes at medium speed of electric mixer. Add eggs and ½ c. flour; beat 2 minutes at high speed.
• Stir in enough flour to make a stiff batter. Spread in well-greased 9-inch square baking pan. Arrange apple slices in 4 rows on top. Sprinkle with Crumb Topping. Cover; let rise until doubled, about 1 hour.
• Bake in moderate oven (375° F.) 45 minutes or until apples are tender. Cool in pan 10 minutes. Remove; cool on rack. Makes 1 coffee cake.

CRUMB TOPPING: Combine ⅓ c. sugar, ¼ c. flour, 2 tblsp. butter and 1 tsp. cinnamon; mix until crumbly.

VANILLA ICING: Combine 1 c. sifted confectioners sugar, 1 tsp. vanilla and 1 tblsp. milk; mix until smooth.
• To freeze, wrap coffee cake as soon as it cools. Label, date and freeze. Recommended storage time: Up to 1 month.
• To serve, let thaw in wrapper at room temperature 1½ to 2 hours. Drizzle with Vanilla Icing before serving.

HERBED SOUR CREAM BREAD

A perfect go-along with spaghetti and a tossed green salad

4¾ c. sifted flour
2 tblsp. sugar
2 tsp. salt
2 pkgs. active dry yeast
1 c. warm dairy sour cream
6 tblsp. soft butter
½ tsp. marjoram leaves
½ tsp. orégano leaves

½ tsp. parsley flakes
½ c. very warm water (120–
    130° F.
2 eggs

• Combine 1 c. flour, sugar, salt, yeast, sour cream, butter, marjoram, orégano, parsley and water in bowl. Beat 2 minutes at medium speed of electric mixer. Add eggs and ½ c. flour; beat 2 minutes at high speed.
• Stir in enough remaining flour to make a soft dough. Cover; let rise until doubled, about 35 minutes.
• Stir down. Turn into 2 well-greased 1-qt. casseroles. Cover; let rise until doubled, about 50 minutes.
• Bake in moderate oven (375° F.) 35 minutes or until done. Cool on rack. Makes 2 loaves.
• Wrap loaves individually as soon as they cool. Label, date and freeze.
Recommended storage time: Up to 1 month.
• To serve, let thaw in wrapper at room temperature 1½ to 2 hours.

RAISIN CINNAMON LOAVES

Great toasted and spread with cream cheese

6 c. sifted flour
¼ c. sugar
2 tsp. salt
1 tsp. ground cinnamon
1 pkg. active dry yeast
1 c. milk
½ c. water
½ c. butter
3 eggs
1 c. raisins

• Mix together 1½ c. flour, sugar, salt, cinnamon and yeast in bowl.
• Combine milk, water and butter in

saucepan. Heat to 120–130° F. Gradually add to dry ingredients; beat 2 minutes at medium speed of electric mixer. Add eggs and ½ c. flour; beat 2 minutes at high speed.
• Stir in raisins and enough flour to make a soft dough. Cover; let rise until doubled, about 50 minutes.
• Stir down. Turn into 2 well-greased 1½-qt. casseroles. Cover; let rise until doubled, about 40 minutes.
• Bake in moderate oven (375° F.) 35 minutes or until done. Cool on rack. Makes 2 loaves.
• Wrap loaves individually as soon as they cool. Label, date and freeze.
Recommended storage time: Up to 1 month.
• To serve, let thaw in wrapper at room temperature 1½ to 2 hours.

RICH EGG BATTER BREAD

Give a loaf of this delicious bread to a new neighbor

6½ c. sifted flour
2 tblsp. sugar
2 tsp. salt
2 pkgs. active dry yeast
2 tblsp. softened butter
2 c. very warm water (120–
    130° F.)
3 eggs

• Mix together 1½ c. flour, sugar, salt and yeast in large bowl.
• Add butter. Gradually add water to dry ingredients; beat 2 minutes at medium speed of electric mixer. Add eggs and ½ c. flour; beat 2 minutes at high speed.
• Stir in enough remaining flour to make a soft dough. Cover; let rise until doubled, about 35 minutes.

• Stir down. Turn into 2 well-greased 1½-qt. casseroles. Cover; let rise until doubled, about 40 minutes.
• Bake in moderate oven (375° F.) 35 minutes or until done. Cool on rack. Makes 2 loaves.

• Wrap loaves individually as soon as they cool. Label, date and freeze.
Recommended storage time: Up to 1 month.
• To serve, let thaw in wrapper at room temperature 1½ to 2 hours.

Freezing Quick Breads

Be sure to use a double-acting baking powder to leaven the quick breads you freeze.

BISCUITS: You can roll out the dough, cut out biscuits and pack the raw rounds in layers separated with freezer paper. Seal, label, date and freeze. Biscuits baked from frozen dough sometimes are tough, with a poorer texture and volume than biscuits baked first and then frozen.
Recommended storage time for frozen dough: 2 weeks.

To serve, bake without thawing on a baking sheet in a hot oven (400° F.) 20 to 25 minutes.

Or bake biscuits, cool, store in rigid containers, seal, label, date and freeze.
Recommended storage time: 3 months.

To serve, heat them in a slow oven (300° F.) 10 to 15 minutes. Many women save leftover biscuits this way.

MUFFINS: Pour the batter into muffin-pan cups lined with paper baking cups. Stack with freezer paper between layers. Overwrap with moisture-vaporproof paper. Seal, label, date and freeze.
Recommended storage time: 12 to 14 days.

To serve, thaw at room temperature for 1 hour. Bake as usual.

Or bake the muffins, cool quickly, package, label, date and freeze.
Recommended storage time: 2 to 3 months.

To serve, reheat in a slow oven (300° F.) about 20 minutes.

DOUGHNUTS: Same as for yeast leavened doughnuts.

COFFEE CAKES: Follow directions for fruit and nut breads.

FRUIT AND NUT BREADS: Use your favorite recipes or the ones in this chapter. Bake, cool, package, label, date and freeze.
Recommended storage time: 2 to 3 months.

To serve, thaw in wrapper at room temperature, 1½ to 2 hours.

Or place unthawed in foil wrapper in slow oven (325° F.) 15 to 25 minutes.

Or pour batter into pans, wrap in moisture-vaporproof material and freeze.
Recommended storage time: Up to 2 months.

To serve, thaw batter at room temperature, but do not let stand longer than necessary. Bake the usual way.

WAFFLES: Bake, cool and freeze in rigid containers with freezer paper between waffles. Seal, label and date. Recommended storage time: 2 to 3 months.

To serve, do not thaw. Reheat in electric toaster, set to "light." Or heat 2 to 3 minutes in a hot oven (400° F.)

PANCAKES: See directions for waffles.

Or pour the waffle or pancake batter into tightly covered containers, seal, label, date and freeze. Recommended storage time: 1 week.

To serve, thaw at room temperature. Many country cooks like to thaw the batter overnight in the refrigerator to bake for breakfast. Add a little milk if the batter is too thick.

FRENCH TOAST: See recipe in this chapter.

Fruited Nut Breads

Many a farm homemaker depends on frozen fruit and nut-filled loaves to make extra-good sandwiches in a hurry. To avoid crumbling, slice nut breads while partially frozen. Some make-ahead cooks slice the frozen loaf, fit the slices back together, wrap and refreeze it. Then it is ready to serve when thawed.

APRICOT/NUT BREAD

Bits of golden fruit show in sliced bread giving it appetite appeal

2¾ c. sifted flour
¾ c. sugar
3 tsp. baking powder
½ tsp. baking soda
½ tsp. salt
2 eggs
¼ c. melted shortening or salad oil
1 c. buttermilk
1½ c. dried apricots, cut in thin strips
1 c. chopped nuts

• Sift together dry ingredients, saving out 2 tblsp. flour to coat apricots and nuts.

• Beat eggs well, add melted shortening and buttermilk. Add this mixture to the dry ingredients. Stir only until dry ingredients are moistened. Coat apricots and nuts with the 2 tblsp. flour; fold into batter.

• Pour batter into greased 9×5×3″ loaf pan. Bake in moderate oven (350° F.) about 1 hour. Cool on rack. Makes 1 loaf.

• Wrap, label, date and freeze. Recommended storage time: 2 to 3 months.

• To serve, thaw in wrapper on rack at room temperature 1½ to 2 hours.

BLACK WALNUT/DATE BREAD

Makes the best bread-and-butter sandwiches you ever tasted

¾ c. black walnuts, finely chopped
1 c. sliced dates
1½ tsp. baking soda
½ tsp. salt
¼ c. shortening
¾ c. boiling water
2 eggs
½ tsp. vanilla
1 c. sugar
1½ c. sifted flour

• Combine nuts, dates, soda and salt in mixing bowl. Add shortening and water. Let stand 15 minutes; blend.

• Beat eggs slightly; add vanilla. Sift in sugar and flour and stir until dry ingredients are moistened. (This is a very stiff mixture.) Add to date mixture, mixing until well blended.

• Grease 4 soup cans. Pour batter into cans, filling two thirds full. Cover with aluminum foil. Bake in moderate oven (350° F.) 25 minutes. Remove foil and bake 10 minutes longer or until center tests done. Cool 15 minutes on rack, remove from cans. Cool.

• To freeze, return to cans, wrap, label and date. Or wrap loaves in foil. Recommended storage time: 2 to 3 months.

• To serve, thaw in wrapper on rack at room temperature 1½ to 2 hours.

LEMON BREAD

Keep this loaf with glazed top on hand for teatime sandwiches

⅓ c. melted butter
1¼ c. sugar
2 eggs
¼ tsp. almond extract
1½ c. sifted flour
1 tsp. baking powder
1 tsp. salt
½ c. milk
1 tblsp. grated lemon peel
½ c. chopped nuts
3 tblsp. fresh lemon juice

• Blend butter and 1 c. sugar; beat in eggs one at a time. Add extract.

• Sift together dry ingredients; add to egg mixture alternately with milk. Blend just enough to mix. Fold in lemon peel and nuts. Turn into greased 8½ × 4½ × 2¾" glass loaf pan.

• Bake in slow oven (325° F.) 70 minutes, or until loaf tests done. (Increase oven to 350° if you use metal pan.)

• Mix lemon juice and ¼ c. sugar; immediately spoon over hot loaf in pan. Cool 10 minutes. Remove from pan; cool on rack. Do not cut for 24 hours. Makes 1 loaf.

• Wrap bread, label, date and freeze. Recommended storage time: 2 to 3 months.

• To serve, thaw in wrapper on rack at room temperature 1½ to 2 hours.

PRUNE/OATMEAL BREAD

Makes really good sandwiches

2 c. sifted flour
3 tsp. baking powder
1 tsp. salt
½ tsp. baking soda
¾ c. sugar
1 c. quick-cooking rolled oats
2 eggs
¼ c. salad oil
1 c. milk
1 c. chopped pitted prunes,
    uncooked

• Sift flour, baking powder, salt, soda and sugar into bowl. Stir in oats.

• Beat eggs, add salad oil and milk. Add to dry ingredients and stir just enough to moisten. Stir in prunes.

• Pour into greased 9×5×3" loaf pan. Bake in moderate oven (350° F.) about 1 hour. Cool on rack. Makes 1 loaf.

• Wrap bread, label, date and freeze. Recommended storage time: 2 to 3 months.

• To serve, thaw in wrapper on rack at room temperature 1½ to 2 hours.

CRANBERRY/ORANGE BREAD

Favorite Christmas bread in homes of Wisconsin cranberry growers—not cake-like, but rich enough

2   c. sifted flour
1   c. sugar
1½ tsp. baking powder
½   tsp. baking soda
1   tsp. salt
½   c. chopped nuts
2   c. fresh cranberries, cut in
      halves
1   orange, juice and grated peel
2   tblsp. melted shortening
Water
1   egg, well beaten

• Sift together flour, sugar, baking powder, soda and salt. Add nuts and cranberries.
• Combine orange juice, grated peel and shortening in measuring cup and add enough water to fill cup three quarters full. Stir in egg. Pour into dry ingredients, stir just to mix.
• Spoon mixture into a greased 9×5×3" loaf pan, spreading evenly, but making the corners and sides a little higher than the center.
• Bake in a moderate oven (350° F.) about 1 hour. Remove from pan; cool on rack. Makes 1 loaf.
• Wrap, label, date and freeze.
Recommended storage time: 2 to 3 months.
• To serve, thaw in wrapper on rack at room temperature 1½ to 2 hours.

HIGH-HAT POPOVERS

They turn golden and puff way up in the oven—a Maine hostess favorite

1   c. sifted flour
¾   tsp. salt
2   eggs, beaten
1   tblsp. melted shortening
1   c. milk

• Sift together flour and salt. Combine eggs, shortening and milk; add to flour. Beat smooth with rotary beater.
• To freeze, grease custard cups generously; fill half full of batter. Set cups on tray and put in freezer.
• To package, dip cups in warm water quickly so frozen batter can be slipped out. Place in plastic freezer bags, seal, label, date and return to freezer. Makes 7 popovers.
Recommended storage time: 1 month.
• To bake, put frozen batter back in greased custard cups. Set cups on baking sheet. Place in a cold oven and heat to 400° F. Bake 1 hour.

FROZEN FRENCH TOAST

Don't waste dry bread slices. Make French toast and freeze it

4   eggs, beaten
2   tsp. sugar
½   tsp. salt
¼   tsp. ground nutmeg
¼   tsp. vanilla (optional)
1½ c. milk
8   slices day-old bread (very dry)
Butter or regular margarine

• Combine eggs, sugar, seasonings, vanilla and milk in large bowl; beat well. Dip each bread slice so it will absorb as much mixture as possible.

• Brown bread on both sides in butter in skillet over moderate heat. Cool.

• To freeze, lay slices on greased baking sheet; put in freezer. When frozen, stack slices with foil or freezer paper between; wrap, label, date and freeze. Makes 8 slices.

Recommended storage time: 2 to 3 months.

• To serve, place frozen bread in toaster until heated through.

CINNAMON/SUGAR DOUGHNUTS

Our best loved cake-style doughnuts

4	eggs, beaten
⅔	c. sugar
⅓	c. milk
⅓	c. melted shortening
3½	c. sifted flour
3	tsp. baking powder
¾	tsp. salt
1	tsp. ground cinnamon
¼	tsp. ground mace or nutmeg
Salad oil	

• Beat eggs and sugar until light. Add milk and shortening. Sift dry ingredients together and add to egg-sugar mixture, mixing well. Cover bowl with plastic wrap, refrigerate about 1 hour.

• Remove half the chilled dough from refrigerator to a well-floured pastry cloth or board. Turn over to coat dough with flour. Roll ⅓" thick, cut with floured (3") doughnut cutter. Let rest, uncovered, 10 minutes.

• Heat salad oil 2" deep in electric skillet to 375° F. Drop 3 or 4 doughnuts at a time into hot oil. As doughnuts rise to surface of fat, turn and continue cooking until doughnuts are golden brown on both sides.

• Lift doughnuts from hot oil, letting each doughnut drain a few seconds. Drain on paper towels. Cool on racks.

• Repeat with remaining dough, rerolling scraps. Makes 22 doughnuts.

• Place in freezer containers, seal, label, date and freeze.

Recommended storage time: Up to 2 months.

• To serve, spread doughnuts in pan and heat, uncovered, in slow oven (300° F.) about 20 minutes. Shake them in paper bag with 1 c. sugar mixed with 1 tsp. ground cinnamon.

Slick Freezing Tricks with Bread
GOOD IDEAS FROM FARM JOURNAL READERS

MUFFINS AND BISCUITS: When there are two or three leftover muffins or biscuits, I store them in the freezer in a plastic bag. When I need a quick topping for a casserole dish, I crumble the frozen muffins or biscuits on top the food, dot it with butter and often sprinkle it with cheese.

ASSORTED BREADS: I buy several different kinds of bread at a time and repackage and freeze slices of different kinds in loaves to fit my family's size. We have a choice of breads at our meals. Everyone approves.

TOASTED PARMESAN ROLLS: Split hard rolls in halves, spread with butter, sprinkle with grated Parmesan cheese and freeze. Wrap in foil or package in plastic freezer bags. To serve, place roll halves, cheese side up, on broiler pan and broil to melt cheese and heat rolls.

BISCUITS: Cut rolled biscuit dough with empty juice can. Fit the rounds of biscuit dough in the washed and dried juice can; seal ends with foil and freezer tape. Freeze. To use, remove foil, push the biscuits out from one end and bake immediately.

HOT BUTTERED HARD ROLLS: These rolls, resembling little loaves of bread, I slice in ½″ slices (but not through the lower crust). I butter the slices liberally, wrap each roll in foil and freeze the small packages. When com-

pany comes, I heat the rolls, wrapped, in a slow oven (325° F.) 30 to 40 minutes or until heated through. Everyone breaks off the hot slices of his roll—easier to eat. When I'm in a hurry, I serve the rolls in the packages and let everyone unwrap his own. They are bound to be hot when left in the foil.

BREAD FOR DRESSING: I start dressing for chicken or other poultry and meats by cutting leftover slices of dry bread in small pieces. Package and freeze it—ready to toast and season for dressing.

CHAPTER 6

Cakes, Cookies, and Pies

Freezing Cakes · Butterscotch Nut Torte, Frozen Cake Roll, Oregon Apple Dapple, Coconut Cupcakes, Wonderful Lard Cakes · Freezing Cookies · Freezing Pies · Ice Cream, Lemon Meringue, Chiffon Pies · Pies Made with Frozen Fruits and Berries · Slick Freezing Tricks

When people want to give praise to cakes, cookies and pies, they say they have a "country-kitchen" taste and look—a tribute to farm women. And country cooks depend on freezing these baked foods to relieve humdrum of meals on busy days, to make entertaining easier.

Of all baked foods, pies and cookies go into farm freezers in the largest quantities. The reasons are simple—they're great farm favorites and time-savers. You can stretch minutes by baking two pies instead of one and doubling the cookie recipe, baking a big batch. When days are crowded, farm homemakers freeze cookie dough to bake when needed. Then they can serve cookies still faintly warm and fragrant from the oven. And if they have several kinds of dough frozen, they can pass an impressive choice of cookies to drop-in guests.

The trend in pies is to freeze fillings to add, unthawed, to pastry at baking time. Fillings frozen in pie pans, then removed, wrapped and put in freezers, are economical of space. Some women prefer to store the fillings in freezer containers or plastic freezer bags, thawing them enough to stir before pouring into pastry-lined pans. If you want to awaken appetites dulled by spring fever, bake Apple/Cranberry Pie—a New Jersey special that in March still has autumn's fresh, tart-sweet taste and rich coloring. You'll have requests for second helpings.

Cakes often go into freezers unfrosted. Farm women find they are easy to wrap and that a frosting quickly spread on before serving gives the cake a freshly made appearance. Be sure to try our Cake Roll, Four Ways, a popular FARM JOURNAL fix-and-forget party dessert introduced to us by a Wisconsin homemaker.

Freezing Cakes

Most cakes freeze successfully—especially angel food, sponge, butter, chiffon, fruit and pound cakes. One-bowl cakes stay moist because they contain high proportions of shortening and sugar. And many homemakers find that their fruit cakes actually improve in flavor during freezer storage.

Cakes, baked and then frozen, have better volume and are of superior quality to those made from frozen batter. Good country cooks usually cut down on the clove measurement because this spice's flavor tends to strengthen during freezing. They use pure vanilla, too, because the artificial flavor often gives the cake an undesirable taste. Recent research indicates that chocolate cakes made with part butter and vegetable shortening freeze best.

To Freeze Baked Cakes: Cool the cake quickly on racks; package and freeze as soon as cool. If you plan to serve it within a couple of days, you need not be as particular about the wrap. For longer storage, you will want to wrap the cake in moisture-vaporproof packaging material. Because few cakes are solid when frozen, freeze the wrapped cake and then place it in a freezer carton or metal container to prevent crushing. Return to freezer. Many women like to bake, freeze and store cakes in covered cake pans.

Frosted Cakes: If you want to frost a cake before freezing it, an uncooked confectioners sugar frosting is the best choice. You can use a candy-type frosting like fudge or penuche, but it may become brittle and crack after a month of storage. Some good cooks like to use boiled frosting on the special-occasion cakes they freeze. It is difficult to wrap such cakes so the frosting will not stick to the wrapping. One way to reduce the annoyance is to insert toothpicks around edge of cake to hold wrapping so it just misses contact with the frosting.

Always freeze frosted cakes before wrapping. You will find them much easier to handle. You can store confectioners sugar frosting in a rigid freezer container with a tight-fitting lid. When you want to use the frosting, thaw it in the covered container at room temperature. If too thick to spread, stir in a little milk.

Unfrosted Cake: An unfrosted cake thaws more quickly than a frosted one and you can use it in more ways.

Cupcakes: Bake as usual, cool, pack in freezer containers and seal or wrap individually in foil and freeze. Individually wrapped cupcakes may be tied "daisy chain" style in stockinet.
Recommended storage time:
Unfrosted cakes, 4 to 6 months
Frosted cakes, 2 to 3 months
Cupcakes, 2 months
To use, *thaw cakes in their wrappers* at room temperature.
Frosted cakes (large), about 2 hours
Unfrosted cake layers, about 1 hour
Cupcakes, 30 minutes

You can reduce defrosting time about one third by thawing wrapped cake in front of an electric fan. Thawing wrapped cake on a rack permits air circulation and speeds up the defrosting.

If you have a microwave oven, follow manufacturer's directions for thawing cake.

Note: Some farm women like to thaw unfrosted cake by putting it in a sealed container or aluminum foil wrap and heating it in a very slow oven (250 to 300° F.) for 20 to 25 minutes. They say the heat redistributes the moisture and gives the cake a fresh taste. The trick is not to heat the cake too long and dry it out. An excellent way to thaw frosted cake is to put it, loosely covered, in refrigerator for 3 to 4 hours.

BUTTERSCOTCH NUT TORTE

Keep rich, nutty layers in freezer to frost quickly. Delicious with coffee

6   eggs, separated
1½ c. sugar
1   tsp. baking powder
2   tsp. vanilla
1   tsp. almond extract
2   c. graham cracker crumbs
1   c. chopped nuts
Whipped Cream Frosting (recipe
    follows)
Butterscotch Sauce (recipe follows)

• Beat egg yolks well, slowly adding sugar, then baking powder and flavorings. Mix well.
• Beat egg whites until they hold stiff

peaks; fold into yolk mixture. Fold in crumbs and nuts. Pour into two 9″ round layer cake pans, greased and lined with waxed paper.
• Bake in slow oven (325° F.) 30 to 35 minutes. Cool 10 minutes, then remove from pans; frost with Whipped Cream Frosting when completely cooled if you wish to use at once. Pour Butterscotch Sauce over top, letting it drizzle down sides of torte. Or freeze unfrosted. Makes 12 servings.

To freeze, wrap layers individually, label and date.

Recommended storage time: Unfrosted torte, 6 months.
• To serve, thaw in wrapper at room temperature about 1 hour. Frost.

WHIPPED CREAM FROSTING: Whip 2 c. heavy cream, slowly adding 3 tblsp. confectioners sugar. Spread frosting between layers and over top of torte.

BUTTERSCOTCH SAUCE

¼ c. water
¼ c. melted butter
1   c. brown sugar, firmly packed
1   tblsp. flour
1   egg, well beaten
¼ c. orange juice
½ tsp. vanilla

• Add water to melted butter; blend in brown sugar and flour.
• Add egg, orange juice and vanilla and mix well.
• Bring to boil and cook until mixture thickens. Cool thoroughly. Pour over torte so it drizzles down sides.

ROCKY MOUNTAIN CAKE

Black walnuts in the frosting make the "rocks" which give cake its name

2   c. sifted flour
1½ c. sugar
1   tblsp. baking powder
1   tsp. salt
1   tsp. ground cinnamon
½   tsp. ground nutmeg
½   tsp. ground allspice
¼   tsp. ground cloves
7   eggs, separated
2   tblsp. caraway seeds
½   c. salad oil
¾   c. ice water
½   tsp. cream of tartar
Rocky Mountain Frosting (recipe
   follows)

• Sift flour, sugar, baking powder, salt and spices together several times.
• Combine egg yolks, caraway seeds, oil and water in large bowl. Add dry ingredients. Beat about 30 seconds at low speed on mixer or 75 strokes by hand. (Add a few drops of lemon extract for different flavor.)
• Add cream of tartar to egg whites. Beat until stiff peaks form.
• Gradually pour egg yolk mixture over beaten whites; gently fold in.
• Pour into ungreased 10″ tube pan. Bake in slow oven (325° F.) 55 minutes, then in moderate oven (350° F.) 10 to 15 minutes.
• Invert pan to cool cake. When completely cool, spread Rocky Mountain Frosting over top and sides of cake.

ROCKY MOUNTAIN FROSTING: In saucepan, blend ½ c. butter with 2½ tblsp. flour and ¼ tsp. salt. Cook 1 minute; do not brown. Add ½ c. milk; cook until thick. While hot, add ½ c. brown sugar, firmly packed; beat well. Add 2 c. confectioners sugar, sifted; beat until thick and creamy. Add 1 tsp. vanilla and 1 c. chopped black walnuts (or other nuts). Spread over top and sides of cooled cake.
• Freeze, wrap, label, date and return to freezer.
Recommended storage time: 2 to 3 months.
• To serve, thaw in wrapper at room temperature about 2 hours.

Note: You can make a double batch of frosting and freeze for that extra cake. If you do not freeze cake, increase amount of cloves to ½ tsp.

FROZEN CAKE ROLL, FOUR WAYS

Basic recipe for cake roll. Look at pictures of cakes in this book

4   eggs, separated
¾   c. sugar
1   tsp. vanilla
¾   c. sifted cake flour
¾   tsp. baking powder
¼   tsp. salt

• Beat egg yolks until light and lemon-colored. Slowly add sugar, beating until creamy. Add vanilla; beat.
• Sift together flour and baking powder; gradually add to egg yolk mixture. Beat only until smooth.
• Whip egg whites with salt until stiff, but not dry. Fold into batter.
• Spread batter evenly in greased 15½ × 10½ × 1″ jelly roll pan lined with heavily greased brown paper.
• Bake in moderate oven (375° F.) 15 minutes, or until top springs back

when lightly touched. Loosen cake edges at once; invert onto clean towel sprinkled with confectioners sugar. Cut off hard edges. Roll up, leaving towel in; cool. Unroll; fill with Strawberry Filling (recipe follows); reroll.

• Wrap, label, date and freeze seam side down.

Recommended storage time; Unfilled rolls, 6 months; filled, up to 1 month.

• To serve, take from freezer, spread top with Strawberry Sauce, slice. Makes 1 roll, or 10 (1″) slices. Repeat recipe for Pineapple and Butterscotch Rolls.

CHOCOLATE CAKE ROLL: Follow recipe, sifting ¼ c. cocoa with flour.

CAKE ROLL FILLINGS, SAUCES

BASIC FILLING: Whip 1 c. heavy cream until it begins to thicken. Gradually add 3 tblsp. sugar and ¼ tsp. vanilla (use almond extract instead of vanilla for Pineapple Roll); beat stiff.

STRAWBERRY ROLL: Fold 1 (10 oz.) pkg. frozen strawberry slices, drained, into whipped cream. Spread on cake; roll. For glaze, bring to boil ¼ c. strawberry jam and ¼ c. light corn syrup; brush on top of roll. Serve with Strawberry Sauce.

STRAWBERRY SAUCE: Mix 1 c. strawberry jam and 1 c. light corn syrup; bring to boil. Cool. Makes 1 pint.

PINEAPPLE ROLL: Fold 1 (8¼ oz.) can crushed pineapple, drained, into whipped cream. Spread over cake; roll. For glaze, bring to boil ¼ c.

apricot jam and ¼ c. light corn syrup; brush on top of roll. Serve with Pineapple Sauce.

PINEAPPLE SAUCE: Mix 1 (8¼ oz.) can crushed pineapple, drained, with 1 c. light corn syrup. Bring to boil and cook until thickened. Makes 1 pint.

BUTTERSCOTCH ROLL: Fold 1 (3 oz.) can chopped pecans into whipped cream. Spread over cake; roll. For glaze, heat ¼ c. light corn syrup and 1 tblsp. melted butter or regular margarine. Brush on top of roll. Sprinkle with ¼ c. chopped pecans. Serve with Butterscotch Sauce.

BUTTERSCOTCH SAUCE: Combine ⅔ c. light corn syrup, 1¼ c. brown sugar, firmly packed, ¼ c. butter and ¼ tsp. salt; boil to heavy syrup; cool. Add 1 (6 oz.) can evaporated milk. Makes about 1 pint.

CHOCOLATE ROLL: Spread whipped cream over cake; roll. Sift ¼ c. confectioners sugar over roll. Serve with Chocolate Sauce.

CHOCOLATE SAUCE: Put 4 squares unsweetened chocolate, ½ c. butter and 2¼ c. evaporated milk in top of double boiler; heat until butter and chocolate melt. Slowly add 3 c. sugar; heat until sugar dissolves. Cool; refrigerate. (Sauce will thicken. If too thick, thin with corn syrup.) Makes 1 quart.

Note: Cake rolls may be filled with different kinds of ice cream, softened just enough to spread, and returned to the freezer. Serve with a sundae sauce.

BOILED RAISIN CAKE

Preferred above all fruit cakes by a Nevada ranch family. Do try it

1½ c. sugar
¼ c. cocoa
2½ c. water
⅔ c. butter
2 c. raisins
2 tsp. ground cinnamon
½ tsp. ground cloves
½ tsp. ground nutmeg
¼ tsp. ground allspice
½ tsp. salt
3½ c. sifted flour
2 tsp. baking powder
1 tsp. baking soda
1 c. chopped walnuts

• Mix sugar and cocoa. Add water, butter, raisins, spices and salt. Boil 4 minutes. Cool to room temperature.
• Sift together flour, baking powder and soda. Add to raisin mixture and mix until well blended. Stir in walnuts. Pour into greased 13×9×2″ pan. Bake in moderate oven (350° F.) 35 minutes or until done. Remove from oven and cool on rack.
• Wrap, label, date and freeze.
Recommended storage time: 6 months.
• To serve, thaw in wrapper at room temperature about 1 to 2 hours.

OREGON APPLE DAPPLE

Two words describe the way this torte tastes—extra-delicious

¼ c. butter or regular margarine
1 c. sugar
1 egg
1 c. sifted flour
1 tsp. baking soda
¼ tsp. salt
1 tsp. ground cinnamon
¼ tsp. ground nutmeg
1 tsp. vanilla
2 c. grated tart apples
½ c. chopped walnuts

• Cream together butter and sugar. Beat in egg.
• Sift together flour, soda, salt and spices. Stir into creamed mixture. Stir in vanilla, apples and nuts.
• Bake in a greased 8″ square pan in moderate oven (350° F.) about 45 minutes. Serve warm with sauce or ice cream. Makes 6 to 8 servings.
• Or freeze. When torte is cool, wrap, label, date and freeze.
Recommended storage time: 2 to 3 months.
• To serve, place unthawed in slow oven (300° F.) 30 minutes. Excellent served with Elin's Sauce, which will keep in the refrigerator 2 to 3 months.

ELIN'S SAUCE: Heat ½ c. butter or regular margarine, ½ c. light cream and 1 c. sugar to boiling point and simmer, stirring occasionally, 20 minutes.

COCONUT CUPCAKES

Coconut serves for frosting, but you may add frosting if you desire

½ c. soft shortening
2 c. sifted cake flour
1 c. sugar
2½ tsp. baking powder
¾ tsp. salt
1 egg
¾ c. milk
1 tsp. vanilla
Flaked coconut

• Stir shortening in bowl to soften. Sift in dry ingredients. Add the egg and half the milk. Beat 2 minutes on low speed of mixer. Add rest of milk and vanilla. Beat 1 minute.

• Place paper liners in medium muffin-pan cups. Fill half full with batter. Top each with 1 tsp. coconut.

• Bake in moderate oven (375° F.) 20 to 25 minutes. Remove from oven; cool on racks. Makes 20 cupcakes.

• Wrap, label and date, or freeze without wrapping, place in plastic freezer bags and return to freezer. Recommended storage time: 2 months.

• To serve, thaw in wrapping at room temperature, about 1 hour.

CARROT/PECAN CAKE

A spiced, nut cake that keeps well—if you hide it in freezer

1¼ c. salad oil
2 c. sugar
2 c. sifted flour
2 tsp. baking powder
1 tsp. baking soda
1 tsp. salt
2 tsp. ground cinnamon
4 eggs
3 c. grated raw carrots
1 c. finely chopped pecans
Orange Glaze (recipe follows)

• Combine oil and sugar; mix well.

• Sift together remaining dry ingredi-

ents. Sift half of dry ingredients into sugar mixture; blend. Sift in remaining half alternately with eggs, one at a time, mixing well after each.

• Add carrots and mix well; then mix in pecans. Pour into lightly oiled 10″ tube pan. Bake in slow oven (325° F.) about 1 hour and 10 minutes. Cool in pan upright. Remove from pan.

• Freeze, wrap, label, date and return to freezer.

Recommended storage time: 2 to 3 months.

• To serve, thaw in wrapper at room temperature 2 hours.

• Split frozen cake into 3 horizontal layers. Spread Orange Glaze between layers and on top and sides.

ORANGE GLAZE

1 c. sugar
¼ c. cornstarch
1 c. orange juice
1 tsp. lemon juice
2 tblsp. butter
2 tblsp. grated orange peel
½ tsp. salt

• Combine sugar and cornstarch in saucepan. Add juices slowly and stir until smooth. Add remaining ingredients. Cook over low heat until thick and glossy. Cool and spread on cake.

Wonderful Lard Cakes

No collection of FARM JOURNAL's cake recipes is complete without including lard cakes made by the meringue method invented in our Coun-

tryside Kitchens. They are treasures in hundreds of recipe files across country. Cakes they produce continue to bring happiness and compliments.

And they are never better than when teamed with the frostings and glazes designed especially to glamorize them. Do try these cakes.

ORANGE LARD CAKE

Dress up this cake by sprinkling coconut on freshly spread frosting

- 2 eggs, separated
- ½ c. sugar
- ⅓ c. lard
- 2¼ c. sifted cake flour
- 1 c. sugar
- 2½ tsp. baking powder
- 1 tsp. salt
- ¼ tsp. baking soda
- ¾ c. milk
- ⅓ c. orange juice, fresh or reconstituted frozen
- ¼ tsp. almond extract

Orange Butter Cream Frosting recipe follows)

- Beat egg whites until frothy. Gradually beat in ½ c. sugar. Continue beating until very stiff and glossy.
- In another bowl stir lard to soften. Add sifted dry ingredients and milk. Beat 1 minute, medium speed on mixer. Scrape bottom and sides of bowl constantly.
- Add orange juice, egg yolks and almond extract. Beat 1 minute longer, scraping bowl constantly.
- Fold in egg white mixture.
- Pour into 2 greased and floured 9″ round cake pans. Bake in moderate oven (350° F.) 25 to 30 minutes.
- Cool layers in pans on racks 10 minutes; then remove from pans. Frost with Orange Butter Cream Frosting. Or freeze unfrosted cake and frost shortly before serving.

- Freeze; wrap, label, date and return to freezer.

Recommended storage time: 2 to 3 months for frosted cake, 4 to 6 months for unfrosted cake.

- To serve, thaw in wrapper at room temperature about 2 hours.

ORANGE BUTTER CREAM FROSTING: Cream ½ c. butter. Gradually add 3 c. sifted confectioners sugar alternately with ⅓ c. orange juice concentrate, thawed, creaming well after each addition until light and fluffy.

YELLOW LARD CAKE

A 2-egg cake with an old-time country taste made in a new way

- 2 eggs, separated
- ½ c. sugar
- ⅓ c. lard
- 2¼ c. sifted cake flour
- 1 c. sugar
- 3 tsp. baking powder
- 1 tsp. salt
- 1 c. plus 2 tblsp. milk
- 1½ tsp. vanilla

Peanut Butter Frosting (recipe follows)

- Beat egg whites until frothy. Gradually beat in ½ c. sugar. Continue beating until very stiff and glossy.
- In another bowl stir lard to soften. Add sifted dry ingredients, ¾ c. milk and vanilla. Beat 1 minute, medium speed on mixer. Scrape sides and bottom of bowl constantly.
- Add remaining milk and egg yolks. Beat 1 minute, scraping bowl constantly. Fold in egg whites.
- Pour into 2 greased and floured 9″ round layer cake pans. Bake in mod-

erate oven (350° F.) 25 to 30 minutes. Cool layers on racks 10 minutes; then remove from pans.

• Brush crumbs from cake while warm for ease in frosting. Frost with Peanut Butter Frosting. Or freeze unfrosted, add frosting before serving.

• Freeze; wrap, label, date and return to freezer.

Recommended storage time: 2 to 3 months for frosted cake, 4 to 6 months for unfrosted cake.

• To serve, thaw in wrapper at room temperature about 2 hours.

PEANUT BUTTER FROSTING: Whip ¾ c. crunchy peanut butter with mixer (or wooden spoon). Gradually beat in ¾ c. cold Basic Sugar Syrup (see recipe in this chapter).

FUDGE PUDDING CAKE

Made with lard, it's chocolate all the way—moist fudgy cake

⅓ c. lard
3 squares unsweetened chocolate
¾ c. sugar
2 tblsp. cornstarch
1 c. milk
2 c. sifted flour
1 c. sugar
1¼ tsp. baking soda
½ tsp. salt
¾ c. milk
2 eggs
Glossy Fudge Frosting (recipe follows)

• Melt lard in 2- or 3-qt. saucepan over low heat. Add chocolate and stir until melted. Remove from heat.

• Mix ¾ c. sugar and cornstarch thoroughly. Stir into chocolate mixture. Add 1 c. milk; stir and cook over low heat until smooth and thick. Remove from heat and cool to room temperature. (Set pan in cold water to speed cooling.)

• Sift remaining dry ingredients together 3 times. Add ½ to chocolate mixture, blend and beat ½ minute (75 strokes). Add ¾ c. milk and blend. Add remaining dry ingredients, blend; beat 1 minute (150 strokes).

• Add eggs, blend and beat ½ minute (75 strokes).

• Bake in paper-lined 13×9×2″ pan in moderate oven (350° F.) about 35 minutes. Cool.

• Frost with Glossy Fudge Frosting.

• Freeze; wrap, label, date and return to freezer.

Recommended storage time: 2 to 3 months.

• To serve, thaw in wrapper at room temperature, about 2 hours.

GLOSSY FUDGE FROSTING

2 tblsp. butter or regular margarine
1 (6 oz.) pkg. semisweet chocolate pieces
⅓ c. Basic Sugar Syrup

• Melt butter over hot (not boiling) water. Remove from heat before the water boils.

• Add chocolate pieces to butter over hot water and stir until melted, blended and thick.

• Add Basic Sugar Syrup gradually, stirring after each addition until blended. Mixture will become glossy and smooth.

• Remove from hot water and cool until of spreading consistency. Spread thinly over cake. Apply quickly by pouring a small amount at a time on top of cake; as it runs down on sides, spread with spatula. Frost top of cake last. Makes frosting for two 8″ layers, or one 13×9×2″ cake.

BASIC SUGAR SYRUP

2 c. sugar
1 c. water

• Boil sugar and water together 1 minute. Pour into jar. Cool, cover and refrigerate. Makes about 2 cups.

Note: Frostings made with Basic Sugar Syrup in higher altitudes (above 3500 feet) take longer and more beating to become thick enough to stay on the cake.

Freezing Cookies

You will kindle delight in your children's eyes if they come home from school to baked cookies from the freezer, or to freshly baked cookies from frozen dough. Take your pick of the two methods, or use both of them at different times, depending on which is more convenient at the moment. Both methods give satisfactory results. Most country cooks prefer to freeze the dough because it takes up less space in the freezer than cookies.

Unbaked Cookies: Here are directions for freezing dough for four types of cookies.

Refrigerator-type:

1. Shape the dough in a roll with a diameter of the desired size. Wrap in plastic wrap or foil; freeze.

To use, thaw just enough to slice with sharp knife, place slices on a lightly greased baking sheet; bake.

2. Or chill the roll of dough in the refrigerator several hours. Slice and store slices in layers, pieces of plastic wrap or aluminum foil between, in freezer containers. Seal, label, date

and freeze. You will reduce the chances of crushing the dough if you hold number of layers to a minimum.

To use, bake slices without thawing.

3. Or pack the dough into empty juice cans, opened at both ends. Seal with foil and freezer tape; label, date and freeze.

To use, remove end wrappings; push out the dough. Slice and bake.

Bar-type:

1. Spread the dough evenly in greased pans; cover with plastic wrap or foil. Seal, label, date and freeze.

To use, bake dough in pan, frozen.

2. Or store dough in wide-topped jar or other freezer container; cover with tight-fitting lid. Seal, label, date and freeze.

To use, thaw dough just enough that you can spread it in a greased pan. Bake as usual.

Drop-type:

1. Store dough in airtight containers and cover or wrap in plastic wrap or foil. Seal, label, date and freeze.

To use, thaw dough at room temperature, just enough that you can drop it from a spoon onto a lightly greased baking sheet. Bake as usual.

2. Or drop dough from spoon onto lightly greased baking sheet and freeze. Remove from freezer, pack unbaked cookies in freezer containers. Cover, seal, label, date and freeze.

To use, bake unthawed cookies.

Rolled-type:

Roll dough, cut and store cutouts in rigid freezer containers with plastic wrap or aluminum foil between layers. Seal, label, date and freeze.

To use, bake unthawed cookies.

Baked Cookies: Cool the cookies quickly when you take them from the oven. Arrange baked cookies in a sturdy box lined with plastic wrap or aluminum foil, separating cookie layers with more plastic wrap or foil. Fold over plastic wrap to cover top of cookies and seal with freezer tape, or press foil over cookies to seal. Close box, label, date and freeze.

To use, let cookies thaw at room temperature in their wrappings. This takes about 10 minutes.

Recommended storage time: Baked and unbaked cookies, 9 to 12 months.

OATMEAL CHIPPERS

All members of the family will like this cookie with chocolate nuggets

½ c. butter or regular margarine
½ c. shortening
1 c. sugar
1 c. brown sugar, firmly packed
2 eggs
1 tsp. vanilla
2 c. sifted flour

1 tsp. baking soda
1 tsp. salt
1 tsp. ground cinnamon
1 tsp. ground nutmeg
2 c. quick-cooking rolled oats
1 (6 oz.) pkg. semisweet chocolate pieces (1 c.)
1 c. chopped walnuts

• Cream together butter and shortening. Add sugars gradually, beating until light and fluffy. Beat in eggs and vanilla. Blend in sifted dry ingredients, mixing thoroughly. Stir in oats, chocolate and nuts.

• Drop by rounded teaspoonfuls 2″ apart onto greased baking sheet. Bake in moderate oven (375° F.) 9 to 12 minutes. Makes 8 dozen.

• Cool cookies quickly on racks; package, label, date and freeze. Recommended storage time: 9 to 12 months.

To serve, let thaw in containers or wrap a few minutes.

FRUIT BARS

Excellent cookies to mail for gifts— they're good keepers

2 c. seedless raisins
1½ c. chopped mixed candied fruit
1 c. chopped walnuts
½ c. orange or pineapple juice
2 tsp. vanilla
1 c. butter or regular margarine
1 c. sugar
1 c. brown sugar, firmly packed
2 eggs, beaten
4½ c. sifted flour
2 tsp. ground cinnamon
2 tsp. baking powder
1 tsp. baking soda

• Rinse raisins in hot water, drain; dry on towel.

• Combine raisins, candied fruit, nuts, juice and vanilla; let stand.

• Cream together butter, sugars and eggs. Sift together dry ingredients and add in thirds to creamed mixture; mix until smooth. Add fruit mixture; blend well. Let stand 1½ hours in refrigerator, or overnight.

• When ready to bake, spread dough in greased 15½ × 10½ × 1″ jelly roll pan. Bake in hot oven (400° F.) 15 to 20 minutes, until lightly browned. Cool in pan. When cool, cut in bars about 3 × 1″. Makes about 4 dozen.

• Package, label, date and freeze. Recommended storage time: 9 to 12 months.

• To serve, thaw cookies in containers or wrap a few minutes.

SIX-IN-ONE COOKIES

You make 18 dozen cookies of six flavors from one batch of dough

2 c. butter
1 c. sugar
1 c. light brown sugar, firmly packed
2 eggs, beaten
1 tsp. vanilla
4 c. flour
1 tsp. baking soda
½ tsp. salt

Dough: Cream butter; gradually add sugars. Cream until light and fluffy.

• Add eggs and vanilla; mix well.

• Sift together flour, soda and salt; gradually add to creamed mixture, beating well after each addition.

Flavors: Divide dough in 6 parts. Add ½ c. shredded coconut to one; ½ c. finely chopped pecans to second; ½ tsp. nutmeg and 1 tsp. cinnamon to third; 1 square unsweetened chocolate, melted, to fourth; ¼ c. finely chopped candied cherries to fifth; leave last portion plain. Chill 30 minutes or longer.

• Shape dough into 6 rolls about 1¾″ in diameter.

• Wrap, seal, label, date and freeze. Recommended storage time for dough: 9 to 12 months.

• When ready to use, slice frozen dough ⅛″ thick. Bake slices on lightly greased baking sheet in moderate oven (375° F.) 10 to 12 minutes. Cool on racks. Makes 18 dozen.

GINGER COOKIES FOR A CROWD

A big recipe to make when you wish to put cookies in the freezer

5½ c. sifted flour
1 tblsp. baking soda
2 tsp. baking powder
1 tsp. salt
¾ tsp. ground ginger
1 tsp. ground cinnamon
1 c. shortening
1 c. sugar
1 egg, beaten
½ tsp. vanilla
1 c. dark molasses
½ c. strong coffee

• Sift together flour, soda, baking powder, salt, ginger and cinnamon.

• Cream shortening; add sugar gradually; beat until light; add egg and vanilla. Add molasses and coffee, then dry ingredients; mix well; chill.

- Roll out on lightly floured board ¼″ thick; cut with round 2″ cutter.
- Place about 2″ apart on greased baking sheet. Bake in hot oven (400° F.) 8 to 10 minutes. Spread on racks to cool. Makes 12 dozen.
- Package, label, date and freeze. Recommended storage time: 9 to 12 months.
- To serve, thaw cookies in containers or wrap a few minutes.

PEPPER NUTS (WITH LARD)

Nibblers like this new version of pfeffernuesse—a German cookie

½	c. lard
1½	c. sugar
1½	c. light corn syrup
¼	tsp. vanilla
6	c. sifted flour
4½	tsp. baking powder
¼	tsp. salt
¼	tsp. ground cardamon
¼	tsp. ground cinnamon
¼	tsp. ground nutmeg
¼	tsp. ground cloves
¼	tsp. ground allspice
¼	tsp. pepper
½	c. milk

- Cream lard; beat in sugar gradually. Blend in corn syrup and vanilla.
- Add dry ingredients, sifted together, alternately with milk. Roll in strips the width of a pencil on a floured surface. Cut in ½″ pieces.
- Place on greased baking sheet and bake in moderate oven (375° F.) about 15 minutes, until browned. Makes 4½ quarts.
- Package, label, date and freeze. Recommended storage time: 6 months if lard contains antioxidant; without, 3 to 4 months.
- To serve, thaw in container or wrap a few minutes. Serve in bowls like nuts or popcorn.

Freezing Pies

Country pie bakers, the best in the world, divide into three groups: those who freeze baked pies, those who freeze pies and bake them later and the up-to-the-minute cooks who fix the fillings and freeze them to combine with pastry at baking time. All agree that the freezer saves time and work because it is almost as easy to make two or three pies or fillings as one.

Many kinds of pies freeze successfully, like those with fruit, vegetable (pumpkin, sweet potato and squash) and mincemeat fillings. Chiffon pies with whipped cream or egg whites folded into the gelatin mixture freeze well and do not "weep" when served. On the other hand, custard pie rarely retains its top quality when frozen.

Pie Shells: You can freeze both pastry and crumb crusts, baked or unbaked. Almost every farm cook has her favorite way of packaging unbaked pastry and here is one system some pie bakers like. Roll pastry the size and shape you will need when baking pie.

Cut a piece of cardboard the same size and cover it with plastic wrap taped in place or with foil. Lay a layer of pastry flat on the cardboard, top with two layers of plastic wrap or foil, then with another layer of pastry. Repeat until you have several layers. Put this stack in a plastic freezer bag, twist to exclude air and fasten with a twist seal.

To use, remove one piece of pastry from the plastic freezer bag, with a sheet of plastic wrap or foil beneath and on top to prevent moisture from condensing on pastry during thawing. Thaw at room temperature 10 to 15 minutes, use like freshly made pastry.

Pie and Tart Shells: Bake pie and tart shells and rounds of pastry in a very hot oven (475° F.) 8 to 10 minutes. Cool; package in rigid containers to prevent crushing and overwrap with moisture-vaporproof packaging material. Or freeze without baking, wrap like baked ones.

Recommended storage time: Baked, 4 to 6 months. Unbaked, 2 months.

To serve, heat baked tart and pie shells and pastry rounds in a slow oven (325° F.) 5 to 8 minutes. Put unbaked pie and tart shells and rounds of pastry in a very hot oven (475° F.) 8 to 10 minutes.

Baked and Unbaked Pies: You can freeze pies before or after baking. Both kinds are satisfactory, although frozen unbaked pies often have a soggy undercrust. We recommend freezing baked pies.

If you bake pies in shiny, lightweight aluminum pans, set them on a baking sheet for baking. *Bake and*

reheat pies on the lower shelf of the oven. And if you use frozen fruit for the filling, thaw it and drain. Also drain canned fruits. Use only a small part of the juice and thicken it either with tapioca or cornstarch, or a combination of tapioca and cornstarch. You can use flour for thickening, but the filling will not be so clear or bright in color.

Unbaked Pies: When using fresh peaches or raw apples, you need to treat them to prevent darkening. Add 1 tblsp. lemon juice or ¼ tsp. powdered ascorbic acid, mixed with 1 tblsp. water to the peaches for a pie. (Handle apricots like peaches.) Dip the apple slices in a solution of ½ tsp. powdered ascorbic acid and 1 c. water (enough for apples to make 5 pies). Or steam apple slices 2 or 3 minutes and cool quickly. Coat sweet cherries and berries with sugar and flour before adding to pie pan. Freeze the pie and then wrap with plastic wrap and seal with freezer tape. Or wrap in aluminum foil and seal with double fold. Place in plastic freezer bag; seal, label, date and freeze.

Recommended storage time: 2 to 3 months.

To serve, unwrap pie and bake on the lower shelf of very hot oven (450° F.) 10 to 15 minutes; complete baking at 375° F. If pie rim browns too quickly, cover with strips of foil.

Baked Pies: Remove the two-crust pies from the oven while crust is light brown. They will brown more when you reheat them. Cool and freeze without wrapping, but be sure the pie is level while freezing. Package like unbaked pies and return to freezer.

Recommended storage time: 4 to 6 months.

To serve, partially thaw baked pie in its original wrap at room temperature, about 30 minutes. Unwrap and place on lower shelf of moderate oven (350° F.) 30 minutes, until warm.

Chiffon Pies: Freeze without wrapping; package like unbaked pies and return to freezer.

Recommended storage time: up to 1 month. (Some cooks freeze chiffon pies 2 to 3 months, but the longer storage often results in rubbery filling.)

To serve, thaw at room temperature about 45 minutes or in refrigerator 2 to 3 hours. Do not heat.

Pie Fillings: An increasing number of farm homemakers freeze pie fillings rather than pies. They take up less space in the freezer. Freeze fruit pie fillings in foil-lined pans, remove filling from pans, overwrap and return to freezer. Place the unwrapped, unthawed filling in pastry-lined pan, add top crust and bake in a hot oven (425° F.) from 10 to 20 minutes longer than for pie with unfrozen filling. Or thaw filling until you can almost stir it before adding to pastry.

Recommended storage time: 2 to 3 months.

Among the favorite mixes containing eggs for freezing are those for pumpkin, squash and sweet potato pies. The usual method is to combine the vegetable, milk, sugar and eggs. Some cooks omit the spices, especially cloves, which increase in strength during storage, and add them at baking time.

Recommended storage time: 5 weeks.

To serve, partially thaw the filling in its container. Add spices if needed. Pour filling into pastry-lined pans and bake pies in the regular way, in a hot oven (400° F.) 45 to 55 minutes, adding 10 to 15 minutes because filling is cold. Test for doneness by inserting a knife 1" from side of filling to see if it comes out clean.

STRAWBERRY CHIFFON PIE

Pie is lovely in color and flavor

Baked 9" pie shell
¼ c. sugar
1 envelope unflavored gelatin
1 (10 oz.) pkg. frozen strawberry
     halves, thawed
3 egg whites
¼ tsp. cream of tartar
Dash of salt
⅓ c. sugar
½ c. heavy cream, chilled

· Stir together in saucepan ¼ c. sugar and gelatin; add strawberries and cook over medium heat, stirring constantly, just until mixture reaches a boil. Remove from heat at once. Set pan in bowl of ice water to chill. Stir occasionally while chilling. Chill until mixture mounds slightly when dropped from a spoon.

· Beat egg whites with cream of tartar, salt and ⅓ c. sugar until mixture is stiff and glossy (do not underbeat). Fold into strawberry mixture.

· Beat cream until stiff; fold into strawberry mixture. Pile into pie shell. Freeze; then overwrap and return to freezer.

Recommended storage time: Up to 1 month.

• To serve, thaw in refrigerator 2 to 3 hours, or at room temperature about 45 minutes. To use without freezing, refrigerate until set, about 3 hours.

PEACH PIE FILLING

Stack several of these in your freezer when peaches are ripe

4   qts. sliced peeled peaches
      (9 lbs.)
1   tsp. powdered ascorbic acid
1   gal. water
3½ c. sugar
½   c. plus 2 tblsp. quick-cooking
      tapioca
¼   c. lemon juice
1   tsp. salt

• Place peaches in large container. Dissolve ascorbic acid in water and pour over peaches. Drain.
• Combine peaches, sugar, tapioca, lemon juice and salt.
• Line four 8″ pie pans with heavy-duty aluminum foil, letting it extend 5″ beyond rim. Divide filling evenly between pans. Makes fillings for four 9″ pies.

To freeze, fold foil loosely over fillings; freeze. Remove from freezer, turn filling from pans and wrap snugly with foil. Return to freezer. Recommended storage time: 6 months.

To bake, remove foil from frozen pie filling and place it, unthawed, in a pastry-lined 9″ pie pan. Dot with butter and if you like, sprinkle on ¼ tsp. ground nutmeg or cinnamon. Adjust top crust; flute edges and cut vents.

Bake in hot oven (425° F.) 1 hour and 10 minutes or until syrup boils with heavy bubbles that do not burst.

APPLE/CRANBERRY PIE FILLING

Taste-testers voted pies made with frozen filling extra-good and pretty

3   c. sliced peeled apples
2   c. cranberries
1¾ c. sugar
¼   c. flour
1   tsp. ground cinnamon
1½ tsp. melted butter

• Combine all ingredients. Arrange in a 9″ pie pan lined with heavy-duty aluminum foil, letting foil extend 5″ beyond rim. Freeze like Peach Pie Filling.
Recommended storage time: 3 months.
• To bake, remove foil and place in pastry-lined 9″ pie pan without thawing. Top with pastry strip lattice.
• Bake in a hot oven (425° F.) 1 hour and 10 minutes.

CONCORD GRAPE COBBLER

What a midwinter treat! It's bound to get a royal table reception

10 c. stemmed and washed grapes
2   c. sugar
2½ tblsp. quick-cooking tapioca
¼   tsp. salt
⅛   tsp. ground cinnamon
2   tblsp. lemon juice
1   tblsp. butter
Biscuit Topping (recipe follows)

• Slip skins from grapes; set aside. Heat pulp to boiling; rub through

coarse sieve or food mill to remove seeds. Discard seeds.
• Combine sugar, tapioca, salt and cinnamon. Add lemon juice and grape pulp. Cook until thickened, stirring.
• Remove from heat; add skins; mix.
• Pour into 8" square foil-lined baking pan; cool and freeze.
• Remove block of filling from pan; overwrap and return to freezer. Recommended storage time: 2 to 3 months.
• To serve, return unthawed filling to pan. Dot with butter. Bake in hot oven (400° F.) about 35 to 40 minutes until bubbling, stir occasionally.
• Remove from oven, cover with Biscuit Topping; return to oven and bake about 20 minutes. Serve warm. Makes 9 servings.

Note: This filling will make two 8 or 9" pies. Pour into foil-lined pie pans; cool and freeze. Remove from pans; overwrap and return to freezer. Recommended storage time: up to 3 months.
• To serve, place unthawed filling in pastry-lined pan; dot with butter; top with lattice crust. Bake in very hot oven (475° F.) 15 minutes; reduce oven to 375° F. and cover pastry rim with foil; bake 40 to 50 minutes.

BISCUIT TOPPING

1½ c. sifted flour
2¼ tsp. baking powder
1   tblsp. sugar
¼   tsp. salt
¼   c. butter or regular margarine
¼   c. milk
1   egg, slightly beaten

• Sift together dry ingredients.
• Cut in butter. Make well in center; add milk and egg all at once; stir with fork until mixture is a soft dough. Turn out on lightly floured pastry cloth and knead dough 10 times. Roll dough into 9×7" rectangle; cut in seven 9" strips. Place 4 strips on hot filling one way, lattice other 3 strips opposite way. Start with center strip. Bake as directed.

ROSY CRAB APPLE PIE

Red, red apples make the prettiest pie

Pastry for 2-crust pie
1   c. sugar
1   tblsp. flour
¼   tsp. salt
6   c. finely chopped unpeeled
       crab apples
1   tsp. vanilla
1½   tblsp. lemon juice
⅓   c. water
1½   tblsp. butter

• Combine sugar, flour and salt; toss together with apples.
• Pour apple mixture into pastry-lined 9" pie pan. Sprinkle with mixture of vanilla, lemon juice and water. Dot with butter. Cover with top pastry; flute edges and cut vents.
• Bake in very hot oven (450° F.) 10 minutes, reduce heat to 375° F. and bake about 45 minutes. Cool.
• Put cooled baked pie in freezer. When frozen, wrap, label, date and return to freezer. Recommended storage time: 4 to 6 months.
• To serve, partially thaw pie in its wrapping at room temperature, about

30 minutes. Unwrap, place on lower shelf of a moderate oven (350° F.) for 30 minutes, or until warm.

You can freeze the unbaked pie; wrap like the baked pie and return to freezer.

Recommended storage time: 2 to 3 months.

• To serve, bake unthawed on lower shelf of preheated very hot oven (450° F.) about 15 minutes; reduce heat to 375° F. and bake about 45 minutes or until done.

Note: Before mixing filling, steam apple bits 1 to 2 minutes and cool quickly to preserve color.

PINK PARTY PIE

FARM JOURNAL'S *famous ice cream pie*

Pink Pastry (recipe follows)
1 qt. strawberry ice cream
1 (10 oz.) pkg. frozen strawberries, thawed and drained
2 egg whites
¼ tsp. cream of tartar
¼ c. sugar
Few drops red food color

• Line a 9" pie pan with Pink Pastry and bake as directed (see recipe which follows).
• Pile softened ice cream into pie shell and spread; freeze overnight.
• You can wrap, label, date and store in freezer.
Recommended storage time: 2 to 3 weeks.
• To serve, heat oven to extremely hot (500° F.). Arrange strawberries on ice cream.

• Beat together egg whites and cream of tartar until frothy. Add sugar, 1 tblsp. at a time, beating until stiff glossy peaks form. Tint a delicate pink. Spread meringue over pie, covering edges so ice cream will not melt.
• *Set pie on wooden bread board* in oven and bake about 5 minutes, or until meringue is lightly browned. Serve at once.

VARIATION

PINK AND WHITE PARTY PIE: Put 2 c. fresh strawberries in baked pie shell, top with 1 pt. vanilla ice cream, 1 c. berries and meringue. Brown.

PINK PASTRY

Good standard pastry recipe—you can of course omit the food color

1 c. sifted flour
½ tsp. salt
⅓ c. lard
3 to 4 drops red food color
2 tblsp. water

• Mix sifted flour and salt; cut in lard with pastry blender. Add food color to water. Sprinkle on the water and mix with a fork until all the flour is moistened. Gather the dough together and press firmly into a ball. Line pie pan with pastry; flute edges and prick pastry. Bake in hot oven (425° F.) 8 to 10 minutes. Cool. Makes enough pastry for 1 (8 or 9") pie shell.

Note: You can substitute ⅓ c. plus 1 tblsp. hydrogenated fat for the lard.

LEMON MERINGUE PIE SUPREME

Meringue billows top nippy, butter-cup-yellow filling—a luscious pie

Baked 9″ pie shell
- 7 tblsp. cornstarch
- 1½ c. sugar
- ¼ tsp. salt
- 1½ c. hot water
- 3 egg yolks, beaten
- 2 tblsp. butter or regular margarine
- 1 tsp. grated lemon peel
- ½ c. fresh lemon juice
- 3 egg whites (room temperature)
- ¼ tsp. cream of tartar
- 6 tblsp. sugar

• Mix cornstarch, 1½ c. sugar and salt in saucepan; gradually stir in hot water. Cook over direct heat, stirring until thick and clear, about 10 minutes.

• Remove from heat. Stir ½ c. hot mixture into yolks; stir this back into hot mixture. Cook over low heat, stirring constantly, 2 to 3 minutes. Remove from heat; stir in butter. Add lemon peel and juice, stirring until smooth. Cool. Pour into pie shell.

• Beat egg whites with cream of tartar until frothy; gradually beat in 6 tblsp. sugar, a little at a time. Beat until meringue stands in firm, glossy peaks. Spread meringue on filling, making sure it touches inner edge of crust all around pie.

• Bake in moderate oven (350° F.) 15 minutes, until lightly browned; cool.

• Place in freezer. When frozen, wrap in plastic wrap or aluminum foil and place in plastic freezer bag or box. Seal, label, date and return to freezer. Recommended storage time: Up to 1 month.

• To serve, remove from freezer 2 to 3 hours before serving.

FLORIDA ORANGE/LIME PIE

Just the pie to end a heavy meal—it's tart-sweet, light airy

- 2 baked 9″ pie shells
- 1 envelope unflavored gelatin
- ⅔ c. orange juice
- 8 egg yolks
- ⅔ c. sugar
- 2 tblsp. grated orange peel
- ½ c. lime juice (or lemon)
- 8 egg whites
- ½ tsp. salt
- 1 c. sugar

• Stir gelatin into orange juice.

• Combine egg yolks, ⅔ c. sugar, orange peel and lime juice in top of double boiler. Stir until smooth. Cook over boiling water, stirring constantly, until thickened.

• Add orange-gelatin mixture; stir well. Remove from heat.

• Beat egg whites with salt added until stiff. Gradually beat in 1 c. sugar. Fold in orange mixture. Pour into pie shells. Chill until set. Makes 2 pies.

• Freeze. When frozen, wrap, label and date. Return to freezer. Recommended storage time: Up to 1 month.

• To serve, unwrap and thaw in refrigerator 1 hour. Spread top with whipped cream.

CHOCOLATE/WALNUT PIE

Rich dessert with snowy coconut halo to serve at your next club luncheon

Baked 9" pie shell
30 marshmallows
1 c. milk
1 tsp. vanilla
⅛ tsp. salt
1 c. heavy cream, whipped
½ c. chopped walnuts
½ c. grated sweet cooking chocolate
¼ c. shredded or flaked coconut
2 tblsp. grated sweet chocolate

• Put marshmallows and milk in top of double boiler. Cook over boiling water until marshmallows are melted, stirring occasionally. Cool.
• Fold in vanilla, salt, whipped cream, walnuts and ½ c. grated sweet chocolate. Pour into pie shell. Sprinkle with coconut and 2 tblsp. chocolate.
• Place in freezer; when frozen, wrap, label, date and return to freezer. Recommended storage time: Up to 1 month.
• To serve, thaw in refrigerator.

Wonderful Pies Made with Frozen Fruits

If you want to rate as the county's best pie baker, use frozen fruits for fillings when the fresh are out of season. Thicken with a mixture of quick-cooking tapioca and cornstarch for a just-right filling—clear, smooth juices, neither too thick nor too runny.

Combine two fruits for new taste adventures. You will have an unforgettable pie.

We developed some recipes for pies made with frozen fruits in our Countryside Kitchens. When you bake them, be sure to notice the proportions of sugar to fruit. The thickenings are adjusted to them.

COUNTRY BLUEBERRY PIE

Tastes as if you rushed berries from the blue thickets into the pie

Pastry for 2-crust pie
3 c. frozen blueberries
Blueberry juice
Water
¾ c. sugar
2 tblsp. quick-cooking tapioca
1½ tblsp. cornstarch
1 tsp. lemon juice

• Thaw berries until most of free ice has disappeared. Drain off juice, measure and add water to make ½ c. liquid; stir into mixture of sugar, tapioca and cornstarch in saucepan. Heat rapidly until thickening is complete. Boiling is not necessary. Cool.
• Add berries and lemon juice to cooled, thickened juice. Pour filling into pastry-lined 9" pie pan. Adjust top crust; flute edges and cut vents.
• Bake in hot oven (425° F.) 30 minutes, or until nicely browned. For a brown undercrust, bake on lowest oven shelf. When pie is cool, put in freezer; when frozen, package, label, date and return to freezer. Recommended storage time: 4 to 6 months.

Note: Blueberries in this recipe are frozen without sugar.

PERFECT CHERRY PIE

Once they taste this pie, you'll freeze cherries just to make it

Pastry for 2-crust pie
- 3   c. pitted tart frozen cherries
- 1   c. tart cherry juice
- 3   tblsp. sugar
- 2   tblsp. quick-cooking tapioca
- 1⅔ tblsp. cornstarch (5 tsp.)
- ⅛   tsp. almond extract

• Thaw cherries until most of the free ice has disappeared. Drain off the juice; measure and stir it into mixture of sugar, tapioca and cornstarch in saucepan. Heat rapidly until thickening is complete. Boiling is not necessary. Set aside to cool.
• Add cherries and extract to cooled, thickened juice. Pour filling into pastry-lined 9″ pie pan. Adjust top crust; flute edges and cut vents.
• Bake in hot oven (425° F.) 30 to 35 minutes, or until nicely browned. For a brown undercrust, bake on lowest oven shelf. When pie is cool, put in freezer. When frozen, package, label, date and return to freezer. Recommended storage time: 4 to 6 months.

Note: Proportions of sugar, tapioca and cornstarch are based on 5 parts cherries frozen with 1 part sugar.

DOUBLE CHERRY PIE

Two kinds of cherries between the same crust make a doubly good pie

Pastry for 2-crust pie
- 2   c. pitted tart frozen cherries
- 1   c. pitted dark sweet frozen cherries
- ⅔   c. tart cherry juice
- ⅓   c. sweet cherry juice
- ¼   c. sugar
- 2⅓ tblsp. quick-cooking tapioca
- 1½ tblsp. cornstarch
- 1   tsp. lemon juice

• Thaw cherries until most of the free ice has disappeared. Drain off juices, measure and stir into mixture of sugar, tapioca and cornstarch in saucepan. Heat rapidly until thickening is complete. Boiling is not necessary. Set aside to cool.
• Add cherries and lemon juice to cooled, thickened juice. Pour filling into pastry-lined 9″ pie pan. Adjust top crust; flute edges and cut vents.
• Bake in hot oven (425° F.) 30 to 35 minutes, or until nicely browned. For a brown undercrust, bake on lowest oven shelf. When pie is cool, put in freezer. When frozen, package, label, date and return to freezer. Recommended storage time: 4 to 6 months.

Note: Proportions of sugar, tapioca and cornstarch are based on 5 parts cherries frozen with 1 part sugar.

GOLDEN PEACH PIE

Winter pie—juicy and luscious with summer's fresh peach taste

Pastry for 2-crust pie
- 3   c. frozen sliced peaches
- 1   c. peach juice
- 1½ tblsp. brown sugar
- 1½ tblsp. sugar

2⅓ tblsp. quick-cooking tapioca
1½ tblsp. cornstarch
⅛ tsp. ground cinnamon
1 tsp. lemon juice

• Thaw peaches until most of free ice has disappeared. Drain off the juice, measure and stir it into mixture of sugars, tapioca, cornstarch and cinnamon in saucepan. Heat rapidly until thickening is complete. Boiling is not necessary. Set aside to cool.
• Add peaches and lemon juice to cooled, thickened juice. Pour filling into pastry-lined 9″ pie pan. Adjust top crust; flute edges and cut vents.
• Bake in hot oven (425° F.) 30 minutes, or until nicely browned. For a brown undercrust, bake on lowest oven shelf. When pie is cool, put in freezer. When frozen, package, label, date and return to freezer. Recommended storage time: 4 to 6 months.

Note: Proportions of sugar, tapioca and cornstarch are based on 5 parts peaches frozen with 1 part sugar.

PEACH/STRAWBERRY PIE

Tune your ears for compliments when forks cut into this lovely pie

Pastry for 2-crust pie
1½ c. frozen sliced peaches
1½ c. frozen strawberries
½ c. peach juice
½ c. strawberry juice
3 tblsp. sugar
2½ tblsp. quick-cooking tapioca
1½ tblsp. cornstarch
1 tsp. lemon juice

• Thaw fruit until most of free ice has disappeared. Drain off the juices and

measure, then stir into mixture of sugar, tapioca and cornstarch in saucepan. Heat rapidly until thickening is complete. Boiling is not necessary. Set aside to cool.
• Add fruit and lemon juice to cooled, thickened juice. Pour filling into pastry-lined 9″ pie pan. Adjust top crust; flute edges and cut vents.
• Bake in hot oven (425° F.) 30 to 35 minutes or until nicely browned. For a brown undercrust, bake on lowest oven shelf. When pie is cool, put in freezer. When frozen, package, label, date and return to freezer. Recommended storage time: 4 to 6 months.

Note: Proportions of sugar, tapioca and cornstarch are based on 5 parts peaches frozen with 1 part sugar; 4 parts strawberries frozen with 1 part sugar.

VARIATION

PEACH/BLUEBERRY PIE: Substitute blueberries for strawberries. If blueberries were frozen unsweetened, use ⅓ c. sugar; add water to the combined fruit juices to make 1 c. liquid.

WONDERFUL STRAWBERRY PIE

The best two-crust strawberry pie you'll ever make or taste

Pastry for 2-crust pie
2⅔ c. frozen strawberries
1⅓ c. strawberry juice
3 tblsp. sugar
2½ tblsp. quick-cooking tapioca
1½ tblsp. cornstarch
1 tsp. lemon juice

• Thaw berries until most of free ice has disappeared. Drain off the juice; measure and stir it into mixture of sugar, tapioca and cornstarch in saucepan. Heat rapidly until thickening is complete. Boiling is not necessary. Set aside to cool.

• Add berries and lemon juice to cooled, thickened juice. Pour filling into pastry-lined 9″ pie pan. Adjust top crust; flute edges and cut vents.

• Bake in hot oven (425° F.) 30 minutes, or until nicely browned. For a brown undercrust, bake on lowest oven shelf. When pie is cool, put in freezer. When frozen, package, label, date and return to freezer.

Recommended storage time: 4 to 6 months.

Note: Proportions of sugar, tapioca and cornstarch are based on 4 parts strawberries frozen with 1 part sugar.

Slick Freezing Tricks for Cakes
GOOD IDEAS FROM FARM JOURNAL READERS

CAKE TO CARRY: Freeze the frosted cake before you tote it to a potluck supper. It is firm and carries well and will be thawed by the time you are ready to serve it.

CHRISTMAS FRUITCAKE: Last year I was chairman of the food committee for our church's big holiday tea. I bought round fruitcakes with holes in the center at the bakery and froze them. Then I sliced each cake into fourths—I cut each wedge in halves to make eighths. I placed each wedge on a board, cut side down, and with a sharp knife, cut thin crosswise slices that resembled triangles—or Christmas trees. At the side end of each, I placed a thin 1″ stick of candy for the tree trunk. At the narrow top, I pressed in a piece of sticky, green gumdrop. The little Christmas trees, arranged on trays, were beautiful—they made a hit.

CAKE CUTOUTS: I freeze unfrosted cake before I attempt to cut it in fancy shapes to frost for parties and birthday cakes. You can make clean, neat cuts and eliminate crumbs.

MARSHMALLOW CAKE FROSTING: Add 10 to 12 marshmallows to your favorite 7-minute cake frosting. They add new flavor and the frosting stays firm when frozen.

CAKE CRUMBS: Freeze leftover cake crumbs. Add a touch of cinnamon and a little brown sugar for a crunchy topping on open-face fruit pies or quick coffee cakes.

CHAPTER 7

Main Dishes

Directions for Cooking, Cooling, Freezing and Serving · Recipes
for Hamburger Mix, Busy-Day Meat Mix, Meat Loaves, Swiss
Steak, Pot Roast, Beef Stroganoff, Stuffed Peppers · Macaroni and
Cheese · Chicken Dishes, Stuffed Crepes and Hot Tamales · Liver
Paste, Ham Combinations, Chop Suey, Sausage Casseroles · Seafood
Scallop, Shrimp Creole · Barbecued Lima Beans · Baked Beans

Listen to a group of farm women discuss frozen main dishes and you'll notice great changes in country cooking. You'll know why meals on the busi-
est days are now so wonderfully good. Homemakers fix them weeks ahead
on less crowded days. These dishes are so hearty that they are almost a
meal by themselves.

Most country cooks like best to freeze slow-cooking food combinations
—those that must bubble lazily over low heat. They couldn't possibly fix
them when they have almost no cooking time. Stewed chicken is one;
baked beans another. And Meat Ball Chowder is our classic—you'll find
the recipe in this chapter. Several farm women keep a supply of it in their
freezers so they will have the right food to tote to community suppers even
if they can spend little time in the kitchen. Such dishes give the impression
the cook wasn't in a hurry—a tribute to her good management.

Of course, you have to reheat main dishes, but most of them take care
of themselves. A Western ranch mother, whose teen-age daughter takes
music lessons and participates in many after-school activities, calls her
specialties shove-in-the-oven dishes. She puts one in the oven to heat for
dinner before she drives to town to bring her daughter home. An excellent
dinner is on the table within 10 minutes after they step into the kitchen.
Magic? No, just intelligent, planned use of the freezer.

Microwave ovens miraculously step up the speed for heating main
dishes. Farm women who own this appliance package stews and casseroles
in single serving portions; in minutes, they can serve a hot meal to anyone
who's in a hurry.

Freezing Main Dishes

Select Ingredients of Good Quality: Especially is this important with the shortening you use. Make certain it is fresh and of high quality, for it has much to do with how successfully dishes freeze. Lard sometimes changes in flavor after being frozen 4 months.

Do Not Overcook: Meat should be tender, but firm. Wait to add vegetables for stews and soups near the end of the cooking period. When you freeze the dish, vegetables should not be tender; they'll finish cooking when you reheat the dish. Undercook macaroni, spaghetti, noodles and rice; otherwise, they'll be soft and mushy when reheated. Parboiled rice is excellent for freezing; cook it only until the hard center disappears, about 20 minutes. Overcooking not only destroys the texture of foods, but it also gives them a warmed-over taste.

Cool Cooked Foods Quickly: This will stop the cooking. Loss of flavor is rapid when foods are held at high temperatures, and spoilage bacteria develop quickly. To hasten cooling, set the saucepan of cooked food in a large pan of cold water (ice water is best). Stir it occasionally with a rubber spatula, using care not to break up or mash the food. When you store a main dish in a Pyroceram or metal container, you can cook and freeze it in the same utensil. A timesaver! If you use cookware designed to move from freezer to oven to table, you can freeze and reheat in the same container—a timesaver. Such utensils must be glass-ceramic if you use a microwave oven.

Pack Food Solidly: A solid pack helps keep out the air. If you pour sauces or gravies over meat, they fill the air spaces and crowd out the air. The meat will taste better.

Thaw in Refrigerator: To thaw main dishes completely, place in refrigerator. There is danger of spoilage which could result in food poisoning if you leave them to thaw at room temperature. You can reheat many of the main dishes in this chapter without thawing.

Reheat Correctly: You can quickly reheat and thaw the food in a saucepan containing a small amount of melted butter or liquid. Or you can set it over low heat, watching and stirring it frequently. Put a covered casserole in a moderate oven (350° F.) or a hot oven (400° F.); it usually heats in about 1 to 1½ hours. Or defrost it in refrigerator overnight before reheating. Reheat foods that scorch easily, like stews and creamed dishes, in a double boiler about 30 to 45 minutes, breaking up the chunks of frozen food carefully. *Time for thawing and reheating is approximate.*

Rotate Frozen Dishes: Keep frozen dishes going to the table and new ones coming into the freezer. Almost all combination dishes lose flavor after 3 to 4 months. It is a good idea to use them within 6 to 8 weeks.

How to Adapt Favorite Recipes

Almost all families have special dishes they like. Often slight changes make them satisfactory for freezing. Here are some facts to keep in mind:

Flavor Changes: Cloves, garlic, black pepper, green peppers, pimiento and celery increase in flavor during freezing. Use them more sparingly. You may wish to add more onion because its flavor often decreases. Salt and chili powder sometimes decrease in strength.

Casserole Toppings: Add bread crumbs and grated cheese at reheating time rather than before freezing.

Rice and Macaroni: Some combination dishes that contain rice, macaroni, spaghetti and noodles may appear dry when taken from the freezer. Add a little milk or water, about ⅓ c., when reheating them.

Chopped Cooked Meats: Freeze them in sauces or gravies to prevent their drying out.

Turkey Fat: It becomes rancid more quickly than chicken fat. Use chicken broth, when available, for making turkey dishes to freeze.

Fat in Sauces and Gravies: It often separates in freezing. Whipping while heating often will recombine it. In making gravy or sauce, use a minimum amount of fat and remove sauce from heat when it starts to thicken. The uncooked flour or cornstarch will finish cooking when reheated.

Beans: Small white beans (navy) are good freezers. Some varieties get mushy.

HAMBURGER MIX

This basic recipe came from Arizona —a busy cook's special

2 c. chopped onions
3 cloves garlic
2 c. chopped celery
¼ c. fat
4 lbs. ground beef
4 tsp. salt
½ tsp. pepper
3 tblsp. Worcestershire sauce
2 (14 oz.) bottles ketchup

• Put onions, garlic and celery in large skillet and cook in fat until soft. (Run toothpick through garlic to aid in removal.) Add ground beef; cook until pink color disappears. Remove garlic and discard.
• Add remaining ingredients; simmer 20 minutes. Cool quickly. Skim off excess fat. Makes 6 pints.
• Pack in moisture-vaporproof containers; seal, label, date and freeze. Recommended storage time: 3 to 6 months.
• To use, partially thaw in container in refrigerator overnight or if container is airtight, under running cold water. Thaw until mixture will slip out of container. Use Hamburger Mix in recipes that follow.

HAMBURGER-FILLED BUNS

Freezer to table with little work

3 c. Hamburger Mix
8 hot buttered buns

• Heat partially thawed Hamburger
Mix. Serve between hot buns, buttered. Makes 8 sandwiches.

VARIATION

STUFFED ROLLS: Spoon 2½ c. thawed
Hamburger Mix generously into split
wiener buns. Wrap in foil or place in
tightly covered pan and heat in moderate oven (350° F.) 30 minutes.
Makes 8 sandwiches.

HAMBURGER/NOODLE CASSEROLE

*Good "dress-up" for leftover corn,
carrots, green and lima beans, peas*

1½ c. noodles, uncooked
  2 c. Hamburger Mix, partly
       thawed
  1 c. canned or cooked vegetables
  ½ c. grated cheese
  1 tsp. chopped parsley

• Cook noodles in boiling salted water
until tender; drain.
• Heat Hamburger Mix in skillet. Stir
in noodles and vegetables. Sprinkle
cheese and parsley over the top.
• Heat, covered, over low heat until
cheese melts. Makes 4 to 6 servings.

BUSY-DAY MEAT MIX

*The best way to salvage roast beef,
lamb or veal—it has many uses*

  4 c. ground cooked meat
  ½ lb. bulk pork sausage
  2 c. bread cubes

  2 eggs, slightly beaten
  1 (4 oz.) can chopped mushrooms
  ¼ c. chopped parsley
  1 tsp. salt
  ¼ tsp. pepper
  2 tsp. Worcestershire sauce
  1 (10¾ oz.) can condensed cream of
       mushroom soup

• Mix all ingredients. Makes 3 pints.
• Pack into freezer containers, seal,
label, date and freeze.
Recommended storage time: 2 to 3
months.
• To use, partially thaw and use to
stuff cabbage leaves, green peppers or
squash, or as a filling for biscuit rolls
or pastry turnovers, which may be
prepared and frozen. Or you can
make individual meat loaves.

INDIVIDUAL MEAT LOAVES: Shape
Busy-Day Meat Mix in ½ c. measure.
Makes 12 individual loaves.
• Freeze on baking sheet; wrap individually in foil or plastic wrap, label,
date and return to freezer.
• To serve, remove wrapping, place
frozen in baking pan. Bake in moderate oven (375° F.) 45 minutes.

TASTY MEAT LOAF

*Well-seasoned loaf is mixed with oat-
meal instead of bread crumbs*

  2 lbs. ground chuck
  ¾ c. quick-cooking oats
  1 c. coarsely shredded carrots
  ½ c. chopped onion
  ½ c. chopped green pepper
  2 tblsp. minced fresh parsley
  ½ c. ketchup
  2 tblsp. yellow mustard
  2 eggs, slightly beaten
  2 tsp. salt

1 tsp. dried thyme leaves
¼ tsp. pepper
¼ c. milk

• Combine all ingredients. Mix lightly, but well. Shape mixture into 10″ loaf. Makes 8 servings.
• Place on baking sheet; freeze. When frozen, remove from freezer, package in foil or plastic wrap, label and date. Return to freezer.
Recommended storage time: 2 to 3 months.
• To serve, bake frozen loaf, uncovered, in shallow pan in a very hot oven (400° F.) 45 minutes. Reduce heat to 350° F. and bake about 15 minutes longer. Let stand 10 minutes before slicing.

COLORADO MEAT LOAVES

Individual frozen meat loaves cook quicker than a single big loaf

2 eggs
1 c. milk
2 c. soft bread crumbs
2 tsp. salt
1 tsp. dry mustard
1 tsp. celery salt
¼ tsp. pepper
6 tblsp. grated onion
2 lbs. ground beef

• Beat eggs slightly in large bowl; blend in milk. Add remaining ingredients. Mix thoroughly.
• Shape mixture into 12 small loaves.
• Place on baking sheet; freeze. When frozen, remove from freezer, package in foil or plastic wrap, label and date. Return to freezer.
Recommended storage time: 2 to 3 months.

• To serve, place frozen loaves, uncovered, in baking pan. Bake in moderate oven (375° F.) 45 minutes.

MEAT LOAF

Good gravy comes with this tasty loaf

2 eggs, slightly beaten
½ c. ketchup
¾ c. warm water
1 (1½ oz.) pkg. onion soup mix
1½ c. soft bread crumbs
2 lbs. ground beef

• Mix eggs, ketchup, water, onion soup mix and crumbs; blend into beef. Shape into loaf. Makes 8 servings.
• Wrap, label, date and freeze.
Recommended storage time: 2 to 3 months.
• To serve, bake frozen loaf, uncovered, in shallow pan in very hot oven (450° F.) 45 minutes; reduce heat to moderate (350° F.) and bake about 15 minutes longer. Transfer loaf to platter; keep warm.
• Make gravy from drippings. Pour over loaf and serve.

SUNDAY SWISS STEAKS

Saucy meat that cooks by itself

½ c. flour
2 tsp. salt
¼ tsp. pepper
6 slices eye of round, 1½″ thick
½ c. thinly sliced onion rings
¼ c. shortening
1 (8 oz.) can tomato sauce
1 c. pizza sauce

• Combine flour, salt and pepper; pound into meat.
• Brown onions lightly in hot fat in

large skillet; remove onions and brown meat. Place onions on top of meat.
· Add tomato and pizza sauces. Cover; simmer 1½ to 2 hours until tender. Serve sauce with meat. Makes 6 servings.
· Package cooked steaks individually, with sauce, in small foil pie pans; wrap with foil. Seal, label, date and freeze.
Recommended storage time: 3 to 6 months.
· To serve, heat in skillet.

RANCHERO POT ROAST

An untended dish—the rich, spicy aroma and flavor mount as it simmers

  2 tsp. salt
  ¼ tsp. freshly ground pepper
  ¼ tsp. paprika
  1 clove garlic, finely chopped
  3 lbs. bottom round roast
  2 tblsp. shortening
  1 c. beef broth
  ½ c. chopped onions
  ½ c. chili sauce

· Rub salt, pepper, paprika and garlic into meat. Cover; refrigerate overnight.
· Next day, brown meat slowly in hot fat to a deep brown.
· Place meat on rack in deep pan. Add broth and onions; cover and simmer 1½ hours (add more broth if necessary). Add chili sauce and cook 30 minutes. Makes 6 servings.
· Cook quickly. Pack into containers, seal, label, date and freeze.
Recommended storage time: 3 to 4 months.
· To serve, partially thaw and simmer until heated and tender, about 1 hour.

BEEF STROGANOFF

Hostess special to fix and freeze a few days before you entertain

  1½ lbs. sirloin steak, cut in
        1 × ¼″ strips
  3½ tblsp. flour
  ¼ c. butter or regular margarine
  1 c. chopped onion
  1 (10½ oz.) can condensed beef
        broth
  1 tblsp. Worcestershire sauce
  1 tsp. salt
  ½ tsp. dry mustard
  2 tblsp. tomato paste
  1 c. dairy sour cream
  1 (6 oz.) can chopped
        mushrooms, drained

· Dredge meat in 1½ tblsp. flour. Brown quickly in melted butter in heavy skillet, turning meat to brown on all sides. Add onion and cook until barely tender, 3 to 4 minutes.
· Remove meat and onion from skillet. Blend remaining flour (2 tblsp.) into drippings in skillet. Add broth and Worcestershire sauce and cook, stirring constantly, until thickened. Stir in salt, mustard and tomato paste. Blend in sour cream. Add meat and mushrooms. Cool. Makes 6 cups.
· Pour into freezer containers, seal, label, date and freeze.
Recommended storage time: 3 to 6 months.
· To serve, partially thaw. Heat in double boiler. Serve over parslied rice, wild rice or noodles.

MEAT BALL CHOWDER

Don't let the number of ingredients scare you—the chowder is simple to fix and there's plenty for two meals

2 lbs. ground lean beef
2 tsp. seasoned salt
⅛ tsp. pepper
2 eggs, slightly beaten
¼ c. finely chopped parsley
⅓ c. fine cracker crumbs
2 tblsp. milk
3 tblsp. flour
1 tblsp. salad oil
4 to 6 onions, cut in eighths
6 c. water
6 c. tomato juice
6 beef bouillon cubes
3 c. sliced carrots (about 6)
3 to 4 c. sliced celery
2 to 3 c. diced potatoes
¼ c. long grain rice
1 tblsp. sugar
2 tsp. salt
2 bay leaves
½ to 1 tsp. marjoram leaves
    (optional)
1 (12 oz.) can Mexicorn

• Combine meat, seasoned salt, pepper, eggs, parsley, cracker crumbs and milk. Mix thoroughly. Form into balls about the size of a walnut (makes about 40). Dip in flour.
• Heat oil in 8- to 10-qt. kettle. Lightly brown meat balls on all sides (or drop unbrowned into boiling vegetables).
• Add remaining ingredients (except add corn last 10 minutes of cooking). Bring to boil; cover. Reduce heat and cook at slow boil 30 minutes, or until vegetables are tender. If dinner must wait, turn off heat at this point. Takes only minutes to reheat. Makes 6 to 7 quarts.
• To freeze, cook until the vegetables are crisp-tender. Cool quickly. Ladle into freezer containers, cover, seal, label and date.

Recommended storage time: 2 to 3 months.
• To serve, partially thaw until chowder softens. Heat until piping hot, about 45 minutes.

PLANTATION STUFFED PEPPERS

Choose plump, crisp peppers to fill with this flavorful meat mix

1 lb. ground beef
1 c. chopped onion
1 clove garlic, chopped
2 tsp. chili powder
1 tsp. salt
½ tsp. pepper
2 (10½ oz.) cans condensed
    tomato soup
½ lb. sharp process cheese,
    shredded or sliced
1½ c. cooked parboiled rice
8 medium green peppers

• Cook ground beef, onion and garlic in skillet until meat is browned. Add seasonings and tomato soup; simmer, covered, 10 minutes. Add cheese. Cook slowly, stirring occasionally until cheese melts. Stir in rice. Cool.
• Cut peppers in halves lengthwise. Remove membranes and seeds. Cook in boiling salted water to cover until barely tender, about 3 minutes. Drain and cool.
• Place peppers on baking sheet. Stuff with rice mixture. Serves 8.
• To freeze, place in freezer until peppers are frozen. Remove, wrap frozen peppers in foil or plastic wrap. Label, date and return to freezer. Recommended storage time: 2 to 3 months.
• To serve, remove wrapping, place partially thawed peppers in shallow

pan. Cover with foil. Bake in hot oven (400° F.) 30 to 45 minutes.

MACARONI AND CHEESE

You couldn't pick a more popular dish to freeze for busy-day meals

1½ c. macaroni, uncooked
2 tsp. salt
2 tblsp. butter or regular margarine
2 tblsp. flour
¼ tsp. pepper
3 c. milk
2 c. grated sharp Cheddar cheese
½ tblsp. grated onion
1 tsp. Worcestershire sauce
½ c. buttered bread crumbs

• Cook macaroni with 1½ tsp. salt in 3 qts. rapidly boiling water; cook about 2 minutes less than for complete doneness. Drain and rinse.
• Make a white sauce of butter, flour, ½ tsp. salt, pepper and milk. Stir cheese into sauce until melted. Add macaroni, onion and Worcestershire sauce. Cool quickly. Place in greased shallow baking dish. Serves 6 to 8.
• Wrap, label, date and freeze.
Recommended storage time: 6 to 8 weeks.
• To serve, thaw 18 hours in refrigerator. Bake, covered with foil, in moderate oven (350° F.) 25 minutes. Remove foil, add crumbs; bake 15 minutes.

TURKEY CURRY WITH RICE

Your guests will be impressed with this colorful, good-tasting meal

1 c. sliced onion
¼ c. sliced carrots
¼ c. sliced celery
1 apple, pared, cored and sliced
1 clove garlic, minced
¼ c. butter
⅓ c. flour
2½ tsp. curry powder
¼ tsp. ground ginger
¼ tsp. ground mace
¼ tsp. pepper
2 c. chicken broth
1 (4 oz.) can mushrooms
3 c. cubed cooked turkey
2 c. dairy sour cream
Hot fluffy rice
Condiments: Chopped onion, raisins, chopped green pepper, chopped tomato, chopped hard-cooked eggs, chopped peanuts, coconut and chutney

• Sauté onion, carrots, celery, apple and garlic in melted butter.
• Slowly blend in flour, curry powder, ginger, mace and pepper. Gradually add chicken broth and undrained mushrooms. Cook over medium heat, stirring constantly, until it comes to a boil. Simmer 5 minutes.
• Stir in turkey. Heat well to blend flavors. Cool quickly; ladle into freezer containers. Seal, label, date and freeze.
Recommended storage time 3 to 6 months.
• To serve, partially thaw in refrigerator. Heat in skillet until bubbly. Stir in sour cream; heat 2 more minutes. Serve with rice. Pass a selection of condiments. Makes 6 servings.

FRICASSEE CHICKEN WITH PARSLEY NOODLES

A good start on Sunday dinner— chicken and homemade noodles

1   (3½ to 4 lb.) stewing chicken,
    cut up
6   c. water
1   branch celery
1   carrot, cut in chunks
1   onion, sliced
1½ tsp. salt
⅛   tsp. marjoram leaves
⅛   tsp. thyme leaves
1   bay leaf

Noodles:
1   tblsp. water
½   tsp. salt
1   egg, beaten
1   c. sifted flour
¼   c. finely chopped parsley

• Simmer chicken in water with vegetables and seasonings until chicken is tender. Strain broth to remove vegetables and herbs. Cool chicken. Reserve broth.
• In the meantime, prepare noodles. Add water and salt to egg. Stir in flour to make a stiff dough, adding parsley with last portion. Knead dough slightly with hands until it doesn't stick. Cut into thirds. Roll one third at a time until very thin on a lightly floured board. Let stand 20 minutes to dry slightly. With long knife cut into ⅛" strips.
• Bring chicken broth to boiling. Add noodles slowly so broth does not stop boiling, and cook 20 minutes, or until tender. Cool. Makes 6 servings.
• Put cooled chicken pieces in freezer containers. Cover with broth, then with cooled noodles and broth. Seal, label, date and freeze.
Recommended storage time: 3 to 6 months.
• To serve, thaw and heat slowly in heavy pan over low heat, until

chicken is heated and sauce is bubbly. Add more water, if needed.

CHICKEN/CORN SUPPER

You can't beat the chicken-corn combination in this old-time special

1   c. butter or regular margarine
¾   c. flour
2   qts. hot chicken broth
2   tsp. salt
½   tsp. pepper
2   tblsp. finely chopped onion
1¾   qts. chopped cooked chicken
     (2 stewing hens)
1½   qts. cooked or canned whole
     kernel corn, or 3 (1 lb.) cans,
     drained
1½   c. grated process cheese
¾   c. chopped pimientos
Buttered bread or cracker crumbs

• Melt butter; blend in flour. Add hot broth and cook over medium heat until bubbling and thickened, stirring. Stir in salt, pepper and onion.
• Combine with remaining ingredients, except crumbs, in a very large bowl. Cool quickly. Makes 4½ quarts.
• Pack in freezer containers, label, date and freeze.
Recommended storage time: 3 to 6 months.
• To serve, thaw food enough to remove from container. Place in a greased casserole and bake, covered, in a hot oven (400° F.) 30 minutes or until food is thawed enough to press into the shape of the baking dish. Sprinkle with buttered crumbs.
• Continue baking, uncovered, until bubbly hot and crumbs are browned. Total baking time, about 1 hour for pints, 1 hour 45 minutes for quarts.

CONNECTICUT BAKED CHICKEN

Easy-to-fix chicken with gourmet flavor and many uses—do try it

2 (2½ to 3 lb.) broiler-fryers, cut
   in serving pieces
1 c. diced celery
2 to 3 sprigs parsley
2 tsp. salt
¼ tsp. pepper
1 tsp. basil leaves
½ c. tomato juice

• Place chicken in Dutch oven or large casserole. Top with remaining ingredients.
• Cover and bake in a moderate oven (350° F.) until chicken is tender but still firm, about 1½ hours. Cool.
• Pack chicken as compactly as possible in freezer containers. Cover with chicken broth. (If more broth is needed, dissolve 1 chicken bouillon cube in 1 c. hot water.) Makes about 2 quarts. Or to save freezer space, cut meat off the bone in as large pieces as possible and pack as directed.
• Seal, label, date and freeze. Recommended storage time: 3 to 6 months.
• To serve, partially thaw and place chicken with bones, covered, in a moderate oven (350° F.) until heated thoroughly. If chicken seems dry, add a little broth while heating. Or use cut-up chicken in casseroles, salads, curry, creamed dishes and chop suey.

CHICKEN PIE SUPREME

FARM JOURNAL'S *famous chicken pie*

Pastry for 2-crust pie
1 (5 lb.) whole stewing chicken
1½ qts. water

2 tsp. salt
1 small onion
1 carrot
1 branch celery
½ c. sifted flour
½ tsp. onion salt
½ tsp. celery salt
Dash of pepper
3½ c. chicken broth
2 or 3 drops yellow food
    color (optional)

• Place chicken in large kettle and add water, 1 tsp. salt, onion, carrot and celery. Simmer, covered, until tender, 3 to 3½ hours.
• Remove chicken and strip meat from bones in large pieces. Refrigerate chicken and broth to cool.
• Combine flour, onion salt, celery salt, pepper and 1 tsp. salt with ½ c. cooled chicken broth. Mix until smooth.
• Put 3 c. chicken broth in skillet; heat and add flour mixture, beating with a wire whip to prevent lumping.
• Cook over medium heat, stirring constantly, until mixture is smooth and thickened. Add food color.
• Add chicken and blend well. Cool.
• Line 9″ deep-dish pie pan with pastry. Fill with cooled chicken mixture. Adjust top crust; cut vents and seal edges. Makes 6 to 8 servings.
• Freeze pie. When frozen, wrap, label, date and return to freezer. Recommended storage time: 3 to 6 months.
• Cool remaining broth and freeze in glass jars or freezer containers.
• To serve, bake frozen pie in hot oven (400° F.) 45 minutes to 1 hour. Chicken filling should be hot, the crust a golden brown. Make gravy with frozen broth to serve with pie.

COUNTRY COMPANY CHICKEN

Do try this treat from kitchens in Kentucky's bluegrass country

1 c. parboiled rice
1 tsp. salt
2 c. boiling chicken broth
1 (10½) oz. can condensed cream of celery soup
1 c. milk
2 c. cooked chicken
1 c. chopped celery
2 tblsp. grated onion or 2 tsp. onion flakes
1 c. slivered, toasted almonds
1 tsp. salt
½ tsp. Worcestershire sauce

• Add rice and salt to broth. Stir until mixture begins to boil. Cover, reduce heat and simmer 15 minutes. (Rice will complete cooking when reheated.)
• Blend together soup and milk. Stir in rice, chicken, celery, onion, ½ c. almonds and seasonings. Cool quickly.
• Line a 2-qt. casserole with aluminum foil. Pour mixture into casserole; sprinkle remaining ½ c. almonds on top. Makes 6 to 8 servings.
• Freeze. Remove from dish, wrap, label, date and store in freezer.
Recommended storage time: 3 to 6 months.
• To serve, grease original casserole. Remove foil from frozen block of chicken mixture and place in casserole. Let partially thaw. Bake, covered, in hot oven (400° F.) 30 to 45 minutes. Remove cover and continue baking until thoroughly heated, about 30 minutes.

RANCH-STYLE CHICKEN

A good Nevada cook gives this baked chicken a savory Italian accent

6 tblsp. shortening or salad oil
½ c. flour
3 tsp. salt
¼ tsp. pepper
2 (2 to 3 lb.) broiler-fryers, cut in serving pieces
1 onion
8 sprigs parsley
1 clove garlic
2 tblsp. olive oil
1 (8 oz.) can tomato sauce
1 (1 lb.) can tomatoes
½ c. pitted, sliced ripe olives and juice
1 (3½ to 4 oz.) can sliced mushrooms
1 tsp. salt
¼ tsp. Italian herb seasoning
⅛ tsp. pepper

• Line two 13×9×2" baking dishes with aluminum foil. Melt 3 tblsp. shortening in each dish.
• Combine flour, salt and pepper in paper bag. Shake chicken pieces to coat with flour. Lightly moisten chicken pieces in melted fat. Arrange in one layer, skin side up in baking dishes. Bake in very hot oven (450° F.) 30 minutes.
• Chop onion, parsley and garlic. Sauté in olive oil. Add tomato sauce, tomatoes, olives, mushrooms and liquid, and seasonings. Pour sauce evenly over the 2 dishes of chicken. Cool. Makes 4 to 6 servings.
• Cover and place in freezer. When frozen, lift foil from baking dish. Wrap in foil or plastic wrap; label, date and return to freezer.

Recommended storage time: 3 to 6 months.
• To serve, remove foil, place in baking dish. Bake, uncovered, in moderate oven (350° F.) 1 hour and 15 minutes.

ARIZONA CHICKEN-STUFFED CREPES

Pancakes that go to company supper or dinner tables and make a hit

1⅓ c. milk
2 tblsp. butter
2 eggs, beaten
½ c. sifted flour
1 tsp. baking powder
½ tsp. salt
1 pt. dairy sour cream
2 whole chicken breasts, cooked and cut in strips
1¼ c. cooked ham, cut in strips
¼ c. canned sweet chili peppers, cut in strips
½ c. grated Gruyère cheese

• Heat 1 c. milk and butter in saucepan until butter is melted; cool.
• Beat in eggs, then flour, baking powder and salt, sifted together; beat until smooth and well blended.
• To bake, lightly grease 4 or 5″ skillet; heat and pour in 3 tblsp. batter for each crepe; tilt skillet to cover evenly. Cook about 1 minute to brown, flip and brown other side.
• Spread one side of all crepes with sour cream (takes about 1 c.); place several strips of chicken, ham and chili peppers on crepes, roll and place in 11×7×1½″ baking dish. Makes 10.
• Wrap, label, date and freeze.
Recommended storage time: 1 month.
• To serve, let thaw in refrigerator.

Mix ⅓ c. milk with 1 c. sour cream, pour over crepes. Top with cheese. Bake, uncovered, in moderate oven (350° F.) 20 minutes, to melt cheese.

Note: If you do not freeze stuffed crepes, let dish stand in refrigerator 24 hours, then bake and serve.

MICHIGAN TAMALE BAKE

Dinner is well on the way when you put this freezer special in oven

Topping:
1 c. yellow cornmeal
1 tsp. salt
1 c. cold water
3 c. boiling water

Filling:
1½ lbs. ground beef
1 c. chopped onion
2¼ tsp. salt
4 tsp. chili powder
2 c. canned tomatoes
1 c. sliced pitted ripe olives
1 c. grated process cheese

• Combine cornmeal, salt and cold water. Stir slowly into boiling water. Cook until mixture thickens, stirring frequently. Cover and continue cooking over low heat 10 minutes.
• Brown beef and onion in large skillet. Pour off excess fat. Add salt, chili powder and tomatoes. Simmer, covered, 20 minutes. Add olives; cool.
• Spread into a 13×9×2″ pan. Top with cornmeal mush. Serves 9.
• Cool, cover with foil, seal, label, date and freeze.
Recommended storage time: 3 to 6 months.
• To serve, bake, uncovered, without thawing, in moderate oven

(350° F.) 1 hour or until hot. Sprinkle with cheese and bake 15 minutes longer.

PENNSYLVANIA LIVER PASTE

Good-for-you pork liver is very tasty fixed this Danish way

1 lb. frozen pork liver
6 slices bacon
1 medium onion
¼ c. flour
1 c. milk
2 eggs
1 tsp. salt
½ tsp. pepper

· Put liver, bacon and onion through food chopper two or three times, using finest blade. Liver is easier to handle if partially frozen. Blend in remaining ingredients.
· Spread mixture in greased 9×5×3″ loaf pan. Bake in slow oven (325° F.) about 1 hour and 35 minutes or until done. Makes 1 loaf.
· Cool, turn out of pan; wrap, label, date and freeze.
Recommended storage time: 6 to 8 weeks.
· To serve, thaw in refrigerator. Makes delicious sandwiches. You also can serve warm like meat loaf.

CREOLE JAMBALAYA ORLEANS

Taste this and you'll know why Louisiana is famed for superior food

1 tblsp. butter or regular margarine
1 tblsp. flour
1 c. cubed cooked ham
¼ c. chopped green pepper
2 c. cooked tomatoes
1¼ c. tomato juice
1 lb. shrimp, cleaned
1 onion, sliced
1 clove garlic, minced
2 tsp. salt
¼ tsp. pepper
1 sprig parsley, chopped, or
　　1 tsp. parsley flakes
⅛ tsp. thyme leaves
1 bay leaf
1 c. parboiled rice

· Melt butter in heavy skillet over medium heat. Stir in flour, ham and green pepper. Cook and stir 5 minutes.
· Add remaining ingredients, except rice. Bring to a boil. Stir in rice and simmer, covered, until rice is almost tender, about 25 minutes. (Rice will complete cooking when reheated.) Remove bay leaf. Makes 2 quarts.
· Cool quickly. Place in freezer containers, seal, label, date and freeze.
Recommended storage time: up to 3 months.
· To serve, partially thaw in refrigerator. Place in a 2-qt. casserole. Pour over ⅓ c. water. Bake, covered, in a hot oven (400° F.) 1 hour.

HOLIDAY HAM RING

Spicy ham loaf for special meals. Serve buttered corn or peas in center

3 lbs. ground smoked ham
1 lb. ground pork
2 eggs
1¾ c. graham cracker crumbs
1½ c. milk
½ tsp. ground allspice
Sauce (recipe follows)

◢ Combine ingredients, except Sauce; mix well. In 15½ × 10½ × 1″ jelly roll pan, shape into ring (or loaf) about 10″ across. Makes 12 to 14 servings.

• Place pan in freezer. When meat is frozen, remove from pan, package, label, date and return to freezer. Recommended storage time: 6 to 8 weeks.

• To serve, partially thaw meat mixture in refrigerator; pour half of Sauce over loaf.

• Bake in moderate oven (350° F.) 45 minutes; pour over remaining Sauce; bake 45 minutes longer, basting several times with sauce.

SAUCE: Combine 1 (10½ oz.) can condensed tomato soup, ½ c. vinegar, ½ c. brown sugar, firmly packed, and 1½ tsp. prepared mustard.

Note: Some cooks like to bake the loaf, cool quickly, wrap, seal, label and freeze. They defrost the loaf in the refrigerator, slice and serve cold.

CASSEROLE OF HAM AND ASPARAGUS

Special for your on-the-go days. It's color-bright and flavorful

    1   c. hot milk
    ¼   lb. process cheese, grated
    ½   c. dry bread crumbs
    ¼   c. ham fat or butter
    ¼   c. chopped pimiento
    1   tblsp. finely chopped onion
    1½   tblsp. chopped parsley
    3   eggs, well beaten
    1½   (10 oz.) pkgs. frozen cut
            asparagus spears

    3   c. diced, cooked ham
    ¾   c. buttered bread crumbs

• Combine hot milk and cheese.

• Blend ½ c. crumbs, fat, pimiento, onion and parsley; add to milk mixture. Blend in eggs. Cool.

• Cook asparagus briefly, boiling only 2 minutes; drain; cool. Combine ham, asparagus and egg mixture. Pour into buttered 1½-qt. shallow casserole. Makes 6 to 8 servings.

• Wrap, label, date and freeze. Recommended storage time: 6 to 8 weeks.

• To serve, thaw in refrigerator and top with buttered crumbs. Bake in moderate oven (350° F.) 40 to 50 minutes.

SKILLET CHOP SUEY

Let company come—with this in the freezer it's easy to get a good meal

    2   c. finely diced fresh pork
    3   tblsp. salad oil
    3   c. chopped celery
    2   c. chopped onion
    2   tblsp. soy sauce
    2   chicken bouillon cubes
    1   c. boiling water
    3   tblsp. cornstarch
    1   tblsp. molasses
    ½   c. water
    1   (1 lb.) can bean sprouts
    1   (4 oz.) can mushroom pieces

• Brown pork in hot oil; add celery, onion and 1 tblsp. soy sauce; cook 10 minutes.

• Dissolve bouillon cubes in boiling water; add to pork. Mix cornstarch with remaining 1 tblsp. soy sauce, molasses and ½ c. water; stir into pork.

Stir in bean sprouts and mushrooms with their liquids. Cook, stirring until thickened. Makes 8 servings. To serve at once, cook 5 minutes longer.

· To freeze, cool quickly, place in freezer containers, seal, label, date and freeze.

Recommended storage time: 2 to 3 months.

· To serve, partially thaw in refrigerator; quickly heat in skillet.

TWIN SAUSAGE CASSEROLES

The unusual seasoning and blend of meats give you tasty variety

1 (6 oz.) can sliced or chopped
   mushrooms
6 chicken bouillon cubes
1 lb. bulk pork sausage
1 lb. ground beef
2 c. chopped celery
1 c. chopped green pepper
½ c. grated Parmesan cheese
1 tsp. salt
½ tsp. marjoram leaves
⅛ tsp. pepper
2 c. long grain rice

· Drain mushrooms and add enough water to liquid to make 5 c. Add bouillon cubes and heat to dissolve.

· Brown meat and drain off fat and liquid. Combine meat with remaining ingredients except bouillon.

· Line two 2-qt. casseroles with aluminum foil. Divide mixture evenly between the 2 casseroles. Pour half of hot bouillon over each casserole. Stir lightly to blend. Cover and bake in moderate oven (350° F.) 15 minutes. Remove from oven, cool slightly and refrigerate before freezing. Each casserole makes 4 to 6 servings.

· Freeze. When frozen, lift foil from casserole. Wrap each in foil; label, date and return to freezer.

Recommended storage time: 1 month.

· To serve, take package from freezer, remove wrappings and place in casserole. Cover and bake in moderate oven (350° F.) 1½ hours.

SHRIMP CREOLE

All-American favorite seafood main dish—it freezes successfully

½ c. chopped onion
½ c. chopped celery
½ c. chopped green pepper
2 cloves garlic, minced
3 tblsp. salad oil
1 (1 lb.) can tomatoes
1 (6 oz.) can tomato paste
1½ tsp. salt
½ tsp. pepper
1 tsp. Worcestershire sauce
⅛ tsp. Tabasco sauce
1 tsp. cornstarch
2 tsp. water
1½ lbs. cleaned raw shrimp

· Cook onion, celery, green pepper and garlic in oil until tender. Add tomatoes, tomato paste and seasonings. Bring to boil, simmer 30 minutes.

· Blend cornstarch and water; stir into sauce. Add shrimp, cover and simmer 15 minutes. Makes 6 cups.

· Cool quickly, pack in freezer container, seal, label, date and freeze.

Recommended storage time: Up to 6 weeks.

· To serve, partially thaw. Heat in skillet. Serve over hot cooked rice.

BARBECUED LIMA BEANS

They'll praise this economical dish at picnics and community suppers

1   lb. dried lima beans
4   c. water
1½ c. chopped onions
1   c. brown sugar, firmly packed
1   c. ketchup
⅔   c. dark corn syrup
1   tblsp. salt
1   tblsp. liquid smoke
9   drops Tabasco sauce
Bacon slices for topping (optional)

• Soak washed beans in water overnight. Do not drain. Add onions, bring to a boil and simmer until beans are almost tender, about 30 minutes.
• Mix remaining ingredients, except bacon. Stir into beans. Makes 9 cups.
• Cool quickly and pack into containers, or place in casserole and wrap; seal, label, date and freeze. Recommended storage time: 6 months.
• To serve, thaw at room temperature until you can remove beans from container. Place in baking pan (or leave in casserole) and put in hot oven (400° F.). Bake, covered, until beans can be pressed into shape of pan or casserole.
• Place bacon strips on top if desired and bake uncovered until beans are bubbly hot and bacon browns, about 1 hour for 1 qt. beans.

INDIANA BAKED BEANS

Brown-beauty beans, hot corn bread and coleslaw make a full meal

2¼ qts. dried navy or pea beans
        (4 lbs.)
4   qts. water

1   c. brown sugar, firmly packed
2   tblsp. salt
3½ tsp. prepared mustard
2   c. ketchup
1   c. molasses
1½ c. chopped onion
¾   lb. salt pork, sliced

• Soak beans in large kettle overnight. Or bring them to a boil and boil 2 minutes. Remove from heat and let them stand, covered, 1 hour.
• Simmer beans in same water until almost tender, about 1 hour.
• Combine sugar, salt, mustard, ketchup, molasses and onion. Add to beans; bring to a boil.
• Pour beans into casseroles or bean pots or leave them in the kettle. Mix some of the salt pork slices into the beans; lay remaining pork on top.
• Bake, covered, in a slow oven (300° F.) 5 hours. Add boiling water during cooking if necessary to keep beans from becoming dry.
• Pack in freezer containers, seal, label, date and freeze. Recommended storage time: About 4 to 6 months. (Storage time depends on freshness of salt pork.)
• To serve, partially thaw in refrigerator; then heat in a saucepan with a little water added or in the top of a double boiler. Or bake beans in a moderate oven (350° F.) until heated, about 45 minutes for a pint, 1 hour for a quart.

Note: Some country cooks like to remove salt pork before freezing because it tends to become rancid. To use part of beans before freezing, cook 1 hour longer than for freezing; uncover the last 30 minutes.

CHAPTER 8

Soups, Salads and Sandwiches

Freezing Soups · Recipes for Beef/Vegetable and Country Bean Soups · Everyday and Fancy Sandwiches · Recipes for Frenched Ham and Ready-to-Go Bologna Sandwiches and Other Sandwiches · Frozen Chicken, Tuna and Fruit Salads

Supper may be the family's stepchild meal in some homes—not in country kitchens. There it is hearty and inviting even if it's often the easiest meal to get. And if you have a freezer, it isn't secondhand food from noontime dinner.

Getting a good country supper often starts when you put a kettle of substantial soup on the range to thaw and then bubble lazily while it heats. You take sandwiches from the freezer to thaw for serving cold or for heating, depending on the weather, and add a frozen salad that doubles for dessert. There's your menu. You may want to bring cookies from the freezer and open a jar of homemade pickles for bonus touches. And you'll pour glasses of milk or a hot beverage.

Not all the sandwiches and salads in this chapter are exclusively for family meals. You'll enjoy splurging with some of them when entertaining your club at luncheon. Chicken/Pineapple Salad, the favorite of a home economist in New Mexico, is a wonderful choice. So is Chicken/Blueberry Salad, made by a treasured New Jersey recipe. Serve either of these salads on crackling-crisp lettuce with hot rolls, muffins or corn sticks and the beverage of your choice. An easy luncheon to get and a compliment winner. Food looks as good as it tastes!

You'll find both everyday and Sunday-best sandwiches in this chapter. Let them inspire you to invent others. And keep a jar of Frozen Tuna Salad on hand for the salad plate or sandwich spreading.

Hearty Country-Style Soups

Many of the best soups require long, slow cooking. Fortunately, country cooks find it takes little more time to make enough for two or three meals than for one. And many of them freeze satisfactorily.

With space in freezers limited, it is best to concentrate the soups from one third to one half. Use less water in making them and add more when you reheat and thaw them.

Among the soups that freeze well are those made with dried beans, split peas, oysters, lobster, chicken and meats with vegetables. Most vegetables, like carrots, celery and soybeans, are good in frozen soups unless overcooked. Do not cook them until tender; they will finish cooking in the reheating. Add them according to the time they require in cooking—carrots before potatoes, for instance. Potatoes tend to become soft when frozen. In soups, as in other dishes when frozen, onions lose flavor, green pepper and garlic flavors intensify.

Cool the soup quickly to check the cooking and avoid the development of bacteria that may cause spoilage. Set the kettle of soup in a big pan containing very cold or ice water and stir occasionally until cool.

SPLIT PEA SOUP

Serve with crunchy French bread

1   lb. green split peas
2   qts. water
2   tsp. salt
¼   tsp. pepper
¼   tsp. dried marjoram leaves
1½ c. chopped onion
1   lb. ham hocks or 1 meaty ham bone
1   c. chopped celery
1   c. sliced carrots
1   c. cubed, pared potatoes

· Wash and pick over peas. Add water, salt, pepper, marjoram, onion and ham hocks. Bring mixture to a boil. Reduce heat and simmer, covered, 1½ hours, stirring occasionally.
· Remove ham. Discard fat and bones. Cut meat in pieces and return to soup mixture. Add celery and carrots; cook 10 minutes. Add potatoes and cook about 15 minutes more (vegetables will not be tender); stir occasionally.
· Cool quickly. Pack into freezer containers. Seal, label, date and freeze. Makes about 2 quarts.
Recommended storage time: 3 months.
· To serve, partially thaw so you can remove soup from freezer container. Add 2 c. water for each quart of soup mixture. Heat.

LENTIL SAUSAGE SOUP

A stick-to-the-ribs soup that men especially like

1   lb. fresh pork sausage links
3   tblsp. water
2   c. chopped onion
1   clove garlic, minced
4   carrots, sliced
¾   c. sliced celery
¼   c. chopped fresh parsley
2   c. lentils, rinsed and drained
1   (1 lb.) can tomatoes, cut up

Take golden, puffed Broccoli Soufflé proudly to the table—everyone will rave over this spectacular vegetable dish and revel in its fine flavor. Prepare it and freeze it; bake weeks later. Recipe in Chapter 2.

Men like the taste of Sunday Swiss Steaks; their wives like to have the frozen steaks heating in a slow oven while they're at church. Ranchero Pot Roast also freezes well. Recipes in Chapter 7.

These three breads—Honey Whole Wheat, Crusty French Bread and Orange Cinnamon Swirl—freeze well. To bring back the fresh-baked taste and come-hither aroma, heat them before serving. Recipes in Chapter 5.

Serve your guests Rocky Mountain Cake, a grand finale for gala meals. Its beauty tempts; its flavor blend of spices, caraway and black walnuts delights. An excellent chiffon cake and a good freezer. Recipe in Chapter 6.

2½ qts. water
1½ tblsp. salt
½ tsp. dried marjoram leaves
¼ tsp. pepper

· Cook sausage in 3 tblsp. water in covered Dutch oven or kettle for 5 minutes. Remove cover; continue cooking until sausage is brown, turning frequently. Drain sausage on paper towels. Cut in chunks; set aside.
· Pour off all but ¼ c. pan drippings. Sauté onion and garlic in drippings until tender (do not brown). Add carrots, celery, parsley, lentils, tomatoes, 2½ qts. water, salt, marjoram and pepper. Bring mixture to a boil. Reduce heat; cover and simmer 25 to 30 minutes or until vegetables are barely tender. Add sausage chunks and heat thoroughly.
· Ladle into freezer containers. Refrigerate until fat congeals. Remove fat. Seal, label, date and freeze. Makes 3 quarts.
Recommended storage time: 1 month.
· To serve, partially thaw so you can remove soup from freezer container. Add 2 c. water for each quart of soup mixture. Heat.

BEEF/VEGETABLE SOUP

Soup's on—your family will rejoice if it's a beef-vegetable combination

1   meaty soup bone (2 to 3 lbs.)
3   c. water
1   tblsp. salt
2½ c. tomato juice
1   c. diced carrots
1   c. diced celery
1   c. diced potatoes
1   medium onion, diced (1 c.)

· Place soup bone, water and salt in kettle. Simmer, covered, 2 hours. Add remaining ingredients and simmer until vegetables are almost tender, about 30 minutes.
· Cool quickly by placing kettle in pan of very cold or ice water. Cut meat in cubes; discard bone. Add meat to soup. Ladle into freezer containers. Place in refrigerator until fat congeals on top. Makes 5 pints.
· To freeze, remove fat from containers; seal, label, date and freeze.
Recommended storage time: 2 to 3 months.
· To serve, partially thaw so you can remove soup from container. To every pint of soup, add 1 cup water. Heat until piping hot.

COUNTRY BEAN SOUP

Thick soup saves freezer space—add the water when reheating it

2   lbs. dried navy beans
2½ qts. water
1   meaty ham bone
2   c. chopped onions
1   tblsp. salt
½ tsp. pepper
1   bay leaf

· Wash and pick over beans. Add water to cover and soak overnight. Or boil beans in water 2 minutes and soak 1 hour.
· Add remaining ingredients. Cover and simmer until beans are tender, 3 to 3½ hours.
· Remove ham bone and bay leaf. Mash beans slightly with a potato masher. Cut meat off bone and add to soup. Makes 4 quarts.

• Pack in containers, seal, label, date and freeze.
Recommended storage time: 3 to 4 months.

• To serve, thaw slightly at room temperature. Add 2 c. water for each quart of soup mixture. Heat.

Sandwiches—Plain and Fancy

If there are lunch boxes to pack, you will save considerable time by freezing sandwiches in advance. Make the sandwiches by the assembly line method—a supply for two weeks at a time. It takes much longer to fix them daily. Lay the bread slices out on the kitchen counter and spread from edge to edge with softened butter or regular margarine to completely cover. Add the fillings to half the slices; then cover with the remaining half.

Wrap as they will be used, individually or in pairs. Or wrap individually in waxed paper and then wrap in pairs in aluminum foil or plastic wrap. Label, date and freeze at once. If there is danger of the sandwiches being crushed in the freezer, you can put them in a box when frozen.

Most women become freehand cooks when they make sandwich fillings. They use imagination and the ingredients they have. But with sandwiches for freezing, certain foods should be avoided. Among them are mayonnaise and salad dressing (don't spread them on the bread instead of butter; like jelly and jams, they soak into the bread). Hard-cooked egg whites often develop off-flavors and become tough. Lettuce, tomatoes, carrots and other raw vegetables lose crispness and color. Natural cheese, sliced, may become crumbly, which

may or may not be objectionable to you in sandwiches. You may prefer to grate the cheese or use process cheese, which does not crumble.

Some of the best fillings for frozen sandwiches are sliced or ground cooked meats and poultry, peanut butter and other nut pastes, commercial cheese spreads like pimiento cheese, process cheese, cooked egg yolk, pickles and olives. Many farm cooks freeze sliced, cooked chicken and beef in broth for sandwiches.

You may prefer to make the fillings in large quantity and store them in small containers. They take up less freezer space and since many of the ingredients are ground, you save time by putting them through the food chopper in big amounts. This cuts down on the daily work, if lunch boxes are a feature of your life.

Dainty, fancy-shaped sandwiches for teas, buffets and other special occasions may be frozen to save time on party day. The crusts will trim off easily if you use frozen bread slices. You can make open-face sandwiches or those with the filling between bread. Arrange them on baking sheets and freeze. Then pack them in layers, with two sheets of freezer paper between, in rectangular boxes. Overwrap the boxes with aluminum foil or plastic wrap; label, date and freeze. Do not pack different kinds of sandwiches in the same box—there may be a transfer of flavors.

To Thaw Sandwiches: Regular sandwiches, left in their wrappings, will thaw in 2 to 3 hours. So sandwiches packed in the lunch box in the morning will be ready to eat by noon.

Thaw the small container of sandwich filling in the refrigerator overnight. It will be ready to spread the next morning.

Thaw small, open-face sandwiches in their original wrapping at room temperature, 10 to 15 minutes. Do not refreeze sandwiches once you thaw them.

Recommended storage time: For sandwiches, about 3 weeks, and for fillings, packed separately, 3 months.

FILLINGS FOR FROZEN SANDWICHES

Add variety to sandwiches with different kinds of bread. When salad dressing is called for, use *only enough* to give filling spreading consistency.

REGULAR SANDWICHES

LIVERWURST mixed with chopped, pimiento-stuffed olives or dill pickles.

PEANUT BUTTER with orange marmalade and honey added for spreading consistency.

BACON, cooked crisp and crumbled, in peanut butter.

CHEESE AND DATES, ground in equal parts, moistened for spreading consistency with orange juice. Point up flavor with a dash of grated orange peel. Add finely chopped nuts for variety.

BROWN BREAD, spread with butter and then with prepared mustard. Top with mashed baked beans, or omit mustard and add finely chopped mustard pickles to the beans.

GROUND COOKED CHICKEN and cream cheese, mixed and moistened with a little lemon juice.

DRIED BEEF, about 6 slices, chopped fine and mixed with 1 (3 oz.) pkg. cream cheese. Season with a bit of prepared horseradish if desired.

SARDINES, canned in oil, drained, bones and skins removed, mashed with hard-cooked egg yolk. Add a dash of lemon juice.

HAM, cooked, chopped and mixed with cream cheese and finely chopped pimiento-stuffed olives or dill pickles.

ROAST BEEF, left over, chopped and mixed with a little chopped pickle and salad dressing.

CHICKEN, cooked, ground, with chopped pimiento added and just enough salad dressing for spreading.

SALMON OR TUNA, canned, drained and mashed with pickle relish and a little salad dressing.

HAM, cooked, chopped and mixed with pickle relish and a little salad dressing.

FANCY SANDWICHES

Use a variety of breads and cut buttered slices into fancy shapes with cookie cutters.

CREAMED CHEESE, softened at room temperature and whipped until very light and fluffy. Use food color to tint cheese delicate pastel colors.

CHICKEN LIVER PÂTÉ, spread on bread and garnished with strips or cutouts of pimientos.

PEANUT BUTTER, mixed with chopped dates and orange marmalade added for spreading consistency.

CHEESE/TUNA, one cup grated Cheddar cheese mixed with ½ c. tuna, drained. Add finely chopped dill pickle and a little salad dressing.

CHEESE/OLIVE, 1 (3 oz.) pkg. cream cheese mixed with 8 ripe olives, pitted and finely chopped. Point up flavors with lemon juice.

DEVILED HAM, mixed with finely chopped nuts and spread on rye bread. Use canned ham. Add a little horseradish if desired.

BACON/CHEESE, crumbled, crisp-cooked bacon mixed into cream cheese with a little grated orange peel added.

FRENCHED HAM SANDWICH

Hearty—a snap to fix when you have French toast in your freezer

> 6 slices baked ham
> 6 slices mild cheese
> 12 slices frozen French toast
> 2 tblsp. butter or regular
>   margarine

• Place 1 slice ham and 1 slice cheese between 2 slices toast; place in shallow baking pan with melted butter. Bake in hot oven (425° F.) 15 to 20 minutes. Turn once. Makes 6 sandwiches.

Note: See directions for freezing French toast in Chapter 5.

READY-TO-GO BOLOGNA SANDWICHES

Tuck them in the oven to heat while you get the rest of the supper

> 1 lb. bologna
> ¾ lb. sharp process cheese
> 3 (3″) dill pickles
> 1 slice onion or 2 tblsp. chopped
>   green onion
> ½ c. salad dressing
> 1 tblsp. prepared mustard
> Soft butter or regular margarine
> 16 wiener buns

• Grind bologna, cheese and pickles in food chopper. If using onion slice, grind and add it to mixture. Or add chopped green onion. Stir in salad dressing and mustard.
• Butter inside of split buns evenly, spreading to the edge. Spread with bologna mixture. Put halves together to make 16 sandwiches.

• Wrap one layer deep in foil packages; label, date and freeze. (They reheat quicker than when stacked.) Recommended storage time: 2 to 3 weeks.

• To serve, if you wish to use some of the sandwiches without freezing, heat in a slow oven (275° F.) until filling is hot and cheese melted, about 30 minutes. Place unthawed frozen sandwiches, unwrapped, in a hot oven (400° F.) and heat 30 to 45 minutes.

Sunday Supper Salads

When company stops by at dusk Sunday afternoon, you often want to say: "Do stay for supper." You'll find the invitation comes easier when there's a good salad in the freezer.

Fruit salads are favorites for freezing, but other kinds are gaining prestige with the discovery of satisfactory ways to handle them. If fresh fruits, like peaches, are used, combine them with citrus fruits, grapefruit and oranges, or add a little lemon juice or powdered ascorbic acid. Often fruit salads contain cream cheese or whipped cream.

Mayonnaise and home-cooked dressings tend to separate when frozen. If you do use them in fruit salads, it is a good idea to mix only a small amount of them with cream cheese or whipped cream. Commercial salad dressing gives good results in chicken, fish and meat salads. The trick is to use a little more commercial salad dressing than usual in salads. In case such salads do separate a little after defrosting, you usually can recombine them by stirring.

Tuna, salmon, shrimp, crab, lobster, ham, pork and beef salads may be frozen. Many farm women like to keep cooked chicken and meat frozen for convenient salad making.

Gelatin salads freeze well if correctly handled. Research shows that it is important to use less liquid than usual—one fourth less. If the recipe calls for 2 cups water or other liquid, reduce it to 1½ cups unless the recipe is especially developed for a frozen salad, like those that follow. Tomato aspic, with the quantity of liquid adjusted, freezes beautifully. Celery, diced, retains some of its texture and color, but few vegetables are satisfactory in frozen salads.

Combinations of fresh fruits, frozen for mixed salads, never go begging. To serve them, partially thaw the fruits and mix in the salad dressing. Add nuts at the same time, if you wish—they sometimes discolor and acquire a bitter taste when frozen in salads. This is especially true if the salad is held more than a few days. Bananas usually are best added at serving time too. Raw apple slices tend to become flabby when frozen.

To Freeze: You can freeze the salad in a large block, wrap it in moisture-vaporproof packaging material and return it to the freezer. Or freeze it in a cylindrical carton if you want to push it out and cut it in round slices for serving. Some women like to slice frozen salads before packaging them, separating the servings with cellophane or plastic wrap. They add an overwrap of moisture-vaporproof paper and place in the freezer.

Or freeze salads in paper cups or individual molds. You can remove them from the cups or molds when they are frozen and wrap individually in foil or plastic wrap. Put them in a box or plastic freezer bag to keep them together in the freezer. If you prefer, freeze the salad in a refrigerator tray with dividers and treat the salad-cubes like individual salads.
Recommended storage time: About 2 to 6 weeks.

To Serve: Never completely thaw frozen fruit salads before serving. Serve them as they come from the freezer or let mellow in refrigerator about 1 hour. You may thaw chicken and tuna salads overnight in refrigerator. Arrange on crisp salad greens and add any last-minute garnishes.

CHICKEN/BLUEBERRY SALAD

Something new and extra-good in chicken salad—a New Jersey favorite

1   envelope unflavored gelatin
¼   c. cold water
1   (10½ oz.) can condensed cream
      of celery soup
½   c. commercial salad dressing
¼   tsp. salt
2   c. finely cut cooked chicken
1   c. diced celery
1   c. fresh or frozen blueberries

• Soften gelatin in cold water. Heat soup over medium heat until boiling. Add to gelatin; stir to dissolve.

• Blend in salad dressing. Add remaining ingredients.
• Pour into a 1½-qt. mold. Chill until firm. Makes 8 servings.
• Wrap, label, date and freeze.
Recommended storage time: Up to 1 month.
• To serve, thaw in refrigerator overnight. Serve on salad greens.

CHICKEN/PINEAPPLE SALAD

A choice recipe from New Mexico where hot corn sticks are a go-with

1½   c. cooked chicken, cut in ½"
      pieces
¾   c. drained crushed pineapple
1   c. chopped pecans
¼   tsp. salt
1   c. heavy cream, whipped
1   c. commercial salad dressing

• Toss together chicken, pineapple, pecans and salt. Blend together whipped cream and salad dressing. Fold into chicken mixture. Pour into a 1-qt. refrigerator tray. Makes 5 to 6 servings.
• Place tray in freezer. When frozen, unmold, wrap, label, date and return to freezer. Or slice and wrap each frozen portion individually in foil. Return to freezer.
Recommended storage time: Up to 1 month.
• To serve, remove from freezer 15 minutes before serving. Serve on crisp greens. Top with salad dressing, if desired, and garnish each serving with a spoonful of cranberry jelly or sauce, drained, or seedless green grapes.

FROZEN TUNA SALAD

Makes excellent sandwich filling

2 (7 oz.) cans tuna
Dash of pepper
2 tsp. prepared mustard
¼ c. pickle relish
½ c. commercial salad dressing
1 tblsp. lemon juice

• Mix together all the ingredients. Divide into 4 custard cups or molds. Makes 2½ cups (or enough filling for 10 sandwiches).
• Wrap, label, date and freeze. Recommended storage time: 6 weeks.
• To serve, partially thaw in refrigerator and unmold on salad greens.

REFRESHING FRUIT SALAD

Not so rich as most frozen fruit salads —doesn't "weep" when served

1 (1 lb.) can fruit cocktail
1 (8¼ oz.) can crushed
      pineapple
1 (1 lb.) can apricot halves
1 banana, diced
3 tblsp. lemon juice
½ tsp. grated lemon peel
1 (3 oz.) pkg. lemon flavor
      gelatin
1 c. heavy cream, whipped

• Drain fruit cocktail, pineapple and apricots, reserving juices. Cut apricot halves in fourths. Gently toss banana with lemon juice and peel.
• Heat 1 c. reserved fruit juices to boiling. Dissolve gelatin in hot juice. Add ½ c. cool, reserved juice. Chill until mixture starts to thicken.
• Whip gelatin mixture with rotary or

electric beater until thick and light. Fold in fruits and whipped cream. Pour into two 9×5×3″ loaf pans or 1-qt. molds. Makes 12 servings.
• Place in freezer; when frozen, remove. Wrap, label, date and return to freezer.
Recommended storage time: About 2 weeks.
• To serve, let partially thaw in refrigerator several hours.

EVER-READY WALDORF SALAD

Apples soften but give salad a pleasing flavor; celery stays crisp

1 (1 lb. 4 oz.) can crushed
      pineapple
3 eggs, slightly beaten
¾ c. sugar
⅓ c. lemon juice
¼ tsp. salt
¾ c. diced celery
3 medium unpeeled apples,
      diced
¾ c. chopped walnuts
1½ c. heavy cream, whipped

• Drain pineapple, reserving ¾ c. juice.
• Combine juice with eggs, sugar, lemon juice and salt. Cook over low heat until thick, stirring constantly. Cool.
• Add pineapple, celery, apples and nuts. Fold in whipped cream. Turn into two 9×5×3″ loaf pans or a 9″ square pan. Makes 12 servings.
• Place in freezer. When frozen, wrap, label, date and return to freezer.
Recommended storage time: Up to 1 month.

• To serve, cut in serving-size portions and place on lettuce-lined plates. Garnish with spoonfuls of salad dressing and insert slices of unpeeled red apples, skin side up, in fan shape.

THREE-WAY FROZEN SALAD

Choose from our trio of colors

2 (3 oz.) pkgs. cream cheese
1 c. heavy cream
⅓ c. mayonnaise
2 tblsp. lemon juice
1 c. miniature marshmallows
1 (10 oz.) pkg. frozen sliced strawberries, thawed and drained
1 (1 lb. 13 oz.) can crushed pineapple, drained

• Whip cheese; slowly add cream, beating until thick. Fold in remaining ingredients.
• Pour into muffin-pan cups lined with cupcake papers. Garnish with nuts, maraschino cherries or coconut. Makes about 18.
• Place in freezer. When frozen, remove from pans; put in plastic freezer bags; seal, and return to freezer. Recommended storage time: Up to 1 month.
• To serve, remove from freezer in the quantity desired; remove paper cups; place on crisp greens.

VARIATIONS

MINT/GRAPE: Substitute 1 c. white seedless grapes for berries; add green food color and ¼ tsp. peppermint extract.

GOLDEN PEACH: Substitute 1 (10 oz.) pkg. frozen peaches, drained, for berries; add yellow food color.

CREAMY FROZEN SALAD

Specialty of a Hoosier hostess—sour cream adds wonderful flavor

2 c. dairy sour cream
2 tblsp. lemon juice
¾ c. sugar
⅛ tsp. salt
1 (8¼ oz.) can crushed pineapple, drained
¼ c. sliced maraschino cherries
¼ c. chopped pecans
1 banana, sliced

• Blend cream, lemon juice, sugar and salt. Stir in other ingredients.
• Pour into 1-qt. mold, or cupcake papers. Makes 8 servings.
• Place in freezer. When frozen, remove from freezer; wrap, label, date and return to freezer. Recommended storage time: About 2 weeks.
• To serve, place on crisp greens.

PART TWO

CHAPTER 9

Preserving Food Safely: New Rules for Home Canning

Acid Foods and Low-acid Foods · News about Tomatoes · Jars and Lids · Boiling Water Canner · Steam Pressure Canner · Step-by-Step Directions for Successful Canning · Storing Canned Foods · Unsafe Canning Practices · Signs of Food Spoilage

Experienced home canners may be tempted to skip this chapter—to get on to the recipes. Don't do it! Some of the recommendations for home canning have changed in recent years. This review of canning procedures will be as useful for the "old hands" as it is for beginners.

Do you find it puzzling, that canning methods and processing times in use for years, handed down from Grandma, are now thought to be unsafe? Well, we wouldn't be Americans if we weren't always trying to solve problems and improve products. When you write us about canning failures— jar lids not sealing, pickles shriveling or changing color, mold forming on jellies and jams—we ask researchers for explanations. In their search for the reasons, scientists and home economists develop better and ever more precise canning directions. You and your family are the winners. While Grandma depended on experience—and a measure of luck—those who follow up-to-date directions and use good equipment can expect good results every time.

Home canning got an undeserved black eye in 1977, when the largest single outbreak of botulism ever recorded in the United States was traced to improperly home-canned peppers served in a Michigan restaurant. Forty-five diners were stricken. No deaths resulted from this outbreak because diagnosis and treatment were swift, but botulism is a wretched illness which is often fatal to the people who eat affected food.

As a home canner, you do not need to risk botulism: the rules for safe canning are explicit. Trouble comes when you don't follow—or don't know—the rules. For people new to gardening and home canning, there's an added danger: some of today's "back to nature" books give incomplete or incorrect canning directions.

How Canning Preserves Food

Food spoilage is caused by the growth of bacteria, yeasts and molds. These microorganisms are found everywhere—in food, in earth, water and air. They grow best in moist, warm food, but they can be destroyed by heat.

If food is to be stored safely in a canning jar, all living forms of these microorganisms must be destroyed, and the jar must be vacuum-sealed to keep air out and prevent recontamination.

This is done in a single process by heat sterilization. The prepared food is packed in a jar, covered with a jar lid and heated in boiling water (or in a steam pressure canner) long enough to kill the microorganisms contained in the food or jar. As the food reaches boiling temperature, steam exhausts the air from the jar. When the jar cools and steam condenses, the lid makes an airtight vacuum seal.

The temperature recommended for sterilizing food depends on whether the food is an *acid food* or a *low-acid food*. Acid foods (fruits, tomatoes, pickles, relishes) can be safely sterilized in a boiling water bath— 212° F. But low-acid foods (vegetables, meats, fish and poultry) must be processed at a temperature of 240° F. —in a steam pressure canner.

Processing time depends on the kind of food, the size of the jar, and whether it was packed raw or hot. There is no guesswork. If you want food safely preserved, you must follow recommended times and temperatures. It takes much longer for heat to penetrate jars of starchy or solid foods such as corn or pumpkin than, say,

jars of green beans. (Note that if you live more than 1,000 feet above sea level, you must also make adjustments in time or pressure—see Index for Altitude Corrections.)

BOTULISM

Ordinary boiling for a reasonable length of time kills most of the microorganisms that spoil food. But the boiling water bath is not hot enough to kill the spores of such heat-resistant bacteria as *Clostridium botulinum*. If you're canning fruits, tomatoes or pickles, there's nothing to worry about —the dangerous *botulinum* doesn't develop in acid foods. That's why it's safe to process acid foods in a boiling water bath.

The spores (the dormant form of *Clostridium botulinum*) are not deadly; in fact, they exist everywhere —in soil, and on the surface of foods which we eat every day. Spores remain dormant in canned acid foods, causing no ill effects.

But these bacteria thrive in low-acid foods and in the moist, low-oxygen environment of an airtight jar. Unless the spores are killed in the jar, they will germinate and grow, producing the deadly toxin responsible for botulism. To can vegetables and other low-acid foods safely, sterilize them at 240° F.—that is, in a steam pressure canner at 10 pounds pressure (more, if you live at higher altitudes—see Index for Altitude Corrections). *If you don't have a pressure canner, don't can vegetables, meats, fish, poultry or anything with milk in it.*

Grandma's Precautions: Even with careful procedures and use of a pressure canner in good working order, Grandma still took no chances with home-canned vegetables and other low-acid foods. Before opening any jar, she examined it carefully for spoilage. Even if the food looked and smelled all right, she still took the precaution of boiling the jar's contents before anyone tasted it. Brisk boiling for ten minutes (pot may be covered or uncovered) is enough to destroy the botulism toxin, if present. Such precautionary boiling is definitely recommended for all home-canned low-acid foods unless you are absolutely sure that your equipment is reliable and that you followed recommended procedures to the letter.

Food showing any signs of spoilage should be destroyed without tasting. Information about spoilage and disposing of contaminated food will be found at the end of this chapter—see Index.

Botulism attacks the nervous system. Early symptoms include nausea, blurred vision and difficulty in swallowing. The patient becomes weak—can't hold his head up. Until medication takes effect, he will need help to breathe; about one-third of the victims of botulism die of respiratory failure.

Acid Foods and Low-acid Foods

Science expresses the quantity of acid in food as a "pH" value—short for "potential of Hydrogen." This means that a chemist can use a testing device on any raw or cooked food and tell you how to classify it. Values of pH range from 0 (acid) to 14 (alkaline). Most foods have a pH value somewhere between 3.0 and 8.0. For canning purposes, the dividing line is 4.6. Acid foods are those with a pH less than 4.6; they're easily and safely canned in a boiling water bath. Any food with pH of or over 4.6 is a low-acid food that must be sterilized at a temperature higher than boiling water —in a steam pressure canner.

All fruits and berries are natural acid foods. This includes tomatoes— botanically, the tomato is a fruit. For canning purposes, properly fermented vegetables, sauerkraut, for example, also are classified as acid foods. During fermentation, their natural sugars are converted to natural acids by bacterial action. Vegetables pickled with vinegar are also considered acid foods, eligible for processing in the boiling water bath. (See Chapter 13, Pickles, Vegetable Relishes and Sauerkraut, for more information about fermentation and vinegar.)

Here is an alphabetical list of acid and low-acid foods. For specific canning directions for each food, see Index.

Acid foods, safe for canning in the boiling water bath:

Apples	Chutney
Apple juice	Conserves
Applesauce	Figs
Apricots	Fruit butters
Berries	Fruit juices
(all varieties)	Fruit purées
Cherries	Gooseberries

Grapefruit	Mangoes	Pickles	Rhubarb
Grapes	Marmalades	Pineapple	Sauerkraut
Guavas	Nectarines	Plums	Tomatoes
Jams, Jellies	Papayas	Preserves	Tomato juice
Ketchup	Peaches	Prunes	Tomato sauce
Loquats	Pears	Relishes	or purée

Low-acid foods, to be canned in a steam pressure canner:

Artichokes	Corn	Peas, Fresh	Spinach and other
Asparagus	Hominy	Black-eye	Greens
Beans, Lima	Meats	Peas, Fresh green	Squash, all varieties
Beans, Snap	(all varieties)	Peppers, Bell,	Sweet Potatoes
Beets	Mushrooms	Green and Red	Tomato Sauce with :
Carrots	Okra	Potatoes	Meat
Celery	Okra and	Poultry	Turnips
Celery and	Tomatoes	Pumpkin	Vegetable-Beef Stew
Tomatoes			Most Soups

New Directions for Processing Tomatoes

Recently, "low-acid" tomatoes have been in the news and home canners ask how they should be processed. Some of the new tomatoes, particularly the pear-shaped and yellow varieties, do, indeed, taste less tart than others. However, when scientists in the USDA Agricultural Research Service and at the University of Minnesota ran pH tests on more than 100 varieties of tomatoes in 1976, they found the acid level in the new varieties to be substantially the same as traditional varieties. Several of the so-called low-acid tomatoes are actually not low in acid but high in sugar, which masks the tart, acid flavor.

Spoilage of home-canned tomato products has been reported, however —including at least two cases of botulism traced to home-canned tomatoes or tomato juice. Since this is supposed to be impossible in an acid food, how could it happen? There are several explanations listed below, and they underscore the need for home canners to follow new directions carefully.

• As tomatoes ripen, they lose acidity. And while all firm prime-quality tomatoes stay safely under the 4.5 pH line, mushy or overripe tomatoes may be in the danger zone with a pH of 4.6. They would be unsafe for canning in a boiling water bath. There are several tomato varieties (listed in Chapter 10) with an acidity low enough to permit the growth of bacteria that causes botulism when the fruits canned are overripe, decayed or picked from dead plants. Home canners have always been cautioned not to can overripe or decayed fruit.

• Open-kettle canning, once permitted for cooked tomato products such as chili sauces and ketchups, is no longer considered safe. The reason? Possibility of mold and bacterial growth. Open-kettle canning means food is cooked in an open kettle, packed boiling hot in sterilized jars and quickly sealed, with no processing after the food is packed in jars.

In transferring food from boiling kettle to sterilized jar, it is impossible to prevent recontamination from microorganisms in the air or in jars and lids. Also, with no processing, air is not exhausted from the jar. Oxygen in head space allows mold organisms to grow. And certain types of mold growth lower the acidity in food. Thus, an acid food like tomatoes can turn into a low-acid food inside the jar, setting up conditions for the growth of *botulinum* bacteria.

• The recommended time for processing all tomato products is longer now than it used to be. Again, the reason is the possibility of growth of molds and flat-sour bacteria. The heat sterilization period must be long enough to kill heat-resistant bacteria which, if they get a start, can contribute to mold growth, even inside a sealed jar.

There's a perfectly logical explana-tion for increasing the sterilization time. It's because so many of us home canners no longer have the ideal place to store our canned food. Jars should be stored in a clean, cool, dark, dry area, preferably between 40° and 50° F.—in the "cold cellar." But many homes no longer have a basement, let alone a cold cellar. When canned food is stored at room temperature, or between 70° and 95° F., it loses quality much faster than at lower temperatures. Under such conditions, errors in canning are even more critical—the potential for spoilage is greater.

Chances of botulism developing in tomato products are extremely rare— but why take any chances? You'll find accurate directions for canning tomatoes in Chapter 10, along with this warning: Moldy canned tomatoes may not always have a rotten odor and the liquid in the jar may not always be cloudy. Don't scrape mold off home-canned tomatoes and then eat the rest in the jar—discard the entire jar, following directions at the end of this chapter.

Your Canning Equipment and Supplies

A Wisconsin farm woman says she starts thinking about her summer canning the first spring evening she hears a cowbird on the pasture fence mock a phoebe in the woods. No doubt you have other signals—the run of smelts, bloom on the dogwood, shooting stars along the path to the garden, purple lilacs brightening bushes. Anything that announces spring is here and summer not far behind!

This means you'll soon be getting the ground ready to plant your garden. So it's time to figure out your family's needs, decide what to plant and calculate how many jars you'll want to fill. In the chapters on canning specific fruits and vegetables, we list a few of the best varieties for canning, but you can also check on good local varieties by asking your County Extension Agent. The charts on the inside back cover of this cookbook will give you approximate yields of many foods when canned.

But your best guide to varieties, yields and the number of canning jars and lids you need will be your own records. In a few seasons, you'll know from experience how much to plant, how much to can, and which varieties you like best.

JARS AND LIDS

Glass jars recommended for home canning—particularly for pressure canning—are the Mason-type jars named after the man who patented the screw-top closure. They are available in half-pint, pint, pint-and-a-half and quart sizes from a number of manufacturers who identify jars with their own brand names. These jars are fully tempered to withstand quick temperature changes without breaking and heat sterilizing temperatures. The top rim of the jar is thick enough to make good contact with the sealing edge of the self-sealing lid, and the neck of the jar is deep enough to permit the lid to be tightened securely, so you get a good seal.

So long as they remain in good condition, these jars can be reused; with careful handling and cleaning, they should last a long time.

Two-quart or half-gallon jars are not recommended for home canning. Unfortunately, some old ones are still in use, and new ones imported from Europe, with glass lids and wire bails, are appearing in some gourmet shops and department stores. Most canners are not designed to hold these huge jars. Also, there is danger that food in the center of the big jars will not be sufficiently heated to kill spoilage organisms.

Before each use, check all jars, used and new. They should be free of cracks and chips, especially on the sealing surface. Old manuals suggest running your finger over the rim of the jar. You can sometimes feel flaws you can't see. Also examine the jars in good light, at eye level. If you wear glasses for close work, wear them when you check over your jars. If there's a chip or nick, or if the rim has a bump, depression or hairline crack, the lid may not seal, or jars may break during processing. Discard faulty jars or set them aside for refrigerator storage; don't use them in the canner.

If jars have any cloudy stains or scaly deposits which harbor bacteria, clean them this way: Fill jars with full-strength vinegar and let stand for two or three hours. Scrub inside of jar with nylon net or stiff brush; do not use a metal scouring pad because it may scratch glass. Reuse vinegar to clean other stained jars; then discard vinegar.

If you have any antique jars or lightning jars (those with glass lids and wire bails), use them for refrigerator preserves or for canisters—but not for canning. Antique jars were designed for shoulder sealing with rubber rings.

Recommended lids for home canning are metal—either steel or aluminum. The two-piece self-sealing lids are most familiar: flat metal disks edged with sealing compound, which are held in place during processing and cooling by screw bands. The sealing edge is shaped in such a way that the lid seats itself on the jar rim without skidding. Screw bands—which may be metal or plastic—are reusable, but the metal lids are not reusable.

Self-sealing one-piece metal caps are relatively new for home canning —they have a sealing compound (gasket) inside the lid. Most have performed well in tests conducted by food scientists at The Pennsylvania State University in 1976 and 1977; such lids have also passed reusability tests.

But note that as of 1977, all plastic lids or lid liners tested at Penn State have proven unsatisfactory in achieving and/or maintaining a seal.

Porcelain-lined zinc caps sealed with a separate rubber ring are also still being used by some home canners, despite the fact they sometimes fail to form seals. Rubber jar rings should be new; they should never be reused.

Before use, inspect all one-piece and two-piece lids to make sure there are no gaps in the sealing compound, or scratches in the enamel lining. Also reject any zinc caps which have chips or cracks in the porcelain. Use only perfect lids for processing. (Use rejects for refrigerator preserves.)

BOILING WATER CANNER

Any large kettle can serve as a boiling water canner, providing it has a tight-fitting cover and is deep enough to hold a rack for the jars, with enough room above jars for water to boil. Measure the height of jars and add 5 inches: 1 inch for the rack, 2 inches for water to cover jar tops and another 2 inches above the water line, so water can bubble vigorously without

boiling over. You can use your pressure canner if it is deep enough. Just set the cover in place but do not fasten it; leave the petcock open for steam to escape.

The rack is necessary to keep jars off the bottom of the canner, so they won't crack and so boiling water can circulate under jars. You can improvise with a round cake rack, but if you do much canning, you will prefer a rack or canning basket designed to keep jars from touching each other and tipping against the sides of the canner. The rack may be of wire or wood, but not resinous wood such as pine.

If you are buying a new canner, you can choose the type best suited to your energy source: one with a ridged bottom provides more heating surface if you cook with gas; one with a flat bottom makes better contact with electric surface units. Canner should be no more than 4 inches greater in diameter than the burner you will put it on. Do not use larger or oval-shaped containers that fit over two burners—some jars might not get enough constant heat.

PRESSURE CANNER

A pressure canner is a deep kettle made of heavy gauge metal—most are aluminum. It is fitted with a rack to hold jars, and it has a tight fitting lid which clamps or locks in place. Most canner lids have a channel for a rubber gasket; with the lid locked in place, steam pressure pushes against the gasket to seal the canner.

There are two general kinds of pressure canners. One has a weighted gauge which actually controls the pressure when the canner is in use. The other has a dial gauge which you read to check on pressure inside the canner. Both types are safe to use. For many decades, pressure canners and cookers have been equipped with a safety valve or fuse—a little round nubbin on the lid which simply blows out if pressure becomes too great. There'll be a spurt of steam as pressure is released—but no explosion.

In addition to the safety fuse, there's an air vent in the lid. On older dial-gauge canners, the vent is called a petcock. When you lock up the canner and apply heat, the interior begins to fill with steam and the air escapes through the vent or petcock. When all the air is exhausted (takes about 10 minutes), you close the vent either by putting a weight over it, or by closing it manually with a flip-type or screw-type valve. With vent closed, steam pressure builds up to the desired 10 pounds, and you can maintain that pressure by controlling the heat under the canner.

If your canner has the weighted gauge, you use the gauge itself to close the vent, setting it for 10 or 15 pounds pressure depending on your altitude. With the right heat under the canner, the gauge will automatically release just enough of the steam buildup to keep pressure at 10 or 15 pounds. Depending on the brand of canner you own, the gauge will jiggle 3 or 4 times a minute, or it will rock steadily and gently, releasing small amounts of steam. If you have too much heat under the canner, the gauge will jiggle more often or release a lot of steam, but will limit pressure to the selected level. These are signals to reduce the heat. If the gauge falls silent, you know you've lost heat and pressure.

If you have a dial-gauge canner, you must keep an eye on the dial and regulate heat as necessary to maintain pressure. Dial gauges can get out of adjustment, so they should be checked for accuracy before the canning season. If the gauge is not giving you a correct reading, the processing will not be accurate and food may be unsafe.

How do you check a dial gauge for accuracy? Try calling your County Extension Home Economist first. Many counties have equipment for testing dial gauges—it's a procedure that takes little time and is often available at home canning "clinics" in the county. The store where you bought the canner may also offer this service.

Otherwise, you can mail the gauge to one of the service centers listed in the instruction book that comes with your canner. Follow directions for removing the gauge carefully, so as not to damage it.

If your gauge is more than 2 pounds off, it should be replaced. If less than 2 pounds, you can make adjustments. For example, if gauge reads high by 1 pound, you would have to process food at 11 pounds pressure to compensate.

Weighted gauges do not get out of calibration and therefore, they do not need to be tested for accuracy.

The gasket will need replacing if it shrinks and fails to seal or if it swells, making the cover difficult to open or close.

If you live at higher altitudes, see Index for Altitude Corrections; you'll need higher pressure to reach the desired temperature.

Do not confuse the pressure canner with the pressure cooker or saucepan. The canner (sometimes called a cooker-canner) is of heavier construction and is big enough to accommodate quart jars. The smallest canner will hold 4 quart jars; the popular medium size holds 7 quart or 10 pint jars. Larger canners are available, but you may not have enough clearance above your range to use them, and they're extremely heavy when loaded.

If you are canning only a few pint or half-pint jars, you may use your pressure saucepan providing you can maintain pressure at 10 pounds. When using the pressure saucepan for canning, follow manufacturer's directions but *do not lower pressure with cold water after processing.*

SMALL TOOLS AND SPECIAL INGREDIENTS

Two extremely useful tools for home canners are a jar funnel, which makes it easier to fill jars without spilling, and a jar lifter, for removing hot jars from boiling water or pressure canners. If you do a lot of canning or freezing, a blancher will be useful, or you can make do with a wire basket to lower food into boiling water and lift it out again. An electric blender or food chopper is essential for relishes, and if you have kitchen scales, colander, sieve or food mill, ladle and long-handled slotted spoon, they'll be handy.

Check recipes to make a list of the herbs, spices, sugar and vinegar you'll need. Use standard vinegar of 5% (or 50 grains) acidity; unmarked or homemade vinegar of uncertain strength should not be used for pickles or relishes. Do not use table salt for canning; buy canning or pickling salt which has no additives. The iodine in most table salt discolors some food and may prevent normal fermentation of pickles or sauerkraut, while the anti-caking additive may cloud the brine. To prevent fruit from darkening, you may want to use ascorbic acid or a commercial antioxidant, which is a combination of ascorbic acid and citric acid.

Steps to Successful Canning

Canning takes some doing, but it's worth your effort when the food inside your jars is prime quality. To ensure good results, plant or buy varieties that can well. Your County Extension Office will have up-to-date information on recommended varieties in your area.

If possible, can fruit and tomatoes on the day you'd like to eat them fresh—when they're firm and ripe. If you buy fruit for canning by the bushel at a roadside stand, sort it carefully and each day, can the fruit at peak ripeness. Can vegetables when they're still young and tender, and preferably on the same day you pick them. Freshness will make a big difference in your success with canning—in the flavor and in the nutrients you save. If you must hold fresh-picked fruits and vegetables for a few hours, store them—unwashed—in a cool, dry and airy place.

Can only freshly killed meats and poultry, chilled at once and kept chilled until canned.

During canning season, nature will have more to say about your daily work schedule than you will. When food ripens, you must take care of it. It helps to be organized: counters clear, jars and lids inspected and ready, and a few hours free so you can concentrate. Pray that company doesn't come or if they do, hope it's someone who knows how to help. Wash and prepare only the amount of food that will fit in the canner at one time—and be realistic about how many canner loads you can handle in a day.

PREPARING JARS AND LIDS

Wash canning jars and screw bands in hot sudsy water and rinse thoroughly. Or wash jars the easy way—in your electric dishwasher. Hold jars in hot water until ready to fill, or drain upside down on a clean towel.

Follow the manufacturer's directions for preparing self-sealing lids. If none are given, wash and rinse lids and scald them by pouring boiling water over them in a pan. Leave them in the hot water until you need them.

If you can with porcelain-lined zinc caps that have been used before, boil them for 15 minutes and leave them submerged in simmering water until used. If caps are new, wash them in hot sudsy water and rinse well. Use new rubber rings; wash and rinse; keep rubbers wet until needed.

It is not necessary to sterilize empty jars which will be filled with food and processed in a pressure canner, or which will be processed for 15 minutes or more in a boiling water canner. Processing sterilizes food, jars and lids all at the same time. But if processing in the boiling water canner is less than 15 minutes, empty jars should be sterilized before filling.

To sterilize jars, put them upright in a large container, add water to cover jars by at least 1 inch, bring to a boil and boil for 15 minutes. Leave jars in hot water until you fill them.

SORTING AND WASHING FRUITS AND VEGETABLES

Sort fruits and vegetables for ripeness or maturity and for size, so they will cook evenly. Discard *all* spoiled, diseased or bruised cherries, berries and plums—don't try to cut out blemishes. After washing large fruits, trim out bruised or decayed areas well below and around the blemish. However, it's sounder practice to set aside blemished fruit for fresh eating. If you're slicing peaches for today's dessert, you can easily slice away bruised or decayed areas. Only the best product is worth preserving. Also reject all moldy or damaged vegetables.

Wash fruits and vegetables thoroughly, small lots at a time, through several changes of water. Wash them even if you'll peel them. Dirt contains bacteria that are difficult to kill; get rid of it! Lift the vegetables and fruits out of the water, so the washed-off soil will stay behind, in the bottom of the sink or pan. Rinse out pans between changes of water. Do not let foods soak in water; they lose nutrients and flavor. Handle gently to avoid bruising.

After washing, remove caps, stems, cores, pits, seeds and skins as individual recipes direct. Potatoes and some fruits will need immediate treatment to prevent darkening.

Prepare only enough jars of food at one time to fill the canner. Charts on the inside back cover of this book tell you how much fresh food it takes to fill quart jars. Recipes give yields.

KEEPING FRUITS FROM DARKENING

Apples, pears, peaches and a few other fruits may turn brown quickly after cutting them because of the action of *enzymes*. Enzymes are specialized proteins in fruits and vegetables that bring about biochemical changes. Responsible for ripening foods, they continue to cause color and flavor changes until food gets too ripe and finally spoils. This ripening-spoiling action speeds up when you peel or cut the food, exposing it to air. Oxygen in the air (also the oxygen dissolved in water) causes oxidation and darkening. To prevent darkening and to preserve the food, you must destroy the enzymes and also remove as much air as possible from food tissues and from jars.

It doesn't take much heat to destroy enzymes: heating food quickly to simmering temperature (185° F.) will do it. But peeled fruits which darken quickly should be pretreated before heating them. You can drop them into a salt-vinegar solution (2 tblsp. each salt and vinegar added to 1 gal. cold water). Or prepare a solution using a commercial antioxidant (a mixture of ascorbic acid and citric acid); follow directions on the label. Cut potatoes into a brine solution (1 tsp. salt to 1 qt. water). Work fast: do not let food stand in solution more than 20 minutes. Drain and rinse before heating or packing in jars.

When you pack jars, add a color protector to peaches, pears, apricots, applesauce and plums to help them keep their bright colors and prevent darkening fruit at the top of the jars. Use the packaged antioxidant mixture following directions on the label. Or sprinkle ¼ tsp. pure powdered or crystalline ascorbic acid (available at drug stores) into each quart jar of fruit before it is sealed.

RAW PACK AND HOT PACK— TWO WAYS TO PACK JARS

The recommended method for packing jars is hot pack: food is brought to a boil or cooked in an uncovered kettle for a specific length of time (see recipes) and then packed hot in jars and processed.

Some foods may be packed raw; low density foods like whole tomatoes will hold their shape better if they're handled as little as possible. And many home canners simply prefer the taste and texture of fruits packed without precooking. Soft berries are always packed raw. The raw pack is also called cold pack, which is somewhat misleading, because after filling the jars, you are usually directed to cover the food with boiling liquid—water, syrup or broth.

There are several reasons why the hot pack usually results in a more satisfactory product. One of the principles of canning is removing air from the jar: to get a vacuum seal, and to reduce the darkening effect of oxygen on food colors. The quickest way to remove air from food tissues is to heat the food in an open kettle. With some of the air removed before food goes into the jar, you'll have less trouble with fruits floating or turning dark near the top of the jar, and fewer unsealed jars.

Preheating food to boiling also reduces the number of bacteria that get into the jar, which is always a good idea. And because food shrinks when it's heated, hot pack allows you to fill more food into each jar.

FILLING JARS, ALLOWING HEAD SPACE

Some foods should be packed loosely, to allow for expansion; others should be tightly packed. Check recipes and follow directions for packing individual foods.

If using hot pack, fill hot jars with food at or near boiling temperatures. Cover with boiling liquid. Pack one jar at a time, leaving recommended head space. Adjust jar lid (see Closing Jars, below) before filling the next jar. Work as fast as you can, and try to keep the sealing edge of the jar clean —a jar funnel helps reduce spillage.

Most raw food is also covered with boiling syrup, water or juice. Covering food with boiling liquid helps prevent darkening.

Head space is specified in recipes for each product to be canned—it means the amount of space to be left between the lid and the top of the food or liquid in the jar. This space allows for expansion—for the bubbling up of food or liquid inside the jar during processing. The amount of head space varies depending on the food, whether it's packed hot or raw, and whether it's processed in boiling water or under steam pressure.

It is important that you leave the correct amount of head space in each jar. If you don't leave enough head space, liquid may be forced out of the jar during processing, perhaps leaving food particles on the sealing edge of the jar, and the lid may not make a perfect seal. If you leave too much head space, there will be too much air inside the jar, which may cause food to discolor or may even prevent the lid from sealing.

When you check head space, do so after you've released air bubbles. These are pockets of air that get trapped inside when you fill jars. To let them out, move a thin plastic spatula gently up and down all around the inside of the jar.

CLOSING JARS

Follow manufacturer's directions for preparing jar lids and have them ready before you start to fill jars.

With Self-sealing Metal Lids and Screw Bands: Wipe off the sealing edge and threads of jars with a clean damp cloth or paper towel. Be sure no food residue remains—even small particles may prevent the lid from making a tight vacuum seal. Put lid on jar with the sealing compound next to the

Altitude Corrections for Boiling Water Bath

All the processing times given with recipes in this book are for altitudes under 1,000 feet. If you live at higher altitudes, simply boil the jars longer, according to this chart:

ALTITUDE	IF PROCESSING TIME IS 20 MINUTES OR LESS	IF PROCESSING TIME IS MORE THAN 20 MINUTES
1,000 feet	add 1 minute	add 2 minutes
2,000 feet	2 minutes	4 minutes
3,000 feet	3 minutes	6 minutes
4,000 feet	4 minutes	8 minutes
5,000 feet	5 minutes	10 minutes
6,000 feet	6 minutes	12 minutes
7,000 feet	7 minutes	14 minutes
8,000 feet	8 minutes	16 minutes
9,000 feet	9 minutes	18 minutes
10,000 feet	10 minutes	20 minutes

For example: Directions for hot pack peaches call for 20 minutes for pints in boiling water canner, and 25 minutes for quarts. At 5,000 feet, pints would be sterilized 20 minutes plus 5 minutes or 25 minutes. Quarts should be sterilized 25 minutes plus 10 minutes or 35 minutes.

glass. Screw the screw band on *tight*, making sure it's also on *straight*. Now . . . how tight? During heat processing, as temperature and pressure increase inside the jar, air is forced out. If the screw band is tightened firmly, the lid permits air to escape but prevents liquids from doing so. But it's possible, if you have a *very* strong grip, to get screw bands too tight— too much torque, the scientists would say. If you see a lot of air bubbles in jars when you remove them from the canner, it's a signal that you may be tightening screw bands too tight. After processing, do *not* tighten screw bands; these lids are self-sealing.

With Self-sealing One-piece Caps: Follow directions above for cleaning rim and threads of jar. Screw cap on

jar. Note that if you tighten this cap too forcibly, it may become distorted in processing. And while it would still make a seal, you would have considerable difficulty in removing the cap.

With Porcelain-lined Zinc Caps and Rubber Jar Rings: Have jar rings ready in a pan of hot water. Wipe off the rim, threads and ledge at base of threads with a clean damp cloth or paper towel. Be sure no food residue remains. Fit a wet rubber ring on the ledge of the jar, below screw threads, stretching it as little as possible. If it won't lie flat, reject it and try another ring. Screw cap down tight and then unscrew it ¼ inch. Loosening the cap this much will permit air inside the jar to escape during processing.

After processing, as soon as you remove jars from canner, retighten the cap. This is necessary to complete the seal when using the rubber rings. In our canning recipes, we refer to this by saying, "Remove jars from canner and complete seals unless closures are self-sealing type."

With Glass Lids and Rubber Jar Rings: Have jar rings ready in a pan of hot water. Wipe off sealing edge of jar with a clean damp cloth or paper towel. Fit a wet rubber ring on the ledge of the jar, stretching it as little as possible. Place glass lid over mouth of jar so that it rests on rubber ring. Put the long wire over the top of the lid so it fits in the groove. Leave short wire up for processing.

After processing, as soon as you remove jars from canner, push short wire down against side of jar to complete seal.

HEAT PROCESSING IN A BOILING WATER CANNER

Use a boiling water canner for these foods: all fruits and fruit juices; jams, jellies and other fruit spreads packed in Mason jars and sealed with lids; pickles and relishes; tomatoes, and fermented vegetables, such as sauerkraut.

The time to process each food depends on the size jar, the kind of food and whether it's raw pack or hot pack —see individual recipes.

To Use Boiling Water Canner: Put the rack in the bottom of the canner and fill it half-full with hot water. Cover and bring water almost to boiling.

As you fill each jar, place it on the rack in the canner. Handle jars carefully; never subject them to abrupt changes in temperature. If jars are raw pack, water in the canner should be hot but not boiling. When canner is full, add enough hot (not boiling) water to raise the water level at least 1 inch, preferably 2 inches above tops of jars. Put on the cover and bring water quickly to a rolling boil.

Start counting processing time as soon as water boils. Adjust heat to maintain a rolling boil throughout the entire recommended time. Add more boiling water if necessary to keep jars covered.

If you lose heat during processing— if the boiling stops—increase the heat until boiling returns. Discount any previous heating, set your timer again, and process the jars again for the entire recommended time. Food will be overcooked, but that's better than spoiled.

HEAT PROCESSING IN A PRESSURE CANNER

Use a steam pressure canner at 10 pounds pressure for all vegetables (except tomatoes), and all meats, poultry, fish, stews, soup and meat sauces.

The time to process each food depends on the size jar, the kind of food and whether it's raw pack or hot pack —see individual recipes.

To Use Pressure Canner: It's easy to forget precision procedures from one canning season to the next. Take time to review the instruction book packed with your canner each time you use it. If you have lost the instruction book

Altitude Corrections for Pressure Canners

Pounds pressure given in canning recipes in this book are for canning at altitudes less than 2,000 feet. Correct as follows for altitudes 2,000 feet and above:

With a *weighted gauge,* use 15 pounds pressure instead of 10.

With a *dial gauge,* add 1 pound pressure for each additional 2,000 feet as shown on chart below:

AT ALTITUDES OF:	MAINTAIN PRESSURE AT:
2,000 feet	11 pounds
4,000 feet	12 pounds
6,000 feet	13 pounds
8,000 feet	14 pounds
10,000 feet	15 pounds

For example: Directions at sea level call for 10 pounds pressure, to achieve a temperature of 240° F. At 6,000 feet, dial gauge must indicate 13 pounds to equal same temperature at sea level. Time remains the same.

for your pressure canner, write to the manufacturer for another copy.

Keep all parts of the pressure canner clean and in good working order. Clean petcock and safety valve by pulling a string through it, or use a pipe cleaner.

Put the rack in the canner and add 2 or 3 inches of hot water (or the amount recommended by manufacturer). Adding 1 or 2 tblsp. vinegar to the water helps prevent mineral stains on canner and jars. Set filled jars on the rack. Fasten cover securely and place canner over heat.

With Dial Gauge: Leave vent open and bring water to a boil; steam will begin to pour out of the vent. Let steam escape for about 10 minutes to exhaust all air in canner. Then close vent and continue heating until pressure climbs to 10 pounds (or more if

you live at higher altitudes—see Altitude Corrections above).

When you reach required pressure, start counting processing time. Regulate heat to maintain needed pressure; do not leave canner untended. If you let pressure fluctuate significantly, liquid may be forced out of the jars.

With Weighted Gauge: To exhaust air from canner, follow directions for your canner. One manufacturer recommends placing the gauge on the canner when you close it—air is automatically exhausted when the control jiggles. Another manufacturer directs you to leave vent open, letting steam escape for about 10 minutes to exhaust all air in canner.

Close vent with the weighted gauge, being sure you set it in the correct position to regulate pressure at 10 pounds (15 pounds at altitudes 2,000 feet or more). Continue heating; the gauge will jiggle (or rock steadily) when it reaches selected pressure. Start counting processing time when it jiggles at least 3 or 4 times a minute (or when it rocks steadily). Regulate heat so the weight continues to jiggle 3 or 4 times per minute (or rocks gently) during entire processing period.

If you lose heat during processing—if the pressure drops—increase heat until pressure returns. Discount any previous heating, set your timer again and process jars again for the entire recommended time.

Cooling Pressure Canners: When processing time is complete, remove canner from heat, but do not open air vent yet. Let canner stand undisturbed until pressure drops to zero—this will take 30 minutes or more. After canner has cooled enough, open the vent or remove the weight. Wait 2 minutes; then open canner, using the lid as a shield to protect your face from steam.

COOLING JARS

With a jar lifter, take the jars from the canner and place them upright on a folded towel or rack, in a place safely shielded from any sudden drafts. Don't set a hot jar on a cold counter; handle jars carefully lest they break. Leave at least 1 inch between jars (preferably more) so they will cool evenly on all sides. Do not touch lids or tighten screw bands of self-sealing jars. But if you are canning with rubber jar rings, immediately tighten zinc caps or push short wires down on glass lids, to complete the seal. Let the jars cool, without moving them, for 12 hours or more.

TESTING FOR SEALS

Self-sealing metal lids may not seal immediately after you remove them from the canner. But you should hear the familiar "ping" within an hour, indicating that air pressure outside and a vacuum inside the jar have worked in concert to make a hermetic (airtight) seal.

After jars are cool, check for seals. With self-sealing lids, look to see if center is down—concave or depressed. Press it. If it's down, or stays down when pressed and doesn't move, the lid is sealed. Some manufacturers suggest tapping the center of the lid with a spoon—if it gives a clear ringing sound, the jar has a good seal. If you're still not sure, turn the jar on its side and roll it; if there's no leakage, the jar is sealed.

If you canned with rubber rings, turn the jar on its side and slowly roll it. If there is no leakage, the jar is sealed.

IF LID DIDN'T SEAL

There are three things you can do if a jar didn't seal. You can refrigerate the food and eat it within a few days. You can pack the food into freezer containers and freeze it; be sure to label it as "canned/frozen" because it won't have the same quality as fresh frozen.

Or you can reprocess the food. Remove the lid and examine the jar. If it was not at fault—that is, if the sealing edge looks all right, adjust head space, clean the jar rim again, scald and put on a *new* lid and a different screw band. If the sealing surface of the jar is defective, transfer food to another lean jar, scald a new lid and screw it ight. Heat process the food again for he full time. Cool and check again or seals. Label the jar to show it's een processed twice.

STORING CANNED FOOD

Most manufacturers of two-piece self-ealing lids suggest that you remove crew bands when jars are cool and ently wash off any food residue that nay have escaped during processing. This is a good idea because metal crew bands may rust in storage and e difficult to remove. Also, residue eft on jars will support mold growth which may penetrate the sealing com-ound and break the seal.

Do not remove screw bands and put nem back on again—the second ghtening may break the seal.

Label jars with information you'll ant: the variety you canned or the ecipe you used, plus the date canned.

The ideal storage place is a clean, ark, dry area with temperature above eezing—between 40° and 50° F. f your storage area is warmer than at, examine stored food more fre-uently—canned food loses quality uch faster when stored at higher mperatures. If your only available orage area exceeds 80° F.—if it's ear hot pipes, furnace or hot water eater—you'd be wiser not to store nned food at all—neither home nned nor commercially canned ods.

If your "cool place" might occa-onally freeze, you can insulate jars wrapping them in newspapers or ankets. Freezing will not cause oilage, but food should not be eaten the seal is damaged or jar broken.

Pretty as they are, your colorful jars of canned food should be stored out of sight, or at least out of light. Fruits are especially prone to turn dark if exposed to light.

UNSAFE CANNING PRACTICES

According to USDA surveys among home canners, many still can jams, preserves, pickles and relishes by the open-kettle method, and process vege-tables in a boiling water bath. Both of these practices are now known to be unsafe, and using them is a form of gambling—playing the "numbers game."

There's no way for you to know the bacterial count on the fresh foods you prepare for canning—bacteria are in-visible. If the numbers are low, if the jars seal and if you store canned food properly, it's unlikely that food will spoil even though you don't process jars as recommended.

But if there's a big load of bacteria in the raw food and you do open-ket-tle canning or follow other unsafe practices, enough bacteria may sur-vive to spoil the food. We now know that heating tomatoes, pickles or pre-serves to boiling in an open kettle is not enough to destroy all spoilage organisms that may be present. Also, it is impossible to prevent microor-ganisms in the air from recontaminat-ing foods while they're being ladled or poured into jars. As for vegetables, the temperature in a boiling water canner is not hot enough to kill heat-resistant bacteria such as *Clostridium botulinum*, which thrive in low-acid foods.

Therefore, for safe canning: Can only fresh, firm food—nothing diseased or decayed or overripe. Be scrupulously clean in washing and preparing food, to keep as few bacteria as possible from getting into your canning jars. Recommended processing times and temperatures are based on a reasonable population of bacteria to be destroyed. Process acid foods in a boiling water bath—*do not do open-kettle canning.* Process low-acid foods in a steam pressure canner—*do not can vegetables or meats in a boiling water canner.*

Also see Chapter 15 for additional information about packing, sealing and processing jellies, jams and preserves.

Some Other Unsafe Canning Methods Should Be Flagged: Since microwave ovens have come into wider use, the idea of oven canning has been "rediscovered." No matter what kind of oven you have, oven canning is just plain dangerous. In the boiling water or steam of a canner, the balance of pressure inside and outside the jars is equalized. In the dry heat of the oven, glass jars may not tolerate the unbalanced pressure—they may explode. Equally important, the center of some jars may never reach the sterilizing point and/or lids may not seal in the oven. Also, metal (in jar lids) should not be used in microwave ovens.

Dishwashers, steamers and slow cookers should not be used for canning; none of these appliances is capable of reaching sterilization temperatures.

Signs of Food Spoilage

Before you open any jar—your own, or a gift jar from a friend—examine it carefully for any signs of spoilage.

A leaking jar or swelling lid are sure signs that food has spoiled and would be dangerous to eat. *Do not taste or serve food from such a jar.*

Early signs of spoiling are not so obvious. So inspect the jar and lid carefully. Is the lid still sealed? Still concave in the center—still drawn down by the vacuum inside the jar? If the lid is flat or slightly rounded, no longer concave, it is beginning to swell.

Also look at the food and liquid in the jar. Bubbles inside the jar are evidence that spoilage yeasts and bacteria are at work—that food is fermenting. Unnatural food colors, cloudy liquid or sediment in the liquid are other signs of spoilage. *Do not taste or serve such food.*

But a sealed jar and normal color do not guarantee that food is safe. Be alert when you open the jar. Spurting liquid is a sure sign of spoilage. If the food smells unnatural—if it smells yeasty, moldy, sour or putrid—it is not safe. If you are not sure, boil the food hard for 10 minutes, smelling it as it boils. Heating makes off-odors more noticeable. Do not taste any questionable food until it has boiled for 10 minutes.

If the food looks soft or mushy or slimy, or if there's a visible white or colored film on the top surface, or a cotton-like mold growth of several possible colors on the surface of the food, it is not safe. *Do not taste or serve any part of foods having these spoilage symptoms.* Destroy entire contents of jar in such a way that children or animals cannot eat it (directions follow).

Until recently, mold growth was regarded more as a nuisance than a danger. Scientists are just beginning to recognize the hazards to humans and animals from eating mycotoxins produced by molds. Moldy jellies and jams should be discarded—not scraped off. By the time you see the cottony growth, molds have sent invisible threads deep into the food.

Also, mold growth inside a sealed jar of acid food may lower the acidity to such a level that there is danger of *botulinum* bacteria germinating. That is why you are cautioned not to taste food or juice with a film on the surface.

To Open Jars: Punch a hole in self-sealing lids and lift up. You should never reuse them anyway. If jars have rubbers, pull the rubber out with pliers to avoid necessity for prying open, which may dent or bend the cap and crack or chip the jar.

HOW TO DISPOSE OF SPOILED FOOD

The Pennsylvania State University Extension Service gives these directions for disposing of spoiled, potentially harmful food:

1. Wear waterproof or plastic disposable gloves when handling. Avoid spilling spoiled food on your skin, clothes or kitchen work surfaces. If it should get on your skin or clothes, wash thoroughly in strongly chlorinated warm water. Clean and sterilize kitchen surfaces with liquid chlorine solution (½ c. chlorine bleach in 1 gal. water).

2. Carefully transfer spoiled food from its container to a vessel in which you can boil it. Boil the food for 20 minutes. Then discard boiled food into a sewer system, or bury it where animals cannot dig it up and in an area which is unplanted.

3. Also boil the contaminated container and lid, your gloves and any other utensils that came in contact with the spoiled food, in a solution of strong soap or detergent. Boil the container even though you will probably not want to use it again for food.

CHAPTER 10

Fruits, Fruit Juices and Tomatoes

Sweetening Canned Fruits · Directions for Canning Fruits · Pear
Mincemeat · Peach Macaroons · Extracting and Canning Fruit
Juices · How to Make Berry Syrups · Canning Tomatoes · Tomato
Sauce · Tomato Juice · V-4 Cocktail · Green Tomato Mincemeat

When plenty spills from almost every vine and bush in the garden, home
canning swings into high gear in country kitchens. Baskets of blushing
peaches and tart-sweet plums and summer apples show up in farm
kitchens in midsummer. Peaches, beans and tomatoes make the "big
three" in country home canning. Farm families like peaches canned and
frozen—"they taste so different."

While the cook fixes fruit for processing, clean glass jars wait in the
electric dishwasher or in a pan of hot water. Shiny, new lids are on the
counter top; the canner is ready. Before sundown, rows of filled jars record the accomplishments of a busy day.

Anytime between spring and frost, farmers say they can tell what week
it is by stepping into the kitchen. They sniff the aroma of the fruit or berries being heated for canning, or they look in bowls or baskets on the
counter top. When cherries, peaches, plums or other fruits are ripe—firm
ripe—the canning kettle is on the stove. At this stage of maturity, fruits
reach their flavor peak; yet they hold their shape when processed.

"Sometimes in the rush of summer," a farm woman confesses, "I swear
I'll never again look at a seed catalog. Then maybe we won't plant so
much. But when I get tasty, economical meals all winter using lots of food
from fruit closet and freezer, I change my mind. I realize that summer's
work pays off in dollars and cents and better meals—that it gives me free
time the other three seasons."

Follow the up-to-date processing times for canning fruits, fruit juices
and tomatoes in this chapter. And don't experiment. Canning is a science
and failure to heed the rules brings waste and disappointment.

Getting Ready to Can Fruits

On Canning Day: Begin by choosing firm ripe fruit. Prepare Sugar Syrup (directions follow) and get jars ready. Wash and sort fruit carefully and prepare according to directions in this chapter for each kind. If necessary, treat to prevent darkening. Pack fruit, hot or raw, into clean hot jars. Cover with boiling syrup or water, leaving recommended head space. Remove air bubbles, wipe sealing edge of jar and adjust lids. Process fruits in the boiling water canner as directed. For complete directions on all canning procedures, see "Steps to Successful Canning" in Chapter 9.

CANNING WITHOUT SUGAR

If there's someone in your family on a special diet or counting calories, you may want to can fruit without sweetening—in its own juice, in extracted juice or in water. The processing time is the same whether or not you sweeten the fruit. Sugar is not necessary to prevent spoilage. Do not use artificial sweeteners in canning because such products may change fruit flavor.

CANNING WITH SUGAR

Most home canners prepare a Sugar Syrup for canning fruit because it helps fruit retain its color and texture. It can be thin, medium or heavy (proportions follow); you'll need about 1 to 1½ cups syrup for each quart of fruit.

SUGAR SYRUPS

	SUGAR CUPS	WATER CUPS	YIELD CUPS
Thin	2	4	5
Medium	3	4	5½
Heavy	4¾	4	6½

Heat sugar and water together until sugar dissolves. Skim if necessary. Fruit juice may be substituted for all or part of the water. To extract juice, heat crushed fruit to simmering point (185 to 210° F.) over low heat. Strain through a jelly bag or cheesecloth.

You may use light corn syrup or mild-flavored honey to replace up to half the sugar in the Sugar Syrup recipes.

Do not use brown sugar or strong-flavored syrups such as molasses; they may darken the fruit and give it a different flavor.

A medium syrup retains best color and shape of fruits, and is specified for most fruits in the directions that follow. However, if you want to save calories, you step down to the thin syrup. You can even make a very light syrup, using only 1 c. sugar to 4 c. water.

If you pack juicy fruit hot, you can add sugar direct instead of making a sugar syrup. Add about ½ c. sugar to each quart raw, prepared fruit. Heat to simmering (185 to 210° F.) and pack fruit in the juice that cooks out.

ALTITUDE CORRECTIONS

All processing times given for the boiling water bath are for altitudes less than 1,000 feet above sea level. See Index for Altitude Corrections if you live above 1,000 feet.

To Can Fruits

APPLES: Wash, peel, core and quarter or slice apples. Drop them into salt-vinegar water (2 tblsp. each salt and vinegar to 1 gal. water) to prevent darkening. Lift out, rinse and boil in thin syrup or water for 5 minutes.

Hot Pack Only: Pack hot fruit to within ½" of jar top. Cover with hot thin syrup or water, leaving ½" head space. Adjust lids. Process in boiling water bath.

Pint jars   15 minutes
Quart jars   20 minutes

Remove jars from canner and complete seals unless closures are self-sealing type.

APPLESAUCE: Make applesauce, sweetened or unsweetened. Heat to simmering (185 to 210° F.). Stir to prevent sticking.

Hot Pack Only: Pack hot applesauce to ¼" of jar top. Adjust lids. Process in boiling water bath.

Pint jars   15 minutes
Quart jars   15 minutes

Remove jars from canner and complete seals unless closures are self-sealing type.

APRICOTS: Use firm, not overripe fruit. Wash. Dip into boiling water for about ½ minute, plunge into cold water and slip off skins. Leave whole or cut in halves and remove pits. To keep from darkening, drop them into salt-vinegar water (2 tblsp. each salt and vinegar to 1 gal. water). Drain and rinse before heating or packing raw.

Raw Pack: Some varieties of apricots hold their shape better raw-packed than hot-packed. If halved, pack in overlapping layers. If whole, pack close. Pack hot jars and cover with boiling medium syrup to ½" of jar top. Adjust lids. Process in boiling water bath.

Pint jars   25 minutes
Quart jars   30 minutes

Remove jars from canner and complete seals unless closures are self-sealing type.

Hot Pack: Prepare apricots as directed for raw pack. Bring to a boil in medium syrup and cook through gently (about 1 to 3 minutes). Pack hot jars with hot fruit and cover with hot medium syrup, leaving ½" head space. Adjust lids. Process in boiling water bath.

Pint jars   25 minutes
Quart jars   25 minutes

Remove jars from canner and complete seals unless closures are self-sealing type.

BERRIES, SOFT (blackberries, dewberries, loganberries, raspberries): Use fully ripe berries. Work with small amounts at a time—1 to 2 qts. Wash with a spray or in water and lift carefully from water. Drain; cap and stem if necessary.

Raw Pack: Pack hot jars with raw berries. Shake down several times while filling jar to pack berries tightly without crushing. Cover with boiling medium syrup, leaving ½" head space. Adjust lids. Process in boiling water bath.

Pint jars 15 minutes
Quart jars 15 minutes

Remove jars from canner and complete seals unless closures are self-sealing type.

Hot Pack: Not recommended for soft berries.

BERRIES, FIRM (currants, elderberries, huckleberries; see separate instructions for blueberries and gooseberries):

Wash and drain as for soft berries. Add ½ c. sugar to each qt. fruit. Cover pan and bring to boil. Shake pan to prevent berries from sticking.

Hot pack only: Pack berries and juice to within ½" of jar top. Adjust lids. Process in boiling water bath.

Pint jars 15 minutes
Quart jars 15 minutes

Remove jars from canner and complete seals unless closures are self-sealing type.

BLUEBERRIES: Can be hot packed but they break down more than other firm berries. Follow this procedure if you want to keep them whole, to use in muffins, etc. Blanch berries first: Put 2 or 3 qts. clean berries in a large square of cheesecloth. Hold cloth by corners and dip berries into boiling water for 15 to 30 seconds. When juice spots appear on cloth, dip berries immediately into cold water;

drain. Pack berries into hot jars, leaving ½" head space. Do not add sugar or liquid. Adjust lids. Process in boiling water bath.

Pint jars 15 minutes
Quart jars 15 minutes

Remove jars from canner and complete seals unless closures are self-sealing type.

CHERRIES, SWEET OR TART: Stem, wash and remove pits, if desired. (Pie cherries should always have pits removed and may be canned in water instead of syrup.) If cherries are left whole, prick skins to prevent splitting.

Raw Pack: Pack hot jars, shaking cherries down several times while filling, for a firm pack. Cover with boiling heavy syrup for tart cherries, thin syrup for sweet cherries. Leave ½" head space. Adjust lids. Process in boiling water bath.

Pint jars 20 minutes
Quart jars 25 minutes

Remove jars from canner and complete seals unless closures are self-sealing type.

Hot Pack: Prepare cherries as directed for raw pack. Add ½ c. sugar to each qt. fruit. Add a little water to prevent sticking while heating to simmering (185 to 210° F.). Ladle into hot jars, leaving ½" head space. Adjust lids. Process in boiling water bath.

Pint jars 15 minutes
Quart jars 15 minutes

Remove jars from canner and complete seals unless closures are self-sealing type.

CHERRIES, SURINAM: Use sound, ripe, freshly picked cherries. Wash and pit.

Precook 2 minutes and use the water from the precooking to make a medium syrup.

Hot Pack Only: Pack cherries in hot jars, shake down and cover with boiling syrup. Leave ½″ head space. Adjust lids. Process in boiling water bath.

Pint jars   15 minutes
Quart jars   16 minutes

Remove jars from canner and complete seals unless closures are self-sealing type.

FIGS: Sort and wash firm, ripe, freshly picked figs. Bring to a boil in hot water. Let stand in hot water 3 to 4 minutes. Drain.

Hot Pack Only: Pack hot in hot jars. Add 2 tsp. lemon juice to each pt. (4 tsp. to each qt.) and cover with hot medium syrup, leaving ½″ head space. Adjust lids. Process in boiling water bath.

Pint jars   90 minutes
Quart jars   90 minutes

Remove jars from canner and complete seals unless closures are self-sealing type.

FRUIT PURÉES: Use sound, ripe fruit. Wash and remove pits if necessary. Cut large fruit in pieces. Simmer until soft, adding a little water to prevent sticking. Put through a food mill or strainer. Add sugar to taste. Heat again to simmering (185 to 210° F.).

Hot Pack Only: Pack hot to within ¼″ from jar top. Adjust lids. Process in boiling water bath.

Pint jars   15 minutes
Quart jars   15 minutes

Remove jars from canner and com-

plete seals unless closures are self-sealing type.

GOOSEBERRIES: Wash and drain. Remove stems and tails.

Hot Pack Only: Prepare 3 c. medium or heavy syrup and bring to boil. Blanch gooseberries, 1 qt. at a time, for 15 to 30 seconds in the boiling syrup. Drain in colander until all berries are blanched. Pack berries closely in hot jars, shaking down to insure close pack. Cover with boiling syrup in which gooseberries were heated. Leave ½″ head space. Adjust lids. Process in boiling water bath.

Pint jars   15 minutes
Quart jars   20 minutes

Remove jars from canner and complete seals unless closures are self-sealing type.

GRAPEFRUIT: Wash thoroughly ripened fruit. With a sharp knife, cut a slice from both ends of the grapefruit, cutting into the flesh. Remove remainder of peel and all of the white membrane by cutting from one end to the other. Loosen sections by cutting as close to membrane as possible on both sides of each section. Discard seeds. Add juice to syrup.

Raw Pack Only: Pack segments firmly into hot jars. Add thin syrup heated to simmering point (185 to 210° F.), leaving ½″ head space. Adjust lids. Process in boiling water bath.

Pint jars   15 minutes
Quart jars   15 minutes

If you pack grapefruit into pre-sterilized jars, process pints and quarts 10 minutes.

Remove jars from canner and complete seals unless closures are self-sealing type.

GUAVAS, SHELLS: Wash, remove stem and blossom and peel. Remove seeds with a spoon.

Hot Pack Only: Drop shells into a medium syrup and cook 2 to 3 minutes. Pack in overlapping layers, adding 1 to 2 tblsp. syrup with each layer; leave ½ " head space. Adjust lids. Process in boiling water bath.

Pint jars   15 minutes
Quart jars   20 minutes

Remove jars from canner and complete seals unless closures are self-sealing type.

LOQUATS, WHOLE: Wash, rinse and drain freshly gathered fruit. Remove stem and blossom end. Peel and remove seeds, if desired.

Hot Pack Only: Cook 2 to 4 minutes in medium syrup and pack hot in hot containers, leaving ½ " head space. Adjust lids. Process in boiling water bath.

Pint jars   15 minutes
Quart jars   20 minutes

Remove jars from canner and complete seals unless closures are self-sealing type.

MANGOES, GREEN: Pick before any color appears and peel as directed for ripe mangoes.

Raw Pack Only: Let stand for 2 minutes in hot medium syrup before packing in hot jars. Cover with boiling syrup, leaving ½ " head space. Adjust lids. Process in boiling water bath.

Pint jars   15 minutes
Quart jars   20 minutes

Remove jars from canner and complete seals unless closures are self-sealing type.

MANGOES, RIPE: Use firm fruit only. Wash and peel from stem to blossom end. Slice, place in hot thin syrup and let stand 1 to 2 minutes.

Raw Pack Only: Pack into hot jars. Boil syrup down 5 to 8 minutes; strain while hot over the raw fruit, leaving ½ " head space. Adjust lids. Process in boiling water bath.

Pint jars   15 minutes
Quart jars   20 minutes

Remove jars from canner and complete seals unless closures are self-sealing type.

NECTARINES: Prepare and process using directions for peaches.

PAPAYAS, GREEN: Wash and peel.

Hot Pack Only: Prepare syrup consisting of 3½ c. sugar, 1 c. vinegar, 1 c. water, ½ oz. piece ginger root and 2 tblsp. cracked cinnamon sticks. Bring to a boil; strain. Simmer green papayas in water 4 minutes; drain. Pack hot in jars. Heat syrup to boiling; pour over fruit, leaving ½ " head space. Adjust lids. Process in boiling water bath.

Pint jars   15 minutes
Quart jars   20 minutes

Remove jars from canner and complete seals unless closures are self-sealing type.

PAPAYAS, RIPE: Choose ripe but firm fruit. Wash, cut and remove seed. Make a medium syrup; after cooling, add juice of one lemon or lime.

Hot Pack Only: Slice fruit; add to the cooled syrup, bringing slowly to a boil. Cook approximately 3 minutes and pack hot in hot jars, adding hot syrup to cover. Leave ½ " head space.

Adjust lids. Process in boiling water bath.

Pint jars   15 minutes
Quart jars   20 minutes

Remove jars from canner and complete seals unless closures are self-sealing type.

PEACHES: Use ripe, firm peaches. Wash. Dip into boiling water for about ½ minute, plunge into cold water and slip off skins. Cut peaches in half and remove pits. Slice if desired. To keep from darkening, drop them into a gallon of salt-vinegar water (2 tblsp. each salt and vinegar to 1 gal. water). Rinse and drain before packing.

Raw Pack: Pack raw fruit (halves) cut side down, edges overlapping, into hot jars. Cover with boiling thin or medium syrup, leaving ½" head space. Adjust lids. Process in boiling water bath.

Pint jars   25 minutes
Quart jars   30 minutes

Remove jars from canner and complete seals unless closures are self-sealing type.

Hot Pack: Heat peach halves or slices to boiling in thin or medium syrup and pack hot in hot jars. (Or, if peaches are very juicy, omit syrup and add ½ c. sugar for each qt. fruit; heat gently, then pack hot jars with hot fruit and juice that has cooked out.) Leave ½" head space. Adjust lids. Process in boiling water bath.

Pint jars   20 minutes
Quart jars   25 minutes

Remove jars from canner and complete seals unelss closures are self-sealing type.

PEARS: Use pears ripened after picking. Can them while still firm. Wash, peel, cut in lengthwise halves or quarters and core. A half-teaspoon measuring spoon and paring knife do a good coring job. Work quickly to prevent darkening or drop into salt-vinegar water (2 tblsp. each salt and vinegar to 1 gal. water); rinse and drain before using.

Hot Pack: Boil pears 3 to 5 minutes in thin or medium syrup. Pack hot, with cups down, to within ½" of jar top. Add hot syrup, leaving ½" head space. (If desired, add 1 tsp. lemon juice to each qt.) Adjust lids. Process in boiling water bath.

Pint jars   20 minutes
Quart jars   25 minutes

Remove jars from canner and complete seals unless closures are self-sealing type.

Note: Pears may be packed raw (without precooking) if they are ripe enough to be quite soft. Pack jars with pears and add syrup to within ½" of jar top. Adjust lids and increase processing time to 25 minutes for pints and 30 minutes for quarts. Most canners consider pears precooked before canning superior to those canned by the raw pack.

PINEAPPLE: Choose sound, ripe fruit. Wash pineapple, slice, peel and remove eyes and core from each slice. Leave slices whole or cut in pieces.

Hot Pack Only: Boil pineapple 5 to 10 minutes in medium syrup. Pack hot in hot jars and cover with syrup, leaving ½" head space. Adjust lids. Process in boiling water bath.

Pint jars   15 minutes
Quart jars   20 minutes

Remove jars from canner and complete seals unless closures are self-sealing type.

PLUMS (including fresh prune plums): Choose ripe, firm fruit just beginning to soften. Greengage and other meaty plums are best for canning. The juicy varieties mush easily. Wash and drain plums. Prick skins with a needle if plums are to be canned whole. This does not prevent skins from splitting, but helps prevent the fruit from bursting. The freestone variety may be cut in half and pitted.
Raw Pack: Pack gently but closely in jar. Cover with boiling syrup, heavy for tart plums, medium for sweet ones. Leave ½″ head space. Adjust lids. Process in boiling water bath.

Pint jars   20 minutes
Quart jars   25 minutes

Remove jars from canner and complete seals unless closures are self-sealing type.
Hot Pack: Prepare as directed for raw pack. Heat fruit in boiling medium syrup. Pack hot fruit to within ½″ of jar top and cover with boiling syrup. Leave ½″ head space. Adjust lids. Process in boiling water bath.

Pint jars   20 minutes
Quart jars   25 minutes
Remove jars from canner and complete seals unless closures are self-sealing type.

RHUBARB: Wash tender, young stalks and cut into ½″ pieces. *Discard leaves.* Can as soon as possible after picking.
Hot Pack Only: Add ½ c. sugar to each qt. rhubarb. Let stand to draw out juice. After juice appears, heat slowly to boiling. Fill jars carefully to prevent mushing; leave ½″ head space. Adjust lids. Process in boiling water bath.

Pint jars   15 minutes
Quart jars   15 minutes

If you pack rhubarb into pre-sterilized jars, process pints and quarts 10 minutes.

Remove jars from canner and complete seals unless closures are self-sealing type.

STRAWBERRIES: Not recommended for canning. Usually the result is not a good product. Plan to freeze them, if possible, or make jam and preserves.

Canned Fruit Specials

If you like to experiment in canning, try these fruit specials of farm women.

SALAD AND DESSERT PEARS: "I like to can pear halves in canned unsweetened pineapple juice instead of syrup. When chilled, they are delicious for salad or dessert."

APPLE/PLUM SAUCE: "When I have both apples and plums, I combine equal amounts of chopped, pitted plums and sliced, unpeeled apples. I cook the mixture until the fruits are soft and put it through the food mill. When sweetened to taste, I can the fruit combination like applesauce."

OLD-TIME PEAR MINCEMEAT

New cooks win praise on pies made with Grandma's pear mincemeat

7 lbs. ripe Bartlett pears
1 lemon
2 (1 lb.) pkgs. seedless raisins
6¾ c. sugar
1 c. vinegar
1 tblsp. ground cloves
1 tblsp. ground cinnamon
1 tblsp. ground nutmeg
1 tblsp. ground allspice
1 tsp. ground ginger

• Core and quarter pears. Quarter lemon, removing seeds. Put pears, lemon and raisins through chopper.
• Combine remaining ingredients in large kettle. Add chopped fruit mixture. Bring to a boil over medium heat; simmer 40 minutes.
• Pack at once in hot pint jars, leaving ½" head space. Adjust lids. Process in boiling water bath 25 minutes.
• Remove jars from canner and complete seals unless closures are self-sealing type. Makes 9 pints.

PEACH MACAROONS

Homely dessert that is distinctive and taste-rewarding—pretty, too

1 egg white, stiffly beaten
¼ c. light corn syrup
½ c. brown sugar, firmly packed
2 c. coarse bread crumbs
16 canned peach halves

• Slowly combine egg white and corn syrup; beat until blended. Add brown sugar and bread crumbs.
• Place 2 tblsp. meringue mixture in

center of each peach half. Arrange in baking pan.
• Bake in moderate oven (350° F.) 20 minutes. Serve warm. Makes 16.

FRUIT COCKTAIL

• Select firm but ripe pineapples, pears and peaches; cut into chunks. Precook separately in thin syrup (1 c. sugar to 2 c. water) 3 to 5 minutes, or until limp. Precook whole seedless white grapes the same way.
• Simmer red maraschino cherries in water 5 minutes to prevent bleeding; drain. Leave whole.
• Fill hot pint jars to within ½" of top with equal amounts of fruit, except cherries (use 3 per pint).
• Cover with strained hot syrup, leaving ½" head space. Adjust lids. Process in boiling water bath 20 minutes. Remove jars from canner and complete seals unless closures are self-sealing type. Cool out of drafts.

WILD PLUMS

• Wash and stem firm, medium-ripe wild plums. Prick the skins and pack cold in hot quart jars to the top.
• Cover with hot heavy syrup (1 c. sugar to 1 c. water) and let set 5 minutes. Drain off syrup; strain and reheat to boiling. The plums will have shrunk to shoulder of jar.
• Cover plums with boiling syrup, leaving ½" head space. Adjust lids; process 25 minutes in boiling water bath. Remove jars from canner and complete seals unless closures are self-sealing type.

Festive Fruit Juices

Fruit juices set the stage for pleasant visiting. One country cook calls them "hostess helpers." She gets the juice (sometimes a blend of juices) ready to pour and puts it in the refrigerator. When guests arrive, her husband and children serve the chilled juice in the living room while she puts the finishing touches on the food.

There are other ways to use luscious juices. Make fresh-tasting jelly from them when the spirit moves you. Tint and flavor cake frostings with them. Combine water and fruit juice for the liquid in making gelatin salads and desserts. Make table syrups to serve hot on pancakes and waffles, cold on ice cream and puddings. Add the juices to fruit cocktails and compotes.

EXTRACTING FRUIT JUICES

Some fruits are heated, to make juice extraction easier; others are simply crushed or ground. Individual directions follow. After heating or grinding, fruit is dripped through a jelly bag. You will obtain clear juice if you press the bag only lightly. Citrus juice is reamed out and strained, but after pressing, most juices should be strained through 3 or 4 thicknesses of cheesecloth that has been washed and rinsed. Note special instructions for clarifying Concord grape juice. Canning directions follow; also see instructions for freezing fruit juices in Chapter 2.

APPLES: Wash fruit, put through food chopper, using coarsest blade. Do not heat. Squeeze juice through strong, clean cloth bag; strain. Add no sugar.

APRICOTS, NECTARINES AND PEACHES: Use firm ripe fruit, wash and remove stems. Drop fruit into boiling water, 1" deep in kettle and boil until tender. Put through colander to remove skins and pits. Strain pulp and juice. Mix equal parts pulp with a thin syrup (1 c. sugar to 4 c. water) or with an equal amount of orange or grapefruit juice.

BERRIES: Wash and crush well-ripened blackberries, boysenberries, loganberries, red or black raspberries or youngberries. Heat to 185° F. Drain and squeeze through a cloth or bag; strain. If you wish to sweeten juice, add 1 c. sugar to 9 c. juice.

STRAWBERRIES: Wash, hull and crush. Heat to 185° F. Drain and squeeze through a cloth bag; strain; add 1 c. sugar to 3 c. juice. If blended with other berry juice, add 1 c. sugar to 9 c. juice.

CHERRIES, RED: Wash, stem, pit and put sweet or tart red cherries through food chopper, or chop. Heat to 185° F. Drain and squeeze through a cloth bag; strain. If you wish to sweeten juice more, add 1 c. sugar to 9 c. juice.

CHERRIES, LIGHT: Wash, stem and pit cherries. Put through food chopper. Do not heat. Drain and squeeze through a bag; strain. Sweeten like red cherries if desired.

CITRUS FRUITS: Have oranges, lemons or grapefruit at room temperature. Navel oranges are not recommended for juice to can or bottle, but if you use them, cut out the navel end before reaming. Ream fruits with any type of reamer you like except the press type; avoid pressing oil from peel and do not try to remove all the pulp. Strain juice through colander. Do not heat juice or add sugar.

CONCORD GRAPES: Wash and stem ripe grapes; crush fruit and heat to 185° F., or until fruit is soft. Drain and squeeze through a cloth bag; strain. Let juice stand 6 to 8 hours, or overnight, to allow sediment to settle to bottom. Carefully pour off clear juice. Add ¼ c. sugar for each qt. juice; mix well.

RED GRAPES: Wash, remove large stems. Put 1 qt. grapes into cheesecloth bag and blanch in 1 gal. rapidly boiling water 30 seconds. Put grapes in bowl, crush and let stand 10 minutes, stirring occasionally. Do not reheat. Drain and squeeze juice through a cloth bag. Strain. Do not add sugar.

WHITE GRAPES: Wash and stem ripe grapes; crush fruit, or put through food mill, but do not crush seeds. Do not heat. Drain and squeeze through a cloth bag; strain. Do not add sugar.

PLUMS: Use firm, rich-flavored, well-colored plums. Wash, crush, and add 1 qt. water to each 2 lbs. plums. Heat at about 180° F. until fruit is soft. Drain and squeeze juice through cloth bag and strain. Add 1 c. sugar to 1 qt. juice, if you want sweet juice.

RHUBARB: Use red varieties. Discard leaves. Wash stalks, cut into 4" lengths. Add 2 qts. water to 10 lbs. rhubarb. Heat until water starts to boil. Drain and squeeze juice through cloth bag. Strain. Add 1 c. sugar to 2 qts. juice.

TO CAN FRUIT JUICE

Take your choice of two ways to process fruit juices. Both ways are effective in preventing spoilage, and both will taste good—not overcooked.

We used to try to protect flavor by processing fruit juice in a simmering (not boiling) water bath. But this processing (pasteurizing) took 30 minutes! The more *rapidly* you heat juice to pasteurizing temperature, the better the flavor. Try one of these methods:

Method 1: Heat prepared juice to a temperature of 140° F. and pour into hot jars, leaving ¼" head space. Adjust lids. Process in boiling water bath.

Pint jars	15 minutes
Quart jars	15 minutes

Remove jars from canner and complete seals unless closures are self-sealing type.

Method 2: Sterilize jars. Heat prepared juice to near-boiling (190 to 210° F.) and pour immediately into hot jars. Adjust lids. Process in boiling water bath.

Pint jars	5 minutes
Quart jars	5 minutes

Remove jars from canner and complete seals unless closures are self-sealing type.

Tasty Table Syrups

Pour boysenberry syrup on a stack of hot, buttered pancakes; strawberry syrup on vanilla ice cream. Visitors to the Pacific Northwest from other areas admire the luscious berry syrups country women often take from their fruit closets or freezers to serve on waffles, pancakes and ice cream—and in punches. "We really like them the way Easterners like maple syrup," an Oregon homemaker says. "We don't have maple groves, but we do have wonderful strawberries, blackberries, boysenberries, red raspberries, blueberries. All make good syrups. Sometimes we blend two or more berries, always including at least one that's tart."

Berry syrups are so delicious and colorful that they are worthy of adoption by country cooks in every state. Gourmet shops in big cities enjoy excellent fruit syrup sales—and pleasing profits. You can make them in your kitchen by directions developed by the Cooperative Extension Service, Oregon State University.

TO MAKE BERRY SYRUPS

1. Start by making a test batch to check the syrup's consistency when cool. Berries vary greatly from one season to another in acid and pectin. If your syrup is too thick, let the rest of the juice stand in the refrigerator overnight. That will destroy some of the pectin, and syrup made will be thinner. If it is too thin, add corn syrup (see note following recipe for Western Berry Syrup).
2. Make a cellulose pulp to mix with crushed fruit or berries—it filters

and clarifies the juice and keeps the jelly bag from clogging. Place 10 unscented white facial tissues in 2 qts. boiling water. Let stand 1 minute. Break tissues into small pieces with two forks and pour through a strainer. Shake, but do not press to remove excess water. Makes 1 c. cellulose pulp.

3. Wash table-ripe berries; cap or stem if necessary. Don't use underripe fruit—its pectin may jell the syrup. Crush berries with a potato masher.
4. For every 3 c. crushed berries or fruit, add 1 c. cellulose pulp. Stir well and heat just to a boil, stirring constantly to prevent sticking. If fruit remains firm, simmer 1 or 2 minutes longer, but don't overcook or you'll destroy fresh fruit flavors.
5. Pour hot fruit into jelly bag or several thicknesses of cheesecloth in a strainer or colander. Let juice drip into bowl. When cool enough to handle, twist the bag or press against the side of the colander to extract juice. Discard dry pulp. Use the basic recipe to make syrup.

WESTERN BERRY SYRUP

Use raspberries, strawberries, boysenberries or any other berries you have

1¼ c. prepared berry juice
1¾ c. sugar

• Combine juice and sugar in a large, heavy kettle. Bring to a full rolling boil; boil 1 minute. Count the time after the mixture comes to a boil you cannot stir down. Remove from heat and skim off foam. Don't boil too long or the syrup may jell.

• Pour into sterilized hot jars, leaving ⅛" head space. Adjust lids. Process in boiling water bath 10 minutes.

• Remove jars from canner and complete seals unless closures are self-sealing type. Makes 1 pint.

Store in a cool, *dark* place. If you prefer, leave 1" head space in the jars and freeze instead of processing in boiling water bath. Once syrup is opened, keep in refrigerator. If the syrup jells when you open it, stir vigorously or place the jar in a pan of warm water, heat until jell dissolves.

Note: For a more tart syrup, add 1 tblsp. lemon juice. For a thicker syrup, use 1½ c. sugar and ¼ c. light corn syrup instead of 1¾ c. sugar.

Tomatoes and Tomato Juice

Home-grown, vine-ripened tomatoes are so superior that most farm families—even when they've given up a big garden—find a sunny spot near the house to set out a few tomato plants. They raise enough for eating fresh all through the season, plus extras for the canning jars. As any home canner knows, tomatoes are one of the easiest foods to can. So long as they are prime quality—that is, red and firm-ripe, but not overripe—they may be safely and easily processed like other acid foods, in a boiling water bath.

But when they become *overripe*, tomatoes lose acidity and such fruit would not be safe for canning in a boiling water bath. This is all fully explained under the heading "New Directions for Canning Tomatoes" in Chapter 9, *Preserving Food Safely.*

KNOW YOUR TOMATOES

There are many varieties of tomatoes developed for home canning. For up-to-date information on what varieties are recommended in your locality, contact your County Agricultural Extension Agent or Home Economist. As of summer 1977, USDA researchers do not recommend the following varieties for home canning: Garden State, Ace, 55VF and Cal Ace. The acid level of these four varieties, they found, moves too close to the borderline of safety when the fruit is overripe, or picked from dead plants.

Do not can any soft, mushy, overripe tomatoes, or any with decayed spots, cracks or growths. And do not use them to make any canned tomato product such as soup, juice or sauce.

If you are uncertain about the variety or quality of your tomatoes, USDA researchers recommend adding 2 tsp. lemon juice per pint, 4 tsp. per quart. Or use powdered citric acid (from the drugstore), ¼ tsp. per pint, ½ tsp. per quart. Do not substitute vinegar—it's not as effective in acidifying the product; besides, it changes the flavor of tomatoes.

Directions for canning tomatoes and tomato juice follow. See Index for recipes for other tomato products—ketchups, sauces, preserves.

TOMATOES: Select sound, ripe tomatoes. Wash. Dip into boiling water ½ minute or until skins crack; dip quickly into cold water. Cut out stem ends, remove cores and slip off skins.

Raw Pack: Leave tomatoes whole or cut in halves or quarters. Pack in hot jars, pressing down gently after each two tomatoes are added, to release juice and fill spaces. Add no water or liquid to cut tomatoes. Add boiling tomato juice to whole tomatoes if necessary. Leave ½" head space. Add ½ tsp. canning salt to pints, 1 tsp. to quarts. Adjust lids. Process in boiling water bath.

Pint jars   35 minutes
Quart jars   45 minutes

Remove jars from canner and complete seals unless closures are self-sealing type.

Hot Pack: Cut peeled, cored tomatoes in quarters or halves. Put in kettle and bring to a boil. Do not add water. Stir to prevent sticking, but use care not to make tomatoes mushy. Fill hot jars with boiling hot tomatoes, leaving ½" head space. Add ½ tsp. canning salt to pints, 1 tsp. to quarts. (Salt may be omitted if you wish.) Adjust lids. Process in boiling water bath.

Pint jars   35 minutes
Quart jars   45 minutes

Remove jars from canner and complete seals unless closures are self-sealing type.

TOMATO SAUCE OR PURÉE: Choose firm, ripe tomatoes. Do not use tomatoes that are mushy or overripe.

Hot Pack Only: Wash, cut out stem ends and trim away small bruised or decayed parts; discard tomatoes with large decayed areas or diseased cracks. Cut into quarters. Simmer until softened, stirring to prevent sticking. Put through a sieve or food mill to remove seeds and skins. Simmer to desired consistency. (Thick sauce is simmered to half its original volume.) Ladle hot sauce into hot jars (use pints or half pints only). Add ½ tsp. canning salt per pint if desired. Leave ¼" head space. Adjust lids. Process in boiling water bath.

Thin sauce   35 minutes
Thick sauce   20 minutes

Remove jars from canner and complete seals unless closures are self-sealing type.

TOMATO PURÉE, SEASONED: Choose firm ripe tomatoes. Prepare about 4 qts. chopped tomatoes at a time.

Hot Pack Only: Wash, chop and simmer tomatoes until soft, following directions for Tomato Sauce or Purée. Prepare seasoning mixture as follows: Chop 3 onions, 3 branches celery and 3 sweet red or green peppers. Simmer in boiling water until soft. Add to tomatoes and put through a sieve. Simmer until pulp is thick (reduced to half the volume). Add ½ tsp. each of salt and sugar for each pint of pulp. Ladle into hot jars, leaving ¼" head space. Adjust lids. Process in boiling water bath.

Pint jars   45 minutes

Remove jars from canner and complete seals unless closures are self-sealing type.

TOMATO JUICE: Select red-ripe, juicy tomatoes. Do not use tomatoes that are mushy or overripe.

Hot Pack Only: Wash, cut out stem ends and trim away small bruised or decayed parts; discard tomatoes with large decayed areas or diseased cracks. Cut into quarters. Cook at once and quickly. Start cooking as soon as you've cut enough tomatoes to cover the bottom of the kettle; cut others into the hot cooking mixture. Stir frequently and cook only until soft—avoid overcooking. (This quick cooking inactivates enzymes and minimizes separation of tomato juice.) Put cooked tomatoes through a sieve or food mill to remove seeds and skins.

Heat strained juice at once to boiling and pour immediately into hot jars.

Add 1 tsp. canning salt per quart (or omit salt if you prefer). Leave ¼" head space. Adjust lids. Process in boiling water bath.

Pint jars	35 minutes
Quart jars	35 minutes

Remove jars from canner and complete seals unless closures are self-sealing type.

Tomato Specialties

By themselves, tomatoes are safe for canning in a boiling water bath. But what about tomato juice and sauce flavored with the addition of onion, celery or other vegetables? We ran pH tests on our Meatless Spaghetti Sauce and V-4 Tomato Juice Cocktail recipes and found them safe for canning in the boiling water bath. Note that the measure of added vegetables is very small in proportion to the tomatoes called for. Do not increase the amount of vegetables—to do so may lower the acidity of the product too much. If you make a half-batch, keep the ratio of vegetable to tomato the same.

Spaghetti sauce with meat added must be processed in a pressure canner; you'll find a good recipe in Chapter 12 (see Index).

V-4 TOMATO JUICE COCKTAIL

Good beginning for any meal

18 lbs. ripe tomatoes (7 qts. juice)
¼ c. chopped onions

1   c. chopped carrots
2   c. chopped celery (5 branches)
⅓ c. sugar
4   tblsp. canning/pickling salt
1   tblsp. celery seeds
⅛ tsp. cayenne pepper

• Dip tomatoes into boiling water for ½ minute to loosen skins. Cool in cold water. Drain. Remove skins and cores.

• Place 12-qt. kettle over heat. Quarter tomatoes directly into kettle a few at a time so that you maintain a boil; this procedure helps minimize separation of juice.

• Gradually add chopped vegetables, always maintaining boil. Add sugar, salt and spices. Simmer 20 minutes, uncovered.

• Put through sieve or food mill. Return juice to pot; bring to boil.

• Pour into 7 hot quart jars, filling to within ½" of jar top. Wipe jar rim; adjust lids.

• Process in boiling water bath 35 minutes. Start to count processing time when water in canner returns to boiling. Remove jars and complete seals unless closures are self-sealing type. Makes 7 quarts.

Note: If you wish to acidify your juice, add 4 tsp. lemon juice to each quart.

MEATLESS SPAGHETTI SAUCE

Heat and serve, or combine with freshly made meat balls if you wish

16 lbs. ripe tomatoes
1 c. chopped onion
4 tblsp. olive oil
1 tblsp. canning/pickling salt
¼ tsp. black pepper
½ tsp. ground bay leaves
2 cloves garlic, minced (or 1 tsp. garlic powder)
1 tsp. basil leaves
1 tsp. orégano leaves
½ tsp. parsley flakes
2 tblsp. brown sugar

· Dip tomatoes into boiling water for ½ minute to loosen skins. Cool in cold water. Drain. Remove skins and cores; quarter.
· In 12-qt. kettle, sauté onions in olive oil until translucent (do not brown). Pour in tomatoes. Bring to boil; simmer, uncovered, for 20 minutes.
· Put through sieve or food mill. Return juice to kettle, adding remaining ingredients. Simmer, uncovered, 1½ to 2 hours or until thick and mixture rounds up on spoon.
· Pour into 3 hot pint jars, filling to within ¼″ of jar top. Wipe jar rim; adjust lids.
· Process in boiling water bath 20 minutes. Start to count processing

time when water in canner returns to boiling. Remove jars and complete seals unless closures are self-sealing type. Makes 3 pints.

TOMATO JUICE COCKTAIL

You can freeze or can it

2 qts. tomato juice, freshly extracted
3 tsp. salt
2 tsp. grated celery
1 tsp. prepared horseradish
3 tblsp. lemon juice
⅛ tsp. Worcestershire sauce
1 tsp. onion juice

· Add seasonings to tomato juice.
· Freeze or can like Tomato Juice. Makes about 2 quarts.

GREEN TOMATO MINCEMEAT

Makes downright delicious pies—the best way to salvage green tomatoes

6 lbs. green tomatoes
2 lbs. tart apples
2 c. raisins
4 c. brown sugar, firmly packed
2 c. strong coffee
1 lemon (grated peel and juice)
2 tsp. grated orange peel
½ c. vinegar
1 tsp. salt
1 tsp. ground nutmeg
1 tsp. ground allspice

· Core and quarter tomatoes and apples; put through food chopper with raisins. Combine all ingredients in large saucepan. Simmer 2 hours, stirring frequently.

• Pack at once in hot pint jars; leave ½″ head space. Adjust lids. Process in boiling water bath 25 minutes.
• Remove jars from canner and complete seals unless closures are self-sealing type. Makes about 10 pints.

Note: The above recipe is safe for canning in a boiling water bath. If your favorite mincemeat recipe includes suet, it must be processed in a pressure canner at 10 pounds pressure for 25 minutes. Leave 1″ head space when filling jars.

CHAPTER 11

Peak-of-the-Season Vegetables

Getting Ready to Can Vegetables · Processing Low-acid Foods in
Pressure Canner · Preparing Vegetables · How to Pack Vegetables
· Adding Salt · Boiling Vegetables Before Tasting or Serving · Di-
rections for Canning

Soon after Christmas, seed catalogs are delivered to RFD boxholders and
confirmed vegetable gardeners start thinking about rows of beans and
corn, squash and peas. What new varieties sound promising? How many
rows of each should be planted this year? Did we can enough last year to
last until the new crop is ready to pick?

This is when home canners bless the records they've kept: how many
rows it takes to fill how many jars . . . and whether or not the "promising
new variety" lived up to its promises in the canning jar.

It's important, blue ribbon canners say, to plant varieties recommended
for canning. If you have doubts about which varieties grow best in your
area, check with your County Agricultural Extension office.

Because of rising energy costs, gardeners are becoming a little more se-
lective about what they freeze, and many are taking a new look at canning
jars. Of course, some vegetables—plump red tomatoes, and green and yellow beans—have always been favorites for home canning. Or as FARM
JOURNAL readers say, "We like them better canned than frozen."

Vegetables for winter eating retain the most food value—the most vitamins—if they're freshly picked and canned the same day. Often, farm
mothers and children visit the garden as soon as the dew had dried, to
collect tender young vegetables before the sun heats them. Before noon, the
day's pick has been processed in the pressure canner, and the jars are
cooling on folded towels.

If you have a pressure canner and a bountiful garden, you have the first
two requirements for good canned vegetables. Add the knowledge of up-
to-date methods and the will to follow them—then you have what it takes
to put up prize winners. Your family will have better meals at a lower cost.

Getting Ready to Can Vegetables

Vegetables (except for tomatoes) are low in natural acids—too low to be safely canned in a boiling water canner. You must process them under 10 pounds pressure in a steam pressure canner (the equivalent of 240° F.) to destroy heat-resistant bacteria capable of growing in low-acid foods.

If you don't have a pressure canner in good working order, don't can vegetables—freeze them. Or, another option is to pickle vegetables, following recipes in Chapter 13. (Vegetables become acid foods—safe for canning in a boiling water canner—when they're properly fermented or when you add the proper concentration of vinegar.)

Before Canning Day: Get out your pressure canner and make sure it's in good working order; if you have a dial-gauge canner, have it checked before you use it, and once again during canning season, if you can a lot. Read through the instruction book, so procedures are fresh in your mind. (See additional information about pressure canners in Chapter 9.)

Also check your supply of canning jars. Use only heat-tempered standard Mason jars for canning in a pressure canner.

When you plan a day for canning vegetables, remember it takes more than the actual processing time to do each canner load: 15 minutes to exhaust air and build pressure; another half hour or more to cool the canner again, before you can open it.

Processing times for some of the vegetables go beyond the setting of a 60-minute timer. For accurate timing, write down the exact time you reach pressure and add the time for processing. That's when you should remove the canner from the heat. For example, if you achieve pressure at 9:35, and your processing time is 80 minutes, you should turn off the heat at 10:55.

On Canning Day: Begin by picking tender young vegetables. Wash jars and get lids ready. Wash vegetables thoroughly, and prepare according to directions in this chapter for each kind. Precook vegetables as directed for hot pack, or pretreat to prevent darkening. Pack into clean, hot jars; individual directions will tell you whether vegetable should be tightly packed or loosely packed. A few starchy vegetables, such as lima beans and peas, should be loosely packed. They absorb a lot of canning liquid in processing; if packed too tight, they'll burst and become mushy. Cover vegetables with boiling water, leaving recommended head space. Remove air bubbles, wipe sealing edge of jar and adjust lids. Process vegetables in the steam pressure canner at a temperature of 240° F. as directed. For complete description of all canning procedures, see "Steps to Successful Canning" in Chapter 9.

ADDING SALT

Most people add salt when canning vegetables to improve flavor, but if you want to omit it, you can; salt is not necessary to preserve food.

ALTITUDE CORRECTIONS

All processing pressures given for the steam pressure canner are for altitudes less than 2,000 feet above sea level. See Index for Altitude Corrections if you live above 2,000 feet.

BOILING VEGETABLES BEFORE SERVING

Spoilage is not always apparent to the eye; even jars with apparent seals can contain hidden toxins. To be on the safe side, boil all low-acid home-canned vegetables for 10 minutes before tasting, even if they are to be served cold. If food foams or smells bad during boiling, destroy it so that it cannot be eaten by people or animals. (See complete discussion of botulism, signs of spoiled food, and how to destroy spoiled food in Chapter 9.)

To Can Vegetables

ARTICHOKES, GLOBE: Wash and trim small artichokes from 1¼" to 2" in length. Prepare a vinegar-water solution (¾ c. vinegar to 1 gal. water). *Hot Pack Only:* Precook artichokes 5 minutes in vinegar solution. Drain and pack hot in hot jars, being careful to not overfill. Prepare a brine by adding ¾ c. vinegar (or lemon juice) and 3 tblsp. salt to 1 gal. water. Pour boiling hot over artichokes, leaving 1" head space. Adjust lids. Process in pressure canner at 10 pounds pressure.

Pint jars   25 minutes
Quart jars   25 minutes

Remove jars from canner and complete seals unless closures are self-sealing type.

ASPARAGUS: Choose young, fresh asparagus. Wash, trim bracts (scales) and tough ends and wash once again. Cut into 1" pieces, or in lengths ¾" shorter than jar and pack in bundles. *Raw Pack:* Pack asparagus very tightly to within 1" of jar top. Add salt (½ tsp. to pints; 1 tsp. to quarts). Cover with boiling water, leaving 1" head space. Adjust lids. Process in pressure canner at 10 pounds pressure.

Pint jars   25 minutes
Quart jars   30 minutes

Remove jars from canner and complete seals unless closures are self-sealing type.
Hot Pack: Put asparagus in blancher basket, submerge in boiling water and boil 2–3 minutes. Pack to within 1" of jar top. Add salt (½ tsp. to pints; 1 tsp. to quarts) and cover with boiling water leaving 1" head space. Adjust lids. Process in pressure canner at 10 pounds pressure.

Pint jars   25 minutes
Quart jars   30 minutes

Remove jars from canner and complete seals unless closures are self-sealing type.

BEANS, FRESH LIMA: Choose only young tender beans for canning. Shell and wash.

Raw Pack: Pack very loosely in jars to within 1" of jar top; do not shake or press down. Add salt (½ tsp. to pints; 1 tsp. to quarts). Fill jars with boiling water, leaving 1" head space. Adjust lids. Process in pressure canner at 10 pounds pressure.

| Pint jars | 40 minutes |
| Quart jars | 50 minutes |

Remove jars from canner and complete seals unless closures are self-sealing type.

Hot Pack: Put shelled beans in blancher basket, submerge in boiling water and boil 3 minutes. Pack hot, loosely in jar, to within 1" of jar top. Add salt (½ tsp. to pints; 1 tsp. to quarts). Fill jars with boiling water, leaving 1" head space. Adjust lids. Process in pressure canner at 10 pounds pressure.

| Pint jars | 40 minutes |
| Quart jars | 50 minutes |

Remove jars from canner and complete seals unless closures are self-sealing type.

BEANS SNAP: Wash beans, trim and string. Cut into 1" pieces or leave whole. (If whole, pack standing on end in jars.)

Raw Pack: Pack tightly to within 1" of jar top. Add salt (½ tsp. to pints; 1 tsp. to quarts). Fill jars with boiling water, leaving 1" head space. Adjust lids. Process in pressure canner at 10 pounds pressure.

| Pint jars | 20 minutes |
| Quart jars | 25 minutes |

Remove jars from canner and complete seals unless closures are self-sealing type.

Hot Pack: Put beans in blancher basket, submerge in boiling water and boil 3 minutes. Pack hot beans loosely

to within 1" of jar top. Add salt (½ tsp. to pints; 1 tsp. to quarts). Cover with fresh boiling water, leaving 1" head space. Adjust lids. Process in pressure canner at 10 pounds pressure.

| Pint jars | 20 minutes |
| Quart jars | 25 minutes |

Remove jars from canner and complete seals unless closures are self-sealing type.

BEETS: Remove beet tops except for 1 to 2" of stem; leave roots on. Wash and sort to uniform size to ensure even cooking. Cover beets with boiling water; boil 15 to 25 minutes; dip in cold water and slip skins, roots and stems. Baby beets (diameter under 1¼") may be left whole, but larger ones should be cubed in ½" pieces or sliced. Large slices should be quartered or halved.

Hot Pack Only: Pack hot beets to within 1" of jar top. Add salt (½ tsp. to pints; 1 tsp. to quarts). Cover with boiling water; add vinegar to retain color (1 tblsp. to pints; 2 tblsp. to quarts). Leave 1" head space. Adjust lids. Process in pressure canner at 10 pounds pressure.

| Pint jars | 30 minutes |
| Quart jars | 35 minutes |

Remove jars from canner and complete seals unless closures are self-sealing type.

CARROTS: Use only young, tender carrots. Wash, scrape or peel. Slice, dice or leave very small carrots whole.

Raw Pack: Pack tightly to within 1" of jar top. Add salt (½ tsp. to pints; 1 tsp. to quarts). Cover with boiling water to within 1" of jar top. Adjust ds. Process in pressure canner at 10 ounds pressure.

| int jars | 25 minutes |
| uart jars | 30 minutes |

Remove jars from canner and complete seals unless closures are self-ealing type.

Iot Pack: Cover carrots with boiling ater and boil 3 minutes. Pack hot arrots to within 1" of jar top. Add alt (½ tsp. to pints; 1 tsp. to quarts). over with fresh boiling water. Leave " head space. Adjust lids. Process in ressure canner at 10 pounds pres-ure.

| int jars | 25 minutes |
| uart jars | 30 minutes |

Remove jars from canner and complete seals unless closures are self-ealing type.

ELERY: Wash and cut branches in " pieces.

Iot Pack Only: Cover with boiling ater and boil 3 minutes. Pack hot to hot jars and add salt (½ tsp. to ints; 1 tsp. to quarts). Cover with oiling water; leave 1" head space. djust lids. Process in pressure canner t 10 pounds pressure.

| int jars | 35 minutes |
| uart jars | 35 minutes |

Remove jars from canner and complete seals unless closures are self-ealing type.

ELERY AND TOMATOES: Wash and hop celery. Wash, core, peel and hop tomatoes.

Iot Pack Only: Use equal measures f the two vegetables. Boil them together for 5 minutes; do not add ater. Fill jars leaving 1" head pace. Adjust lids. Process in pressure anner at 10 pounds pressure.

| Pint jars | 30 minutes |
| Quart jars | 35 minutes |

Remove jars from canner and complete seals unless closures are self-sealing type.

CORN, CREAM STYLE: Harvest in small quantities (2 to 3 doz. ears at a time), if convenient. *Immediately prepare for processing.* Cut ends from ears, peel off husks and silk. Trim blemishes from ears. Cut kernels from cob about ⅔ the depth of the kernels. Scrape cob to remove the remaining corn, but not any of the cob. *Use pint jars only.*

Raw Pack: Pack very loosely to within 1" of jar top, do not shake or press down. Add ½ tsp. salt to each pint and fill with boiling water, leaving 1" head space. Adjust lids. Process in pressure canner at 10 pounds pressure.

| Pint jars | 95 minutes |

Remove jars from canner and complete seals unless closures are self-sealing type.

Hot Pack: Add 2½ c. boiling water to each quart of prepared corn and boil 3 minutes. Pack hot corn loosely into pint jars to within 1" of jar top. Add ½ tsp. salt to each pint jar. Adjust lids. Process in pressure canner at 10 pounds pressure.

| Pint jars | 85 minutes |

Remove jars from canner and complete seals unless closures are self-sealing type.

CORN, WHOLE KERNEL: Prepare as directed for hot pack cream-style corn, but do not scrape the cob. Use only the whole kernels.

Hot Pack Only: Add 1 pt. boiling water to each qt. of prepared corn and heat to boiling. Pack corn loosely in jars to within 1″ of jar top and cover with boiling water. Leave 1″ head space. Add salt (½ tsp. to pints; 1 tsp. to quarts). Adjust lids. Process in pressure canner at 10 pounds pressure.

Pint jars   55 minutes
Quart jars   85 minutes

Remove jars from canner and complete seals unless closures are self-sealing type.

HOMINY: Shell 2 qts. dry white or yellow field corn. Make a solution of 8 qts. water and 4 tblsp. household lye (be sure it's the kind suitable for use with food) in an enameled or stainless steel pan or kettle—*do not use aluminum, iron, copper, tin or zinc*. Add the dry corn to this brine and boil hard for 30 minutes, or until hulls come loose. Let stand for about 20 minutes, then rinse corn thoroughly with several hot water rinses, followed by cold water rinses to cool the corn sufficiently for handling and to remove all of the lye. Work hominy, using your hands, to remove dark kernel tips. The corn and the tips are more easily separated if placed in a coarse sieve and floated off by water.

Next, add about 1″ of water to cover hominy and boil for 5 minutes; drain; cover with fresh water and boil again. Repeat three more times for a total of five 5-minute boilings in fresh water. Then cook the corn kernels in fresh water about 30 to 45 minutes (or until soft) and drain. Makes about 6 quarts.

Hot Pack Only: Fill jars with hot hominy to within 1″ of jar top. Add salt (½ tsp. to pints; 1 tsp. to quarts).

Fill with boiling water, leaving 1″ head space. Adjust lids. Process in pressure canner at 10 pounds pressure.

Pint jars   60 minutes
Quart jars   70 minutes

Remove jars from canner and complete seals unless closures are self-sealing type.

MIXED VEGETABLES: Choose combinations you like—carrots, green beans, celery, lima beans—and prepare each vegetable according to individual directions.

Hot Pack: Mix vegetables together. Cover with boiling water and boil 3 minutes. Drain. Pack hot vegetables to within 1″ of jar top. Add salt (½ tsp. to pints, 1 tsp. to quarts). Cover with boiling water, leaving 1″ head space. Adjust lids. Process in pressure canner at 10 pounds pressure; for processing time, use the numbers of minutes for pints and quarts needed for the vegetable requiring the longest processing time.

MUSHROOMS: Select fresh mushrooms at peak maturity; can promptly. Soak in cold water for 2 minutes; agitate water vigorously, several times, to remove adhering soil. Trim stems and discolored parts. Wash and rinse again in clean water. Can small mushrooms whole; cut larger ones in halves or quarters. *Use half-pint or pint jars only.*

Hot Pack Only: Cover mushrooms with boiling water and boil 5 minutes, or steam 5 minutes. Pack hot mushrooms to within 1″ of jar top. Add salt (¼ tsp. to half-pints, ½ tsp. to pints) and, for better color, crystalline ascorbic acid (1/16 tsp. to half-pints,

⅛ tsp. to pints). Cover with boiling water, leaving 1″ head space. Adjust lids. Process in pressure canner at 10 pounds pressure.

Half-pint jars   45 minutes
Pint jars   45 minutes

Remove jars from canner and complete seals unless closures are self-sealing type.

OKRA: Wash and trim only tender pods, being careful not to remove the "cap" unless okra will be used for soup. For soup, slice across pods in ½ to 1″ lengths. If canning whole, do not cut into the pod. Cook in boiling water for 1 minute.

Hot Pack Only: Pack hot okra in jar to within 1″ of jar top. Add salt (½ tsp. to pints; 1 tsp. to quarts). Cover with boiling water, leaving 1″ head space. Adjust lids. Process in pressure canner at 10 pounds pressure

Pint jars   25 minutes
Quart jars   40 minutes

Remove jars from canner and complete seals unless closures are self-sealing type.

OKRA AND TOMATOES: Prepare okra as directed for that vegetable. Leave small pods whole, or cut in 1″ lengths. Combine with peeled, quartered tomatoes. Add salt (½ tsp. to pints; 1 tsp. to quarts). Heat to boiling.

Hot Pack Only: Pack hot seasoned vegetables in either pint or quart jars, leaving 1″ head space. Adjust lids. Process in pressure canner at 10 pounds pressure.

Pint jars   25 minutes
Quart jars   35 minutes

Remove jars from canner and com-

plete seals unless closures are self-sealing type.

PEAS, FRESH BLACK-EYED (Cowpeas, Black-eyed Beans): Shell and wash, discarding the too-mature peas.

Raw Pack: Pack raw peas loosely in jars to within 1″ of jar top; do not shake or press down. Add salt (½ tsp. to pints, 1 tsp. to quarts). Cover peas with boiling water, leaving 1″ head space. Adjust lids. Process in pressure canner at 10 pounds pressure.

Pint jars   35 minutes
Quart jars   40 minutes

Remove jars from canner and complete seals unless closures are self-sealing type.

Hot Pack: Cover peas with boiling water and boil 3 minutes. Drain. Pack hot peas loosely in jars to within 1″ of jar top. Add salt (½ tsp. to pints; 1 tsp. to quarts). Cover with boiling water, leaving 1″ head space. Adjust lids. Process in pressure canner at 10 pounds pressure.

Pint jars   35 minutes
Quart jars   40 minutes

Remove jars from canner and complete seals unless closures are self-sealing type.

PEAS, FRESH GREEN: Harvest at the best stage for eating. Can immediately after picking to assure sweet-flavored freshness. Wash, shell peas and wash again.

Raw Pack: Fill jars very loosely to within 1″ of jar top; do not shake or press down. Add salt (½ tsp. to pints; 1 tsp. to quarts). Cover with boiling water, leaving 1″ head space. Adjust lids. Process in pressure canner at 10 pounds pressure.

Pint jars   40 minutes
Quart jars   40 minutes
Remove jars from canner and complete seals unless closures are self-sealing type.

Hot Pack: Put shelled peas in blancher basket, submerge in boiling water and boil 3 minutes. Pack hot peas loosely in jar to within 1" of jar top. Add salt (½ tsp. to pints; 1 tsp. to quarts). Cover with boiling water, leaving 1" head space. Adjust lids. Process in pressure canner at 10 pounds pressure.

Pint jars   40 minutes
Quart jars   40 minutes
Remove jars from canner and complete seals unless closures are self-sealing type.

PEPPERS—BELL, GREEN AND RED: Wash, remove stem, core inside partitions, remove seeds.

To peel: cook in hot water 3 minutes or until skins will slip easily. Or heat in a hot oven (450° F.) 8 to 10 minutes until skins blister and crack. Do not let burn. Remove skins with slender knife blade. Chill peppers immediately in cold water.

Hot Pack Only: Flatten peppers and pack carefully in horizontal layers. Fill hot jars to within 1" from top; cover with boiling water to within 1" of jar top. Add ½ tsp. salt and ½ tblsp. lemon juice (or 1 tblsp. vinegar instead of lemon juice) to pints; 1 tsp. salt and 1 tblsp. lemon juice to quarts (2 tblsp. vinegar if substituted for lemon juice). Vinegar and lemon juice improve flavor. Adjust lids. Process in pressure canner at 10 pounds pressure.

Pint jars   35 minutes
Quart jars   45 minutes

Remove jars from canner and complete seals unless closures are self-sealing type.

Note: Canning peppers is not making the best use of your time. The product is soft and unappetizing. The better way to preserve an excess of peppers is to dice and freeze them, or choose relish recipes that call for peppers.

POTATOES, NEW: Wash, scrape, and rinse small freshly dug new potatoes. Boil 10 minutes in water; drain.

Hot Pack Only: Pack hot to within 1" of jar top. Add salt (½ tsp. to pints; 1 tsp. to quarts). Cover with fresh boiling water. Leave 1" head space. Adjust lids. Process in pressure canner at 10 pounds pressure.

Pint jars   30 minutes
Quart jars   40 minutes

Remove jars from canner and complete seals unless closures are self-sealing type.

POTATOES, CUBED: Select firm, mature boiling potatoes. Scrub, pare and cut into ½" cubes. To prevent darkening, cut directly into brine made with 1 tsp. salt dissolved in 1 qt. water. Drain and rinse.

Hot Pack Only: Put potato cubes in blancher basket and submerge in boiling water; boil 2 minutes. Pack hot to within 1" of jar top. Add salt (½ tsp. to pints, 1 tsp. to quarts). Cover with boiling water, leaving 1" head space. Adjust lids. Process in pressure canner at 10 pounds pressure.

Pint jars   35 minutes
Quart jars   40 minutes

Remove jars from canner and complete seals unless closures are self-sealing type.

PUMPKIN OR WINTER SQUASH, STRAINED: Wash firm, fully ripe pumpkin or squash. Cut in pieces, remove seeds and peel. Steam until tender, about 25 minutes. Put through a food mill or purée in blender. Simmer until heated through, stirring to prevent sticking.

Hot Pack Only: Pack hot pulp in hot jars to within 1″ of jar top. Press down to remove air bubbles. Add no liquid or salt to strained pumpkin. Adjust lids. Process in pressure canner at 10 pounds pressure.

Pint jars   65 minutes
Quart jars   80 minutes

Remove jars from canner and complete seals unless closures are self-sealing type.

SPINACH AND OTHER GREENS: Pick fresh, tender spinach and wash thoroughly. Cut out tough stems and midribs. Put in blancher basket and submerge in boiling water; heat just until wilted.

Hot Pack Only: Pack hot, loosely in jar, to within 1″ of jar top. Add salt (¼ tsp. to pints, ½ tsp. to quarts). Cover with boiling water, leaving 1″ head space. Adjust lids. Process in pressure canner at 10 pounds pressure.

Pint jars   70 minutes
Quart jars   90 minutes

Remove jars from canner and complete seals unless closures are self-sealing type.

SQUASH, BANANA: Prepare as directed for Pumpkin.

SQUASH, SUMMER: Wash, trim ends, remove any imperfect portions, but do not peel. Cut uniform pieces, quarters, halves or ½″ slices.

Raw Pack: Pack tightly to within 1″ of jar top. Add salt (½ tsp. to pints; 1 tsp. to quarts). Fill jar with boiling water to within 1″ of top. Adjust lids. Process in pressure canner at 10 pounds pressure.

Pint jars   25 minutes
Quart jars   30 minutes

Remove jars from canner and complete seals unless closures are self-sealing type.

Hot Pack: Prepare as directed for raw pack. Add water to cover squash and bring to a boil. Pack hot squash loosely to within 1″ of jar top. Add salt (½ tsp. to pints; 1 tsp. to quarts). Cover with boiling water, leaving 1″ head space. Adjust lids. Process in pressure canner at 10 pounds pressure.

Pint jars   30 minutes
Quart jars   40 minutes

Remove jars from canner and complete seals unless closures are self-sealing type.

SWEET POTATOES: Scrub and sort for size. Boil or steam 20 to 30 minutes to facilitate slipping the skins. Do not stick with fork while removing skins. Cut in uniform pieces if large, or leave smaller ones whole.

Dry Pack: Pack hot sweet potatoes tightly, pressing gently to fill air spaces to within 1″ of jar top. Do not add salt or liquid. Adjust lids. Process in pressure canner at 10 pounds pressure.

Pint jars   65 minutes
Quart jars   95 minutes

Remove jars from canner and complete seals unless closures are self-sealing type.

Wet Pack: Pack hot sweet potatoes to within 1" of jar top. Add salt (½ tsp. to pints; 1 tsp. to quarts). Cover with either boiling water or medium Sugar Syrup (see Index), as you prefer. Leave 1" head space. Adjust lids. Process in pressure canner at 10 pounds pressure.

Pint jars   55 minutes
Quart jars   90 minutes

Remove jars from canner and complete seals unless closures are self-sealing type.

TURNIPS: Prepare as directed for carrots. Process in pressure canner at 10 pounds pressure. The time for *pint* jars is 5 minutes longer than for carrots.

Pint jars   30 minutes
Quart jars   30 minutes

Remove jars from canner and complete seals unless closures are self-sealing type.

VEGETABLE SOUP MIXTURE: Any mixture of vegetables may be used. Prepare them as directed in previous instructions for canning each individually. Combine vegetables and boil 5 minutes with water to cover. Peeled, cut-up tomatoes may be used for part of the liquid.

Hot Pack Only: Pour hot vegetable mixture into jars. Cover with boiling liquid. If enough tomatoes are used, no additional water will be necessary. Add salt (½ tsp. to pints; 1 tsp. to quarts). Leave 1" head space. Adjust lids. Process in pressure canner at 10 pounds pressure.

Pint jars   60 minutes
Quart jars   70 minutes

Remove jars from canner and complete seals unless closures are self-sealing type.

VEGETABLE/BEEF STEW: Cube and combine the following ingredients:

2   qts. stewing beef (1½" cubes)
2   qts. potatoes (½" cubes)
2   qts. carrots (½" cubes)
3   c. celery (¼" pieces)
1¾ qts. small whole onions (1"
       or less in diameter)

Raw Pack Only: Brown meat cubes in a small amount of fat, if you wish. Pack jars loosely with meat and vegetables to within 1" from top. Add salt (½ tsp. to pints; 1 tsp. to quarts). Do not add liquid. Adjust lids. Process in pressure canner at 10 pounds pressure.

Pint jars   60 minutes
Quart jars   75 minutes

Remove jars from canner and complete seals unless closures are self-sealing type.

CHAPTER 12

Meat, Poultry and Game

Methods for Canning Beef, Veal, Pork, Lamb, Big Game, Ground
Beef, Sausage, Chickens, Recipes for Blue Ribbon Mincemeat and
State-of-the-Union Mincemeat • Turkey, Other Poultry, Giblets,
Rabbits

You'd think from the freezers and lockers full of meat that freezing has
just about replaced meat and poultry canning on farms. But not quite!
Good country cooks say there's a place in their kitchens for both the frozen and canned. "I can step to the cupboard, take a couple of jars of
home-canned beef or chicken off the shelf and put them on to heat," says
a farm woman. "It's an old farm custom," she adds, "and my salvation
when I look out the window and see a car driving in near mealtime."

Canned meat and chicken also provide a change of pace from frozen
and cured meats, homemakers say. And they take no thawing time. Meats
and poultry in jars are invaluable when freezer and locker space vanishes.
But clever cooks also plan to keep jars of meat and chicken on hand just
to please the people at the table with their old-fashioned taste.

Most cooks have pet ways (but few recipes) for dressing up canned
meats and chicken—often heirloom methods their grandmothers and
mothers used. One woman puts chicken on to heat while she cooks noodles in chicken broth she canned. She combines them to serve. Another
homemaker heats canned beef, adds barbecue sauce and spoons it over
fluffy rice or toasted buns. Two hearty main dishes for 30-minute dinners!

A Kansas farmer and his wife, when driving to Colorado for a family
reunion, took along 52 quarts of canned beef to make cooking easier and
more economical in their rented cottage. They also wanted to treat their
grown children, now scattered in homes of their own, to what-mother-
used-to-make. Their reward—many compliments.

Getting Ready to Can Meat and Poultry

To can any meat, poultry or game safely, you need a pressure canner in good working order. Meats and meat broths are extremely low-acid foods which require processing under 10 pounds pressure in a steam pressure canner (the equivalent of 240° F.) to destroy heat-resistant bacteria capable of growing in low-acid foods. If you have a dial-gauge canner, have it checked before you use it. (See additional information about pressure canners in Chapter 9.) Use only heat-tempered standard Mason jars for canning in a pressure canner. For meat, wide-mouth jars with straight sides are better because the meat will come out more easily.

You'll need sharp knives for cutting and trimming meat, and a pencil-type thermometer if you pack meat raw.

HOT PACK AND RAW PACK

Although the two methods of packing meat are described as *hot pack* and *raw pack,* in both cases the meat is heated before you cap the jars.

For hot pack, you precook meat to medium doneness in a pan or in the oven, pack it hot in jars, cover it with boiling liquid (broth, or meat juices plus boiling water) and adjust lids—the same procedures you follow for hot pack vegetables.

If you pack meat raw, you must preheat it in the jar without a lid—a step called *exhausting.* The open jars of meat are placed in hot water and heated slowly until a thermometer in one of the jars registers 170° F.

Whichever method you use—

precooking or exhausting—such heating drives air out of meat tissues so that a vacuum will be formed in jars after processing. This gives you a better seal; it also protects meat flavor.

The addition of salt is optional. In the small amounts used for seasoning, salt does not help preserve the meat.

Control of head space is very important. If less than designated, it is likely to cause seal failures.

Before adjusting lids, wipe jar tops and threads carefully with a *dry* paper towel. Any fat left on sealing surface could prevent the lid from sealing. For complete description of canning and procedures, see "Steps to Successful Canning" in Chapter 9.

KEEP WORK SURFACES AND UTENSILS SUPER-CLEAN

At room temperature, bacteria grow rapidly in meat, so it's important to follow good procedures. Work with small quantities while cutting and trimming meat and poultry and *work fast.* Keep the rest of it as cold as you can, in the refrigerator or a meat cooler.

Wash knives and other utensils and rinse well with boiling water before and after each use.

Use a wood cutting board, but keep it scrupulously clean. The cutting board (or any tabletop where meat has rested) needs special treatment to keep bacteria under control. Scrub it well with hot soapy water, rinse with boiling water and disinfect it as follows: Mix ¼ c. chlorine bleach (ordinary household laundry bleach) with

1 qt. water; leave this solution on the board for at least 15 minutes; rinse off with scalding water. Do this before and after handling meat.

Wipe-up cloths should be rinsed with cool water to remove meat juices, boiled in soapy water and rinsed in the chlorine solution.

BOILING MEAT BEFORE SERVING

Spoilage is not always apparent to the eye; even jars with apparent seals can contain hidden toxins. When you open jars, smell the food. If there is any off-odor or off-color, any mold or softening of the meat—these are signs of spoilage, and food should be destroyed without tasting.

As a further safety precaution, boil all home-canned meat/meat mixtures, poultry—and broths—for 10 minutes before tasting, even if they are to be served cold. Add water if necessary, stirring so that all food is thoroughly boiled. Heating intensifies any off-odors; if food smells queer, destroy it without tasting. (See complete discussion of botulism, signs of spoiled food and how to destroy spoiled food in Chapter 9.)

ALTITUDE CORRECTIONS

All processing pressures given for the steam pressure canner are for altitudes less than 2,000 feet above sea level. See Index for Altitude Corrections if you live above 2,000 feet.

To Can Meats and Large Game

Select clean, fresh meat that has been well bled. Chill immediately after slaughtering until all the animal heat is gone. Can or hold several days at 40° F. or lower. You will find chilled meat easier to handle.

Do not wash meat. Wipe it off with a damp cloth before cutting. Washing draws out the juices.

Carefully cut meat from bones and trim out all visible fat. Cut large pieces across the grain in lengths to fit jars, with grain of meat running the height of jar. Cut less choice pieces of meat in cubes for stew. Use bones to make broth for canning or soup stock. *Raw Pack:* Pack raw meat loosely in jar, leaving 1¼″ head space. Set open filled jars on rack in kettle of hot water. Water level should be 2″ below

jar tops. Put thermometer in center of center jar, cover kettle and heat slowly until thermometer registers 170° F. This takes about 70 minutes.

Remove jars from kettle; remove thermometer. Add salt (½ tsp. to pints, 1 tsp. to quarts). Do not add liquid—meat makes its own broth during processing. Wipe jar rims with dry paper towel to remove any trace of fat. Adjust lids. Process in pressure canner at 10 pounds pressure.

Pint jars   75 minutes
Quart jars   90 minutes

Remove jars from canner and complete seals unless closures are self-sealing type.

Hot Pack: Put meat pieces in a large

shallow pan. Add a little water to prevent sticking. Cover and heat until medium done (center of meat pieces faintly pink); stir occasionally so meat will heat evenly.

Pack hot meat loosely in clean hot jars. Add salt (½ tsp. to pints, 1 tsp. to quarts). Cover with boiling meat juices and/or boiling water, leaving 1¼″ head space. Wipe jar rims with dry paper towel to remove any trace of fat. Adjust lids. Process in pressure canner at 10 pounds pressure.

Pint jars   75 minutes
Quart jars   90 minutes

Remove jars from canner and complete seals unless closures are self-sealing type.

CANNING GROUND MEAT

1. Select and grind small pieces of meat or meat from the less tender cuts. Make certain all the meat is fresh. Keep it cold. Discard any pieces of questionable freshness. Don't use chunks of fat.

2. Add 1 tsp. salt to each pound of ground meat. Mix well.

3. Form meat into fairly thin, small patties that can be packed into glass jars without breaking.

4. Put meat patties into cooking pan and precook in a moderate oven (350° F.) until medium done. The red color should be almost gone from center of patties.

5. Pack hot patties into clean hot jars, leaving 1¼″ head space.

6. Skim fat from drippings, add water and heat to boiling.

7. Cover patties with boiling liquid (or boiling broth). Leave 1 to 1¼″ head space.

8. Remove air bubbles. Readjust head space.

9. Wipe jar rims with dry paper towel. Adjust lids.

10. Process at once in a pressure canner at 10 pounds pressure.

Pint jars   75 minutes
Quart jars   90 minutes

Remove jars from canner and complete seals unless closures are self-sealing type.

CANNING PORK SAUSAGE

Use your favorite sausage recipe (or see recipe in chapter on Curing Meat, Poultry, Fish), omitting the sage. It often gives canned sausage a bitter taste. Use a light hand with other herbs, spices, onion and garlic because flavors change in processing and storage. You may wish to add more seasoning when you heat the sausage for serving. Use ¾ lean meat to ¼ fat to make the best sausage. Process like ground meat.

CANNING SOUP STOCK

If you use the bones of beef or chicken to make broth, skim off the fat and remove all pieces of bone, but do not strain out bits of meat or sediment. Fill hot jars with boiling stock, leaving 1″ head space. Wipe jar rims with dry paper towel. Adjust lids. Process at once in a pressure canner at 10 pounds pressure.

Pint jars   20 minutes
Quart jars   25 minutes

Remove jars from canner and complete seals unless closures are self-sealing.

CANNING CORNED BEEF

Remove corned beef from brine when it's cured (see directions in Chapter 16, Curing Meat, Poultry and Fish). Wash, drain and cut into pieces or strips to fit jars. Cover meat with cold water and bring to a boil. If broth tastes very salty, drain, cover with fresh water and bring again to a boil. Pack meat hot into hot jars, to within 1¼" of jar top. Cover with boiling broth or boiling water, leaving 1" head space. Remove air bubbles; readjust head space. Wipe jar rims with dry paper towel. Adjust lids. Process at once in a pressure canner at 10 pounds pressure.

Pint jars   75 minutes
Quart jars   90 minutes

Remove jars from canner and complete seals unless closures are self-sealing type.

SPAGHETTI SAUCE WITH MEAT

Easy dinner: open a jar, simmer 10 minutes, serve on spaghetti

12 lbs. ripe tomatoes
4   (18 oz.) cans tomato paste
3½ c. chopped onions
4   (4 oz.) cans sliced mushrooms
    (optional)
5   cloves garlic, minced
¼   c. olive oil
3   lbs. lean ground beef
¼   c. olive oil
3   tblsp. canning/pickling salt
¼   c. brown sugar, firmly packed
2   tsp. black pepper
2   tsp. orégano leaves
1½ tsp. ground paprika
2   tsp. basil leaves

• Dip tomatoes into boiling water for ½ minute to loosen skins. Cool in cold water. Drain. Remove skins and cores; cut in quarters. Boil 20 minutes, uncovered, in 5-qt. kettle. Put through food mill or sieve. Combine with tomato paste.
• In small skillet sauté onions, mushrooms if desired, and garlic in ¼ c. olive oil until translucent.
• In 12-qt. kettle, sauté meat in ¼ c. olive oil. Add onion and tomato mixtures, and salt, sugar and spices. Bring to boil; simmer, uncovered, until mixture is very thick, about 45 minutes.
• Ladle into hot jars, filling to within 1" of jar top. Wipe jar rim; adjust lids. Process in pressure canner at 10 pounds pressure.

Pint jars   60 minutes
Quart jars   75 minutes

• Remove jars and complete seals unless closures are self-sealing type. Makes 7 quarts.

BLUE RIBBON MINCEMEAT

Farm homemaker won first place with this at Colorado State Fair

6   c. ground beef
12 c. chopped apples
6   c. seedless raisins
1   c. apple cider
1   tblsp. ground cinnamon
1   tblsp. ground allspice
1   tblsp. ground nutmeg
3½ c. sugar

• Cook beef well, but do not brown.
• Put meat and apples through food chopper, using medium blade.
• Combine all ingredients in large kettle. Simmer 30 minutes.
• Pack at once in hot *pint* jars, leav-

ing 1" head space. Wipe jar rim; adjust lids. Process in pressure canner at 10 pounds pressure.

Pint jars　　　75 minutes

• Remove jars from canner and complete seals unless closures are self-sealing type. Makes 8 pints.

STATE-OF-THE-UNION MINCEMEAT

Old recipe from Alaska where bear meat used to substitute for beef

3　lbs. lean beef
1　qt. water
1　c. chopped dates
1　c. washed suet, finely
　　chopped
3½　lbs. apples
1　lb. seedless raisins
1　lb. white raisins
4　c. orange marmalade
2　qts. apple cider
2　tblsp. ground cinnamon
1　tsp. ground cloves
1　tsp. ground nutmeg
3　tblsp. salt

• Simmer beef in water until tender (add more water if needed). Drain. Trim away bone and gristle. Put meat through food chopper, using medium blade.

• Combine all ingredients in large kettle. Mix well. Bring to a boil; reduce heat, simmer 1½ hours; stir often.

• Pour at once into hot *pint* jars, leaving 1" head space. Wipe jar rim; adjust lids. Process in pressure canner at 10 pounds pressure.

Pint jars　　　75 minutes

• Remove jars from canner and complete seals unless closures are self-sealing type. Makes about 10 pints. Store in cool, dark place.

HAMBURGER SAUCE MIX

Good between toasted buns—or use in many substantial tasty dishes

2　lbs. lean ground beef
2　large onions, chopped (3 c.)
2　(6 oz.) cans tomato paste
1⅓　c. water
2　tsp. salt
½　tsp. pepper

• Cook beef and onions in large container until meat browns. Pour off fat.

• Add remaining ingredients, bring to a boil and simmer 5 minutes.

• Pack at once into hot pint jars, leaving 1" head space. Wipe jar rim; adjust lids. Process in pressure canner at 10 pounds pressure.

Pint jars　　　75 minutes

Remove jars from canner and complete seals unless closures are self-sealing type. Makes about 3 pints.

Note: If you double the recipe, process quarts 90 minutes.

PIZZABURGERS

All taste-testers made the same comments—delicious and satisfying

1　pt. Hamburger Sauce Mix
1　tsp. orégano leaves
6　to 8 buns
6　to 8 slices sharp process cheese

• Blend Hamburger Sauce Mix and orégano. Bring to a boil and simmer 10 minutes.

• Spoon between hot, buttered buns. Put a slice of cheese in every bun. Serve at once. Makes 6 to 8.

CHILI AND BEANS

From start to finish, this dish takes 20 minutes of your time

1 pt. Hamburger Sauce Mix
½ c. tomato juice or water
1 (15½ oz.) can kidney beans
1 tblsp. chili powder
½ tsp. cumin seed (optional)
2 cloves garlic

· Combine Hamburger Sauce Mix, tomato juice, kidney beans, chili powder and cumin seed. Place garlic cloves on toothpicks; add to mixture. Bring to a boil and simmer 10 minutes. Remove garlic. Makes 4 servings.

Note: Chili and beans is a hot dish. Decrease chili powder if you like a milder seasoning.

TAMALE AND CHILI BAKE

Serve this easy South-of-the-Border dish with pineapple salad

1 pt. Hamburger Sauce Mix
½ c. water
1 tblsp. chili powder
1 clove garlic
2 (15 oz.) cans tamales
¼ lb. sliced sharp process cheese

· Combine Hamburger Sauce Mix, water and chili powder. Place garlic on a toothpick and add to mixture. Bring to a boil and simmer 10 minutes. Remove garlic.
· Unwrap tamales and place in a greased 2-qt. casserole. Top first with Hamburger Sauce Mix, then cheese.
· Bake in moderate oven (350° F.) until bubbling hot, about 30 minutes. Makes 6 servings.

HURRY-UP SPAGHETTI

You can fix a wonderful spaghetti supper in half an hour

1 pt. Hamburger Sauce Mix
1 c. tomato juice
1 tsp. basil leaves
1 tsp. parsley flakes
½ tsp. orégano leaves
¼ tsp. salt
2 cloves garlic
1 (12 oz.) pkg. spaghetti, cooked
Cheese

· Combine Hamburger Sauce Mix, tomato juice, herbs and salt. Stick a toothpick through garlic cloves; add to mixture. Bring to a boil and simmer, covered, 15 minutes. Remove garlic.
· Place hot, drained spaghetti in serving dish. Pour sauce mixture over spaghetti. Sprinkle with grated Parmesan cheese or other hard cheese. Serve at once. Makes 4 servings.

BEST-EVER PIZZA

Tasty version of the popular American-style pizza. Seasoning is mild

1 (13¾ oz.) pkg. hot roll mix
1 pt. Hamburger Sauce Mix
1 tsp. orégano leaves
1 c. shredded Mozzarella cheese

· Prepare dough from hot roll mix by package directions, but omit rising. Divide dough in half. Roll on lightly floured surface to make two 14″ circles.
· Place on ungreased baking sheets; turn up edges to form 13″ circles.
· Combine Hamburger Sauce Mix

and orégano. Spread on dough circles.
• Bake in very hot oven (450° F.)
10 minutes; remove from oven and
top with cheese. Bake about 10 minutes longer or until crust is done.
Serve at once. Makes 2 pizzas.

To Can Poultry and Small Game

Follow these directions for canning chicken, turkey and other kinds of poultry and game birds.

Can only healthy, plump chickens. The texture and flavor of young birds is not as good as mature birds for canning. (See directions for killing and dressing chickens in Chapter 3. Refrigerate dressed chicken promptly and keep chilled until you're ready to can it.)

Cut up chickens as you would for table use and divide into three groups —meaty pieces, bony pieces and giblets.

To make broth, cover bony pieces with cold water and bring to a boil; simmer until meat is tender. Drain broth and skim off fat. (Strip meat from bones and can for chicken á la king, if you wish.)

Bone the chicken breasts; saw drumsticks off short; leave bones in other meaty pieces. Trim off fat.

Hot Pack, with Bone: Place chicken in a kettle, cover with hot broth, cover kettle and simmer until meat is medium done (barely pink in center).

Pack hot drumsticks and thighs in hot jars, placing skin next to glass. Fit breasts in center and add small pieces where needed; fill to within 1¼″ of jar top. Add salt (½ tsp. to pints; 1 tsp. to quarts). Cover meat with boiling broth, leaving 1″ head space. Wipe jar rim with dry paper

towel. Adjust lids. Process in pressure canner at 10 pounds pressure.

Pint jars   65 minutes
Quart jars   75 minutes

Remove jars from canner and complete seals unless closures are self-sealing type.

Hot pack, without Bone: Follow directions for hot pack with bone, except remove all bones, either before or after cooking. Leave skin on. Note that chicken without bone must be processed for a longer period of time.

Process in pressure canner at 10 pounds pressure.

Pint jars   75 minutes
Quart jars   90 minutes

Remove jars from canner and complete seals unless closures are self-sealing type.

Raw Pack, with Bone: Do not precook chicken. Pack raw drumsticks and thighs into clean jars, placing skin next to glass; fit breasts in center, filling to within 1¼″ of jar top.

Set open filled jars on rack in kettle of hot water. Water level should be 2″ below jar tops. Put thermometer in center of center jar, cover kettle and heat slowly until meat is steaming hot and thermometer registers 170° F. This takes about 75 minutes.

Remove jars from kettle; remove thermometer. Add salt (½ tsp. to pints, 1 tsp. to quarts). Do not add liquid. Wipe jar rim with dry paper towel. Adjust lids. Process in pressure canner at 10 pounds pressure.

Pint jars   65 minutes
Quart jars   75 minutes

Remove jars from canner and complete seals unless closures are self-sealing type.

Raw Pack, without Bone: Follow directions for raw pack with bone, except remove all bones; leave skin on. Note that chicken without bone must be processed for a longer period of time.

Process in pressure canner at 10 pounds pressure.

Pint jars   75 minutes
Quart jars   90 minutes

Remove jars from canner and complete seals unless closures are self-sealing type.

HOW TO CAN RABBITS

Skin rabbits, remove entrails and cut out the waxy glands under the front legs where they join the body. Wash the carcass in salty water; cut in pieces. Can like chicken.

HOW TO CAN GIBLETS

It is best to can livers separately, to protect the delicate flavor. Gizzards and hearts may be canned together.

Put hearts and gizzards in sauce-pan; cover with broth made from bony pieces, or hot water. Cover pan and simmer until medium done, stirring occasionally. Precook livers same way.

Pack giblets hot in pint jars, to within 1″ of top of jar. Add ½ tsp. salt. Cover with boiling broth, leaving 1″ head space. Remove air bubbles; wipe rim of jar with dry paper towel. Adjust lids.

Process at once in pressure canner at 10 pounds pressure.

Pint jars   75 minutes

Remove jars from canner and complete seals unless closures are self-sealing type.

HOW MUCH WILL IT MAKE?

The amount of fresh meat you can pack into a canning jar depends on size of pieces and (in the case of chicken) whether you pack with or without bone.

For a quart jar, allow about this much untrimmed meat:

Beef round, 3 to 3½ lbs.
Beef rump, 5 to 5½ lbs.
Pork loin, 5 to 5½ lbs.

For a quart jar, allow about this much ready-to-cook (dressed and drawn) chicken:

Canned with bone, 3½ to 4¼ lbs.
Canned without bone, 5½ to 6¼ lbs.

CHAPTER 13

Pickles, Vegetable Relishes and Sauerkraut

Pickling Methods, Ingredients and Equipment • How to Can Pickles • Recipes for Dill, Whole Sweet, Sliced, Chunk and Mixed Cucumber Pickles • Brine-Cured Cucumbers • Crock Sauerkraut • Dilled Beans and Okra • Beet, Zucchini and Onion Pickles • Corn Relish • Green Tomato Pickles • Pepper Pickles and Relishes

When a farm woman goes to the garden every morning to look at the cucumber vines, she's daydreaming. She sees sparkling jars lined up on the shelves of her fruit closet, filled with all kinds of pickles. And she imagines how she will tie ribbons on the fancy packs at Christmas time and nest them in baskets for gifts. She can almost hear guests around her table repeat, "Please pass the pickles!"

Between garden and table, there are many kitchen hours invested in making the old-time greats—fresh-pack dills, bread-and-butters, piccalilli, chow-chow, icicle pickles and others with fanciful names. Certain that her efforts will pay off next winter, the country cook watches for the signs. When sprawling vines yield basketfuls of cucumbers and the dill heads turn green-gold on their feathery stalks, it's time to make sours—the name Pennsylvania Dutch women gave to pickles.

We asked FARM JOURNAL readers for their best pickle recipes. Thousands of these treasures came to our Test Kitchens in friendly response and from them—after months of sorting, testing and tasting—we picked a hundred or so to include in the first edition of our FREEZING & CANNING COOKBOOK, published in 1963. Dozens of these recipes have since become country classics: made up year after year by women whose families clamor for them.

Since 1963, the *art* of pickling has given in a bit to the *science* of pickling, as researchers have identified some of the reasons why even experienced home canners sometimes reported failures. High on the list of errors is open-kettle canning. Scientists are unanimous today in insisting that all pickles and relishes be processed in a boiling water bath to seal the jars and protect the contents from spoilage. It is equally important that each recipe formulation be sufficiently acid to register a pH of 4.2 or below (but never above 4.5).

In this chapter and the next, we reprint our favorite recipes with renewed confidence that your experience with them will be successful. We have retested every recipe—to verify the processing time that will protect crispness and flavor, and also to verify the acidity of the product.

Now—it's up to you. We'll repeat advice from the earlier book: Regardless of what kind of pickles you choose to make, there are rules to heed for success. So start your pickling by reading.

How to Succeed with Pickles

There are two kinds of cucumber pickles—brine-cured (also called fermented) and quick-process or fresh-pack.

In the brine cure, vegetables are held in brine until fermentation is complete (about 4 to 6 weeks); they can wait in the brine even longer. Homemakers who champion the brine cure say it's easy to put cucumbers down in brine as they come from the garden, to make pickles as needed or on days when they are not so rushed. Experts agree these pickles—if well made—are worth the time and attention they require, they are so crisp and good tasting.

But many quick-process pickles are delicious and our FARM JOURNAL readers have shared some superlative recipes with us. About 99 percent of the reader recipes were for the fast-fix kind. Some quick-process pickles are canned immediately after a brief simmer in a vinegar solution; but most (not so quick) are soaked in a brine for a few hours, or overnight—or even for a few days. Some recipes call for treatment with ice water or boiling water before the cucumbers are covered with pickling syrups.

Recipes for quick-process cucumber pickles and relishes come first in this chapter. Following them, you will find directions for cured pickles and for brining cucumbers and sauerkraut. The chapter ends with recipes for other vegetable pickles—dilled beans, green tomatoes, pickled peppers—plus all the wonderful mixed pickles that farm women make up from garden remnants.

START WITH RIGHT INGREDIENTS

Cucumbers: It pays to grow your own cucumbers for pickling and to plant a variety developed for pickling—you'll find this information on the seed packet or in the seed catalog. Slicing cucumbers, grown for salads and fresh eating, do not make good pickles; waxed cucumbers from the supermarket are totally unacceptable.

Immature cucumbers make the best pickles; use them as soon as possible after picking. If you cannot begin the pickles within an hour or two, store them without washing in the refrigerator. *Do not use cucumbers that have been picked more than 24 hours* —you'll simply be wasting your time. Cucumbers that wait too long may make pickles that are hollow, shriveled or tough. Some cucumbers, because of growing conditions, are hollow to start with. Watch for them when you wash cucumbers—usually they float in water. Set them aside to chop for relishes.

Sort cucumbers for size and degree of maturity; use appropriate recipes. (You'll find a special collection of recipes calling for ripe yellow cucumbers in this chapter.)

Onions: For pickling, use onions that have been cured. Onions are cured by drying. Onions pulled fresh from the garden do not pickle well.

Other vegetables: Use fresh-picked vegetables, free of blemishes. Sort for size and maturity and can as soon as possible after picking. To make attractive relishes, chop vegetables into uniform-sized pieces.

Vinegar: Use a clear vinegar with specified acid strength of 5 percent or 50 grains (look on bottle label). Do not use homemade vinegar because its acid content varies greatly; often it is too weak to prevent spoilage, or it may be strong enough to cause shriveling. White distilled vinegar gives pickles the best color and should always be used with light-colored vegetables and fruits. Cider vinegar may darken pickles, but cooks like it for its flavor and aroma.

Salt: Use pure granulated salt (pickling or canning salt) for pickles and for brining. The additives in table salt to prevent caking will make brine cloudy; iodized salt may darken pickles and retard fermentation.

Canning/pickling salt is a fine-granulated salt. If you use a coarse granulated pure salt (Kosher or flake salt), measure half again as much—

that is, use 1½ c. coarse salt for each 1 c. fine-granulated salt.

Spices: Use fresh spices for the best flavor. Whole spices help produce pickles of superior color. Tie them in cheesecloth bags to boil with pickling liquids *and remove them before canning.* If poured into the jar, they may darken pickles and cause an off-flavor.

Sugar: Use granulated white sugar—cane or beet sugar—unless the recipe specifies some other sweetener. Some recipes call for brown sugar for added flavor in products where the darker color is not objectionable.

Water: Use soft water for making a pickling brine; the minerals in hard water interfere with the curing process. If you don't have soft or softened water on tap, boil the hard water for 15 minutes. Let it stand for 24 hours. Skim off any film and ladle water from the container, being careful not to disturb the mineral sediment on the bottom. Before using boiled water, add 1 tblsp. vinegar to each gallon water.

No additives: Although several of the recipes in this book originally called for alum to give pickles "icicle crispness," we did not use alum when we retested recipes. It's an old-fashioned ingredient which most authorities today consider unnecessary. As you try the updated recipes, you will see there are other ways to insure crispness. Using fresh-picked vegetables, absolutely sanitary methods in brining and soaking, and accurate processing times, are all important. Follow recipe directions carefully—

don't take shortcuts. You'll have the satisfaction of making superior pickles entirely from natural ingredients— plus your own skill and know-how.

USE RIGHT EQUIPMENT

Container for Brining: Use stoneware crocks made in America (some foreign crocks have glazes high in lead), or any glass, stainless steel, enamelware or food-grade plastic such as polyethylene. Do not use any container that is chipped or cracked. Do not use wood or any other metal container for brining. For a cover, use a heavy plate or glass lid big enough to cover vegetables but small enough to fit inside the container. You'll also need a weight to hold the cover down and keep vegetables submerged in brine—a good one is a glass jar filled with water.

Pickling Kettle: Best choice is stainless steel, unchipped enamelware or flameproof glass in a size large enough to prevent food boiling over. You can use aluminum, although it may react with pickling liquids to give food a metallic taste. Do not cook pickles in iron, unlined copper or galvanized metal containers. Iron darkens pickles; the other two combine with pickling acids and salts to form very small quantities of harmful substances.

Other Equipment: For description of canning jars, lids, boiling water canner and other small equipment, see Chapter 9.

To Can Pickles and Relishes

Choose your recipe first. Follow directions for washing and preparing vegetables. After washing cucumbers, trim $\frac{1}{16}''$ from blossom end. (The blossom sometimes contains enzymes which may cause pickles to soften.)

Weigh vegetables and measure vinegar accurately. Do not reduce the amount of vinegar or change the proportion of vinegar to vegetable in the recipes that follow. The correct proportion of vinegar to vegetable is necessary to produce a product with enough acidity to be safely processed by the boiling water bath. If the pickle is too sour for your taste, increase the sugar the next time you make it. Or try another recipe.

Prepare jars and lids. Jars should be sterilized if recipe calls for processing less than 15 minutes.

Pack pickles into hot jars. Cover with boiling water or brine, leaving recommended head space. Remove air bubbles, wipe sealing edge of jar with a clean, moist paper towel and adjust lids. Process pickles and relishes in the boiling water canner as directed. For complete directions on all canning procedures, see "Steps to Successful Canning" in Chapter 9.

Open-kettle canning is no longer considered safe for canning pickles and relishes. There is always danger of spoilage organisms entering the food when it is transferred from kettle to jar. Wild yeasts and molds, harmless in themselves, may start to grow in a jar of unprocessed pickles. Their growth reduces the acidity of the product. Botulinum spores, if present, could germinate in the resulting low-acid product and produce the deadly poison that causes botulism. Processing jars after filling destroys such yeasts and molds, and also helps achieve a good seal.

ALTITUDE CORRECTIONS

All processing times given for the boiling water bath are for altitudes less than 1,000 feet above sea level. See Index for Altitude Corrections if you live above 1,000 feet.

HOW TO MEASURE CHOPPED FRUITS AND VEGETABLES (WATER DISPLACEMENT METHOD)

The way you measure ingredients for relishes will make a difference in yields. In testing recipes, we measured *before* chopping, using the water displacement method. We measured all chopped ingredients this way, except tomatoes.

If your recipe calls for 4 c. chopped apples, for example, put 1 qt. water into a 2-qt. bowl. Quarter apples and remove cores; add apples to bowl (pressing down if they float) until water reaches 2-qt. line. Drain and chop.

If your recipe calls for ⅔ c. chopped onion, put 1⅓ c. water into a 2 c. measure. Add enough peeled onion to bring water line to 2 c. Drain and chop onion.

If you use the blender for chopping, measure water and solids in blender container, chop and drain in a colander.

STORE PICKLES PROPERLY

After jars have cooled, check all the seals. Label, date and store in a cool, dark, dry place. Jars that are not sealed should be refrigerated and used. Keep opened jars refrigerated and make sure liquid covers the pickles.

Discard, without tasting, any pickles that have a bad odor, or that appear moldy, mushy or gassy when opened.

Quick Process Pickles

FRESH-PACK DILLS

Taste Panel's favorite recipe for fresh-pack dills. Rated Excellent

9	lbs. (3–5″) pickling cucumbers
1½	c. canning/pickling salt
2	gals. water
¾	c. canning/pickling salt

¼	c. sugar
9	c. water
6	c. 5% acid strength cider vinegar
2	tblsp. pickling spices
24	heads fresh dill
16	tsp. mustard seeds

• Wash cucumbers; cut $\frac{1}{16}''$ off blossom end. Place in 3-gal. crock. Cover with brine made by adding 1½ c. salt to 2 gals. water. Cover. Let stand 12–18 hours in cool place. Drain.

• Combine ¾ c. salt, sugar, 9 c. water and vinegar in 5-qt. kettle. Tie pickling spices in cheesecloth bag; add to pot. Bring to boil. Let stand while you pack jars.

• Pack cucumbers into 8 sterilized hot quart jars. Place 3 dill heads and 2 tsp. mustard seeds in each jar. Reheat syrup to boiling; remove spice bag. Pour liquid over cucumbers, filling to within ¼" of jar top. Wipe jar rim; adjust lids.

• Process in boiling water bath 10 minutes. Start to count processing time when water in canner returns to boil. Remove jars and complete seals unless closures are self-sealing type. Makes 8 quarts.

DILL PICKLES

Dill seeds do the seasoning

4   lbs. (4") pickling cucumbers
    (36 to 40)
½   c. dill seeds
24   whole peppercorns
2   c. 5% acid strength cider
    vinegar
4½   c. water
⅓   c. canning/pickling salt

• Wash cucumbers; cut $\frac{1}{16}''$ slice off blossom ends. Pack raw into 8 sterilized hot pint jars. Add 1 tblsp. dill seeds and 3 peppercorns to each jar.

• Combine vinegar, water and salt in 2-qt. saucepan; bring to boil. Pour

over cucumbers, filling to within ¼" of jar top. Wipe jar rim; adjust lids.

• Process in boiling water bath 10 minutes. Start to count processing time when water in canner returns to boil. Remove jars and complete seals unless closures are self-sealing type. Makes 8 pints.

EXTRA GOOD SWEET DILLS

Perfect blend of dill and sweet flavors

4 lbs. (4") pickling cucumbers
Ice water
6 tblsp. dill seeds
1 medium (2") onion, peeled and
    thinly sliced
2 c. 5% acid strength cider
    vinegar
1 c. water
1 c. sugar
2 tblsp. canning/pickling salt

• Wash cucumbers; cut $\frac{1}{16}''$ slice off blossom end. Cover with ice water. Let stand in refrigerator 3 to 4 hours. Drain; quarter lengthwise.

• Add 1 tblsp. dill seeds and 2 slices onion to each of 6 sterilized hot pint jars. Pack cucumbers into jars.

• Combine vinegar, 1 c. water, sugar and salt in 2-qt. saucepan. Bring to boil. Pour over cucumbers, filling to within ¼" of jar top. Wipe jar rim; adjust lids.

• Process in boiling water bath 10 minutes. Start to count processing time when water in canner returns to boiling. Remove jars and complete seals unless closures are self-sealing type. Makes 6 pints.

Sliced Cucumber Pickles

Pickles with sandwiches once were the rule. Now it's often pickles in sandwiches. Some of the recipes that follow call for thin slices, others for thick slices and chunks. Some are sweet and some are sweet-sour. But in all of them you see signs of an imaginative cook behind the scenes.

SWEET PICKLE SLICES

Superior flavor and nice green color

4 lbs. 3–5" pickling cucumbers
4 c. 5% acid strength cider vinegar
¼ c. sugar
3 tblsp. canning/pickling salt
1 tblsp. mustard seeds
2 c. 5% acid strength cider vinegar
3 c. sugar
2¼ tsp. celery seeds
1 tblsp. whole allspice

• Wash cucumbers; cut ⅛" slice off both ends. Cut crosswise in ⅛ to ¼" slices, making about 4 qts.
• Combine 4 c. vinegar, ¼ c. sugar, salt and mustard seeds in 6-qt. kettle; bring to boil. Add cucumber slices. Bring to boil; simmer until all slices turn slightly yellow, stirring often, about 8 minutes. *Do not overcook.* Drain well.
• Combine 2 c. vinegar, 3 c. sugar, celery seeds and allspice in 2-qt. saucepan. Bring mixture to boiling.
• Pack cucumber slices into 5 sterilized hot pint jars. Pour boiling liquid over cucumbers, filling to within ¼" of jar top. Wipe jar rim; adjust lids.
• Process in boiling water bath 15 minutes. Start to count processing

time when water in canner returns to boiling. Remove jars and complete seals unless closures are self-sealing type. Makes 5 pints.

CRISP-AS-ICE CUCUMBER SLICES

Always gets compliments. Crisp, good in sandwiches or with roast beef

4 lbs. 3–4" pickling cucumbers
8 small onions (1½" diameter), thinly sliced
2 green peppers, seeded and cut in strips
½ c. canning/pickling salt
Ice cubes
4 c. sugar
4½ c. 5% acid strength cider vinegar
1½ tsp. ground turmeric
½ tsp. ground cloves
3½ tsp. mustard seeds

• Wash cucumbers; cut ⅛" slice off both ends. Cut crosswise in ⅛ to ¼" slices, making about 4 qts. Layer cucumbers, onion, green pepper and salt in 2-gal. crock or glass bowl. Cover with ice cubes. Let stand 3 hours in refrigerator, adding more ice if needed. Drain well.
• Combine sugar, vinegar, turmeric, cloves and mustard seeds in 6-qt. kettle; bring to boil. Add vegetables. Heat over low heat to scalding (do not boil). Stir mixture frequently so it heats evenly.
• Ladle into 5 sterilized hot pint jars, filling to within ¼" of jar top. Wipe jar rim; adjust lids.
• Process in boiling water bath 5 min-

It's easy to mix and bake these No-Knead Batter Breads for your freezer. Heat them before serving—the aroma will make everyone hungry. After reheating, drizzle icing on the Apple Crumb Kuchen. Recipes in Chapter 5.

Select the reddest crab apples you can find to make Rosy Crab Apple Pie. This old-fashioned dessert will have rich color, luscious flavor. Freeze it baked or unbaked. Recipe in Chapter 6.

Cool as snowy mountain peaks, these creamy dessert squares will win compliments on summer evenings. Recipes for Velvety Lime Squares, Vanilla Almond Crunch and Raspberry Swirl in Chapter 4.

You can make your family happy all winter long by filling jars with luscious jams, jellies and preserves made from sun-ripened fruits. Taste-testers selected recipes in Chapter 15 from thousands of farm favorites.

utes. Start to count processing time when water in canner returns to boiling. Remove jars and complete seals unless closures are self-sealing types. Makes 5 pints.

OLIVE OIL PICKLES

If you like salty pickles, try these

24 small (1 to 1½" diameter)
    pickling cucumbers (7 lbs.)
1 c. canning/pickling salt
4 c. 5% acid strength cider
    vinegar
½ c. olive oil
¼ c. sugar
¼ c. mustard seeds
¼ c. celery seeds

• Wash cucumbers, peel and cut crosswise in ⅛ to ¼" slices. Layer cucumbers and salt in 1-gal. crock or glass bowl. Cover. Let stand 12–18 hours at room temperature. Drain. Rinse well with cold water.

• Combine vinegar, olive oil, sugar and spices; boil 1 minute. Add cucumbers; simmer until cucumbers change color, about 2 minutes.

• Pack cucumbers into 5 sterilized hot pint jars. Cover with hot syrup, filling to within ¼" of jar top. Wipe jar rim; adjust lids.

• Process in boiling water bath 5 minutes. Start to count processing time when water in canner returns to boiling. Remove jars and complete seals unless closures are self-sealing type. Makes 5 pints.

Note: Usually you wipe jar rims with a moist paper towel. Because of oil in this syrup, wipe sealing edge of jar with a clean *dry* paper towel.

CURRY PICKLES

If you like curry, you'll love this

16 (5") pickling cucumbers (3 lbs.)
10 medium onions (1½ lbs.)
2 large sweet red peppers
½ c. canning/pickling salt
Cold water
4 c. sugar
2 c. 5% acid strength cider
    vinegar
1 c. water
3 tblsp. pickling spices
2 tblsp. celery salt
1 tsp. curry powder

• Pare cucumbers; cut ⅛" slice off both ends. Peel onions; seed peppers; cut all vegetables in thin slices; layer with salt in 1-gal. crock or glass bowl. Add enough cold water to cover. Let stand overnight, covered, in cool place. Drain well.

• Combine sugar, vinegar, 1 c. water, pickling spices (tied in cheesecloth bag), celery salt and vegetables in 4-qt. kettle. Bring to boil; cover; boil 15 minutes or until cucumbers are translucent. Stir in curry powder.

• Ladle into 6 sterilized hot pint jars, filling to within ¼" of jar top. Wipe jar rim; adjust lids.

• Process in boiling water bath 5 minutes. Start to count processing time when water in canner returns to boiling. Remove jars and complete seals unless closures are self-sealing type. Makes 6 pints.

BEST-EVER
BREAD-AND-BUTTERS

The ice cubes help keep the thin cucumber slices from breaking

32 (4") pickling cucumbers
   (6 lbs.)
½ c. canning/pickling salt
Ice cubes
1 qt. 5% acid strength vinegar
4 c. sugar
2 tblsp. mustard seeds
1 tblsp. celery seeds
1 tblsp. ground ginger
1 tsp. ground turmeric
½ tsp. white pepper
2 qts. sliced onions (about 3 lbs.)

· Wash cucumbers; cut ⅛" slice off both ends. Slice crosswise, no more than ⅛" thick (the thinner the better), making about 4 qts. Layer cucumber slices and salt in 1-gal. crock. Cover with ice cubes. Let stand 3 hours in refrigerator or until cucumbers are crisp and cold. Add more ice if needed. Drain well.
· Combine vinegar, sugar, mustard seeds, celery seeds, ginger, turmeric and pepper in 8-qt. kettle. Bring mixture to boil; boil 10 minutes. Add cucumbers and onions; bring back to a rolling boil.
· Immediately pack into 8 hot pint jars, filling to within ¼" of jar top. Wipe jar rim; adjust lids.
· Process in boiling water bath 15 minutes. Start to count processing time when water in canner returns to boiling. Remove jars and complete seals unless closures are self-sealing type. Makes 8 pints.

CELERY/CUCUMBER PICKLES

Wonderful with cold cuts

3 lbs. large green pickling
   cucumbers
Ice water
9 branches celery

2 medium (2") onions, peeled
   and sliced
4 c. 5% acid strength cider
   vinegar
1 c. sugar
½ c. water
⅓ c. canning/pickling salt

· Wash cucumbers; cut 1/16" slice off blossom end. Soak in ice water in 1-gal. crock for 5 hours in refrigerator.
· Cut each cucumber lengthwise into 4 to 6 strips. Pack closely in 3 sterilized hot quart jars, adding 3 branches celery and 3 slices onion to each jar.
· Combine vinegar, sugar, water and salt in 2-qt. saucepan. Bring to rolling boil. Pour over cucumbers, filling to within ¼" of jar top. Wipe jar rim; adjust lids.
· Process in boiling water bath 5 minutes. Start to count processing time when water in canner returns to boiling. Remove jars and complete seals unless closures are self-sealing type. Makes 3 quarts.

SWEET DILLS

Crisp dill with sweet-sour flavor

4 lbs. (3–5") pickling cucumbers
3 medium (2") onions, peeled
   and thinly sliced
16 heads fresh dill
6 c. 5% acid strength cider
   vinegar
6 c. sugar
6 tblsp. canning/pickling salt
1½ tsp. celery seeds
1½ tsp. mustard seeds

· Wash cucumbers; cut ⅛" slice off each end. Cut cucumbers crosswise in ⅛" to ¼" slices.

• Place 2 slices onion and 1 dill head in each of 8 sterilized hot pint jars. Pack raw cucumber slices. Place 1 slice onion and 1 dill head on top.
• Combine vinegar, sugar, salt, celery seeds and mustard seeds in 4-qt. kettle. Bring to boil. Pour over cucumbers, filling to within ¼" of jar top. Wipe jar rim; adjust lids.
• Process in boiling water bath 10 minutes. Start to count processing time when water in canner returns to boiling. Remove jars and complete seals unless closures are self-sealing type. Makes 8 pints.

EASY SWEET DILLS

Use some of your dill pickles to make these on a leisurely winter day

2   qts. sliced processed dill
        pickles
4   cloves garlic, crushed

2⅓ c. 5% acid strength cider
        vinegar
2   tblsp. whole allspice
1   tblsp. whole peppercorns
4   c. sugar
1   c. brown sugar, firmly packed

• Drain dill pickles.
• Combine garlic, vinegar, spices and sugar in 4-qt. kettle. Bring to boil; simmer 5 minutes. Add pickles. Bring to boil.
• Pack cucumbers into 4 sterilized hot pint jars; cover with syrup, filling to within ¼" of jar top. Wipe jar rim; adjust lids.
• Process in boiling water bath 5 minutes. Start to count processing time when water in canner returns to boiling. Remove jars and complete seals unless closures are self-sealing type. Makes 4 pints.

Yellow-Ripe Cucumber Pickles

Grandmother knew what to do with those big cucumbers that ripened under the leaves before she saw them. She turned them into relishes to please her family and friends. She made firm, spicy, transparent pickles with them. And many old-fashioned country cooks considered them perfect for chicken dinners—along with fluffy mashed potatoes and gravy.

Today's farm cooks treasure their grandmothers' recipes. Every summer they fill jars with tempting pickles to use in the frosty days ahead. Here are a few recipes they prize.

YELLOW CUCUMBER PICKLES

Chunks of vegetables make these special—perfect with baked beans

12   lbs. yellow pickling
        cucumbers (12 large)
4½   lbs. onions, peeled
12   branches celery
½   c. canning/pickling salt
3   c. sugar
4   c. 5% acid strength cider
        vinegar
2   tblsp. mustard seeds
1   c. prepared mustard

• Peel cucumbers; cut in half lengthwise; scrape out seeds and soft centers with a spoon. Cut cucumbers, onions and celery in small ¼" cubes. Layer vegetables and salt in 2-gal. crock. Mix well. Let stand 2 hours. Drain and rinse.

• In 8-qt. kettle, combine sugar, vinegar, mustard seeds and prepared mustard. Blend with wire whisk. Bring to boil; add vegetables; return to boil, stirring constantly.

• Ladle immediately into 13 sterilized hot pint jars, filling to within ¼" of jar top. Wipe jar rim; adjust lids.

• Process in boiling water bath 5 minutes. Start to count processing time when water in canner returns to boiling. Remove jars and complete seals unless closures are self-sealing type. Makes 13 pints.

SPICY RIPE CUCUMBER PICKLES

Long, slow cooking retains cucumbers' delightful crispness

12 lbs. yellow pickling
       cucumbers (12 large)
½ c. canning/pickling salt
Water
5 c. sugar
2½ c. 5% acid strength vinegar
2 tblsp. whole cloves
4 sticks cinnamon

• Peel cucumbers, cut in half lengthwise; scrape out seeds and soft centers with spoon. Cut into 1–2" squares. Layer into 2-gal. crock with salt. Cover. Let stand 12–18 hours at room temperature. Drain. Place into 8-qt. kettle.

• Cover with 3 qts. cold water. Gradually heat to boiling. Cover. Reduce heat so cucumber slices do not boil but stay steaming hot (about 180° F.) for 2 hours. Drain. Chill in ice water for 2 hours. Drain.

• In 3-qt. saucepan, combine sugar, vinegar and spices (tied in cheesecloth bag). Bring to boil. Pour over cucumbers. Bring cucumbers to boil; simmer, uncovered, for 15 minutes. Remove spice bag.

• Pour into 1-gal. crock or glass bowl. Cover. Let stand in refrigerator for 12–18 hours. Drain syrup into 3-qt. saucepan; bring to boil.

• Pack cucumber slices into 8 sterilized hot pint jars; cover with syrup, filling to within ¼" of jar top. Wipe jar rim; adjust lids.

• Process in boiling water bath 5 minutes. Start to count processing time when water in canner returns to boiling. Remove jars and complete seals unless closures are self-sealing type. Makes 8 pints.

COLORING PICKLES

While all the pickles look fine in their natural color, country cooks sometimes like to color them red or green with food coloring, to look festive on Christmas tables, or for gifts. But now red dye No. 2 is banned. In retesting recipes, we used the only red food coloring now permitted. The color is not as pretty. It tends to fade and it made our pickles look rusty instead of red. We have left mention of it in the recipes, but have marked it "optional." You may prefer to leave your pickles natural color—or use green food coloring if you want some to look Christmasy.

CHEERFUL SWEET PICKLES

These will cheer up your family on a wintry day

9 lbs. yellow pickling
    cucumbers (9 large)
½ c. canning/pickling salt
3½ qts. water
3½ c. sugar
2 c. 5% acid strength cider
    or white vinegar
1 tsp. whole cloves
2 sticks cinnamon
1 (4 oz.) jar maraschino cherries
2 tsp. red food color (optional)

• Peel cucumbers; cut in half lengthwise and scrape out seeds and soft centers with a spoon. Cut cucumbers into ½" cubes; layer with salt in 2-gal. crock. Cover with water. Cover. Let stand 12–18 hours in cool place.
• Pour cucumbers and liquid into 10-qt. kettle. Heat to boiling; drain well. Return to crock.
• In 2-qt. saucepan, combine sugar, vinegar and spices (tied in cheesecloth bag). Bring to boil; pour over cucumbers in crock. Cover. Let stand 12–18 hours.
• Pour cucumbers, spice bag and syrup into 5-qt. kettle. Simmer until cucumbers are tender when pricked with fork, about 15 minutes. *Do not overcook.* Remove spice bag. Add undrained cherries and food color if desired.
• Ladle into 6 sterilized hot pint jars, filling to within ¼" of jar top. Wipe jar rim; adjust lids.
• Process in boiling water bath 5 minutes. Start to count processing time when water in canner returns to boiling. Remove jars and complete seals unless closures are self-sealing type. Makes 6 pints.

GOLDEN CUCUMBER RELISH

You couldn't serve your company a relish that would get more praise

12 lbs. yellow pickling cucumbers
    (12 large)
2 lbs. onions, peeled (12 medium)
¾ c. canning/pickling salt
2 branches celery
3 sweet red peppers
3 green peppers
4 c. 5% acid strength cider
    vinegar
4 c. sugar
2 tblsp. mustard seeds

• Peel cucumbers, slice lengthwise; scrape out seeds and soft centers with a spoon. Dice cucumbers and onions into ¼" cubes. Place in 1-gal. crock. Add salt; mix well. Cover. Let stand 10–12 hours at room temperature. Drain well. Dice celery and peppers in ¼" cubes.
• Combine vinegar, sugar and mustard seeds in 8-qt. kettle; add vegetables and boil gently, uncovered, over low heat, stirring often, until celery, onions and cucumbers are translucent, about 30 minutes.
• Ladle into 8 hot pint jars; cover with syrup, filling to within ¼" of jar top. Wipe jar rim; adjust lids.
• Process in boiling water bath 15 minutes. Start to count processing time when water in canner returns to boiling. Remove jars and complete seals unless closures are self-sealing type. Makes 8 pints.

CHRISTMAS PICKLES

Makes a splendid holiday gift

**10 lbs. yellow pickling cucumbers
(10 large)**
½ c. canning/pickling salt
**3 large red sweet peppers,
seeded and cut in strips**
**3 large green peppers, seeded
and cut in strips**
4 c. diced celery
1⅓ c. diced onions (2 medium)
5 c. sugar
4 c. 5% acid strength vinegar
2 tsp. mustard seeds
1 c. water

• Peel cucumbers and dice in ½"
cubes into 1-gal. crock (do not remove seeds). Sprinkle uniformly with
salt. Cover. Let stand 12–18 hours
in cool place. Drain well. Rinse.

. Place in 8-qt. kettle. Add peppers,
celery, onions, sugar, vinegar, mustard
seeds and water. Heat to boiling; cook
slowly, uncovered, stirring occasionally until cucumbers are translucent, about 30 minutes. *Do not overcook.*

• Ladle into 8 hot pint jars; cover
with liquid, filling to within ¼" of jar
top. Wipe jar rim; adjust lids.

• Process in boiling water bath 15
minutes. Start to count processing
time when water in canner returns to
boiling. Remove jars and complete
seals unless closures are self-sealing
type. Makes 8 pints.

Cured Pickles (*8 to 14 Days*)

Investing a week or two in curing cucumbers before you pack them in jars
results in superior eating. The pickle
maker who shares her recipe for
Whole Sweet Pickles considers it the
best one she ever used to make small
sweet pickles. Other homemakers sent
this recipe with words of praise. But
some excellent cooks had failures
making pickles this way and they
wonder why.

There is no "trick" to success, according to the expert who retested this
recipe—it's purely a matter of *daily
care.* She suggests that you pretend
you are a pediatrician and treat your
cukes like babies. Make up a chart for
their daily care, and use the most sanitary practices. Wash and scald the
crock, plate and weight each time you
change the water. If you tend the babies on schedule, she says, every 24
hours, you'll succeed with this "pas-
teurization" cure. But if you are care-
less and forget to change the water
even one day, you are in trouble!

These crisp pickles are well worth
your attentiveness. You may want to
try the boiling water cure on your
dills, too.

WHOLE SWEET PICKLES

*Be sure to use freshly picked cucum-
bers for these party pickles*

**1 gal. small (2") pickling
cucumbers**
Boiling water
⅓ c. canning/pickling salt
1½ gals. water

4 c. 5% acid strength cider
   vinegar
1 c. water
8 c. sugar
¼ c. mixed pickling spices

• Small cucumbers tend to be hairy and sometimes gritty; wash them well, scrubbing gently with a soft brush. Trim ¹⁄₁₆″ off blossom ends.
• Wash and *scald* a 2-gal. crock, a heavy plate (to cover cucumbers) and a glass jar filled with hot water (to serve as a weight.)
• Place cucumbers into crock; cover them with boiling water, add the scalded plate and weight and cover all with a clean terry towel; tie with twine. Count this as Day 1.
• On the next 5 days (same time each day), drain water from crock, remove cucumbers, wash and *scald* crock, plate and weight. Rinse cucumbers and return to crock; cover with boiling water, add plate and weight. Re-cover with towel.
• On the 7th day, make brine by dissolving salt in 1½ gals. water. Drain cucumbers; wash and *scald* crock, plate and weight. Return cucumbers to crock; cover with brine; add plate and weight. Re-cover with towel.
• On 8th day, drain cucumbers. Pierce each in several places with a fork to prevent shriveling when syrup is added.
• Combine vinegar, 1 c. water, sugar and spices (tied in a cheesecloth bag) in a 4-qt. kettle. Boil 20 minutes, uncovered. Remove spice bag.
• Pack cucumbers into 8 sterilized hot pint jars. Cover with syrup, filling to within ¼″ of jar top. Wipe jar rim; adjust lids.
• Process in boiling water bath 5 min-

utes. Start to count processing time when water in canner returns to boiling. Remove jars and complete seals unless closures are self-sealing type. Makes 8 pints.

CRISP PICKLES

These pickles are so great, everyone will want to know your secret. It isn't magic—it's the daily care!

4 lbs. (4–5″) pickling cucumbers
   (about 20)
Boiling water
8 c. sugar
4 c. 5% acid strength cider
   vinegar
5 tblsp. canning/pickling salt
2 tblsp. mixed pickling spices
2 tsp. celery seeds

• Wash cucumbers well and cut ¹⁄₁₆″ slice off blossom ends. Wash and *scald* a 1-gal. crock, a heavy plate (to cover cucumbers) and a glass jar filled with hot water (to serve as weight).
• Place cucumbers into crock; cover them with boiling water, add the scalded plate and weight and cover all with a clean terry towel; tie with twine. Count this as Day 1.
• On the next 5 days (same time each day), drain water from crock, remove cucumbers, wash and *scald* crock, plate and weight. Rinse cucumbers and return to crock; cover with fresh boiling water, add plate and weight. Re-cover with towel.
• On the 7th day, drain cucumbers and cut crosswise in ¼″ to ½″ slices. Return to *scalded* crock.
• Combine sugar, vinegar, salt and spices (tied in cheesecloth bag) in 3-qt. saucepan. Bring to rolling boil.

Pour over cucumbers. Add *scalded* plate and weight. Re-cover.
- On the 8th day, drain syrup with spice bag into 3-qt. saucepan. Lift cucumbers out of crock. Wash and *scald* crock and weights. Return cucumbers to crock. Bring syrup to rolling boil; pour over cucumbers.

Add weights and re-cover.
- On the 9th day, repeat.
- On the 10th day, repeat.
- On the 11th day, repeat.
- On the 12th day, drain syrup into 3-qt. saucepan. Pack cucumbers into 5 sterilized hot pint jars. Bring syrup to boil, remove spice bag; pour syrup over cucumber slices, filling to within ¼" of jar top. Wipe jar rim; adjust lids.
- Process in boiling water bath 5 minutes. Start to count processing time when water in canner returns to boiling. Remove jars and complete seals unless closures are self-sealing type. Makes 5 pints.

EXPERT'S SWEET PICKLES

Crisp, sweet-tart and less work to make than you think

8	lbs. (3–3½") pickling cucumbers (about 75)
1½	c. canning/pickling salt
Water	
10	c. sugar
5	c. 5% acid strength cider vinegar
18	drops oil of cinnamon (or 3 sticks cinnamon)
18	drops oil of cloves (or 2 tblsp. whole cloves)
2	tblsp. celery seeds

- Wash cucumbers; cut 1/16" slice off blossom ends. Place in 3-gal. *scalded* crock.
- Combine salt and 1 gal. water in 6-qt. kettle; bring to boil. Pour over cucumbers. To keep cucumbers submerged in brine, cover them with a *scalded* heavy plate that fits inside the crock. Put a weight on top of the plate; a glass jar filled with hot water, capped and *scalded*, makes a good weight. Brine should cover plate by 2" for easy skimming; if necessary, make additional brine (3 tblsp. salt to 1 qt. water) and pour over cucumbers. Cover crock with a clean terry towel; tie with twine.
- Let stand 7 days at room temperature; EACH DAY, uncover crock and remove any yeast film or mold with scalded stainless steel spoon. If mold forms on plate or weight, wash and *scald* them. Re-cover with clean towel.
- On 8th day, drain; empty cucumbers into sink. Wash thoroughly with cold water. Cut lengthwise or into chunks and return to freshly washed and *scalded* crock. Bring 1 gal. fresh water to boil; pour over cucumbers. Weight with *scalded* plate and jar. Re-cover with clean towel.
- On 9th day, drain. Remove cucumbers from crock; wash and scald crock. Bring 1 gal. fresh water to boil; pour over cucumbers. Weight with scalded plate and jar; re-cover.
- On 10th day, repeat.
- On 11th day, drain; remove cucumbers. Wash and *scald* crock, plate and jar. Return cucumbers to crock. Combine sugar, vinegar, cinnamon and cloves (if using whole spices, tie in cheesecloth bag) and celery seeds in 4-qt. kettle. Bring to boil; pour over

cucumbers. Weight with plate and glass jar. Re-cover with clean towel.

• On 12th day, drain syrup into 4-qt. kettle; bring to boil; pour over cucumbers. Weight with plate and glass jar; re-cover.

• On 13th day, repeat.

• On 14th day, drain syrup into 4-qt. kettle. Pack cucumbers into 12 sterilized hot pint jars. Bring syrup to boil. Pour over cucumbers, filling to within ¼" of jar top. Wipe jar rim; adjust lids.

• Process in boiling water bath 5 minutes. Start to count processing time when water in canner returns to boiling. Remove jars and complete seals unless closures are self-sealing type. Makes 12 pints.

14-DAY SWEET PICKLES

Adaptation of an heirloom recipe long treasured in country kitchens

3½ qts. (2") pickling cucumbers
¾ c. canning/pickling salt
Water
6⅔ c. 5% acid strength cider
        vinegar
4 c. sugar
2 tsp. celery seeds
5 sticks cinnamon
2 c. sugar

• Wash cucumbers; cut ¹⁄₁₆" slice off blossom ends. Cut in lengthwise halves or leave whole. Place in 2-gal. *scalded* crock.

• Combine salt and 2 qts. water in 3-qt. saucepan; bring to rolling boil. Pour over cucumbers. To keep cucumbers submerged in brine, cover them with a *scalded* heavy plate that fits inside the crock. Put a weight on top of the plate; a glass jar filled with hot water, capped and *scalded,* makes a good weight. Brine should cover plate by 2" for easy skimming; if necessary, make additional brine (3 tblsp. salt to 1 qt. water) and pour over cucumbers. Cover crock with a clean terry towel; tie with twine.

Let stand 7 days at room temperature; EACH DAY, uncover crock and remove any yeast film or mold with scalded stainless steel spoon. If mold forms on plate or weight, wash and *scald* them. Re-cover with clean towel.

• On 8th day, drain; remove cucumbers. Wash and *scald* crock, plate and weight. Rinse cucumbers; return to crock. Pour 2 qts. fresh boiling water over cucumbers. Add plate and weight. Re-cover.

• On 9th day, repeat.

• On 10th day, repeat.

• On 11th day, drain; remove cucumbers. Wash and *scald* crock, plate and jar. If cucumbers are whole, prick each in several places with a fork to prevent shriveling when syrup is added. Return cucumbers to crock. Combine vinegar, 4 c. sugar, celery seeds and cinnamon in 3-qt. saucepan. Bring to rolling boil; pour over cucumbers. Weight with plate and jar. Re-cover with clean towel.

• On 12th day, drain syrup into 3-qt. saucepan, adding ⅔ c. sugar; bring to a boil and pour over cucumbers; replace plate and jar. Re-cover.

• On 13th day, repeat, adding ⅔ c. sugar.

• On 14th day, drain syrup into 3-qt. saucepan, add remaining ⅔ c. sugar. Pack cucumbers into 6 sterilized hot pint jars. Bring syrup to boil, remove cinnamon sticks. Pour over cucumbers, filling to within ¼" of jar top. Wipe jar rim; adjust lids.

• Process in boiling water bath 5 minutes. Start to count processing time when water in canner returns to boiling. Remove jars and complete seals unless closures are self-sealing type. Makes 6 pints.

CINNAMON CUCUMBER RINGS

Prepare these candied pickles for a holiday treat—children love them

12 lbs. (5–6") green pickling
    cucumbers (not more than
    2" in diameter)
 2 c. canning/pickling salt
 8 qts. water
 1 c. 5% acid strength cider
 1 tsp. green food color (optional)
Water
 6 c. 5% acid strength cider
    vinegar
 2 c. water
12 c. sugar
 4 sticks cinnamon

Wash cucumbers; slice ⅛" off each end. Cut crosswise in ½" slices. With melon baller, take out seeds so slices become rings. (You should have 8 qts.) Place in 3 gal. *scalded* crock.

• Dissolve salt in 8 qts. cold water, pour over cucumbers. To keep cucumbers submerged in brine, cover them with a *scalded* heavy plate that fits inside the crock. Put a weight on top of the plate; a glass jar filled with water, capped and *scalded* makes a good weight. Brine should cover plate by 2" for easy skimming; if necessary, make additional brine using original proportions and pour over cucumbers.

Cover crock with a clean terry towel; tie with twine.

• Let stand 5 days at room temperature; EACH DAY, uncover crock and remove any yeast film or mold with scalded stainless steel spoon. If mold forms on plate and weight, wash and *scald* them.

• On 6th day, drain. Combine 1 c. vinegar and food color in 10-qt. kettle. Add cucumber rings and water to cover. Gradually bring to boil; cover; reduce temperature so rings do not boil, but stay steaming hot (about 180° F.) for 2 hours. Drain. Place rings in 1-gal. *scalded* crock.

• In 3-qt. saucepan, combine 6 c. vinegar, 2 c. water, sugar and cinnamon. Bring to boil and pour over rings. Weight with *scalded* plate and jar. Cover with towel. Let stand 12–18 hours at room temperature.

• On 7th and 8th days, drain syrup into 3-qt. saucepan. Heat to boiling; pour over rings. Weight and re-cover.

• On 9th day, drain syrup into 3-qt. saucepan. Pack rings into 9 sterilized hot pint jars. Bring syrup to rolling boil; remove cinnamon sticks; pour syrup over rings, filling to within ¼" of jar top. Wipe jar rim; adjust lids.

• Process in boiling water bath 5 minutes. Start to count processing time when water in canner returns to boiling. Remove jars and complete seals unless closures are self-sealing type. Makes 9 pints.

ICICLE PICKLES

The color of these sweet pickles is exceptionally good—so is the taste

 8 lbs. (4–5") pickling cucumbers
    (4 qts. sliced)

2   c. canning/pickling salt
Boiling water
2½ qts. 5% acid strength cider
      vinegar
5   lbs. sugar (10 c.)
2   tblsp. whole allspice

• Wash cucumbers; cut ⅟₁₆″ slice off blossom end. Cut in quarters lengthwise. Place in *scalded* 2-gal. crock. Dissolve salt in boiling water; pour over cucumbers. Place a heavy, *scalded* plate over cucumbers; it should fit snugly inside crock. Weight with *scalded*, capped jar filled with hot water. Cover crock with terry cloth towel; tie with twine.
• EACH DAY for 7 days skim brine with *scalded* stainless steel spoon to remove yeast or mold film. Replace towel with clean one.
• On 8th day, drain brine and remove slices from crock. Wash and *scald* crock and weights. Return cucumbers to crock. Cover with fresh boiling water. Weight down. Cover. Let stand 24 hours.
• On 9th day, drain. Repeat as for 8th day.
• On 10th day, drain. Remove cucumbers. Wash and *scald* crock and weights. Return slices to crock. Make syrup by combining vinegar, 2½ c. sugar and allspice in 4-qt. kettle. Boil 20 minutes. Pour over slices. Weight down. Cover. Let stand 24 hours.
• On 11th day, drain syrup, add 2½ c. sugar, bring to boil. Pour back over slices. Weight down with clean weights. Cover. Let stand 24 hours.
• On 12th day, repeat as for 11th day.
• On 13th day, repeat as for 12th day.
• On 14th day, drain syrup, heat to boiling. Pack slices into 5 sterilized hot pint jars; cover with syrup, filling to within ¼″ of jar top. Wipe jar rim; adjust lids.
Process in boiling water bath 15 minutes. Start to count processing time when water in canner returns to boiling. Remove jars and complete seals unless closures are self-sealing type. Makes 5 pints.

FERMENTED DILL PICKLES

Unsurpassable olive-green dills

20   lbs. (3½–5½″) pickling
       cucumbers
½   c. mixed pickling spices
2   to 3 bunches fresh dill
2   c. 5% acid strength vinegar
1½ c. canning/pickling salt
2   gals. water
20   cloves garlic, peeled

• Wash cucumbers, handling gently to avoid bruising. Cut ⅟₁₆″ off blossom ends. Drain on rack or wipe dry.
• Place half the pickling spices and a layer of dill in a 5-gal. *scalded* crock or stone jar. Fill with cucumbers to 3 or 4″ from top of crock. Mix vinegar, salt and 2 gals. water and pour over cucumbers. Place a layer of dill and remaining spices over top of cucumbers. Add 10 cloves garlic.
• To keep cucumbers submerged in brine, cover them with a *scalded* heavy plate that fits inside the crock. Put a weight on top of the plate; a glass jar filled with water, capped and *scalded,* makes a good weight. Brine should cover plate by 2″ for easy skimming; if necessary make additional brine using original proportions and pour over cucumbers. Cover crock with a clean terry towel; tie with twine.

· Keep pickles at room temperature. EACH DAY, uncover crock and remove any yeast film or mold with *scalded* stainless steel spoon. Do not stir pickles around in jar but be sure they are completely covered with 2" of brine. If necessary, make additional brine, using original proportions. Wash and scald plate and weight if mold forms on them. Change towel weekly.

· In about 3 weeks the cucumbers become olive-green and any white spots inside the fermented pickles will be eliminated by processing.

· Drain cucumbers. Strain and reserve brine in 10-qt. kettle. Bring to boil. Pack pickles, along with some of the dill, into 10 sterilized hot quart jars.

Add 1 clove garlic to each jar. Pour boiling brine over cucumbers, filling to within ¼" of jar top. Wipe jar rim; adjust lids.

· Process in boiling water bath 10 minutes. Start to count processing time when water in canner returns to boiling. Remove jars and complete seals unless closures are self-sealing type. Makes 10 quarts.

Note: Garlic gives dill pickles the flavor of those sold in delicatessens. The original brine is usually cloudy as a result of yeast development during fermentation, but this cloudiness will settle in the jar. Fermentation brine is generally preferred for added flavor; it is safe to use.

Brine-cured Pickles (*The Long Cure*)

You're less likely to have problems with pickles spoiling if you know what happens when you submerge cucumbers in brine for the long cure.

The salt in the brine draws juices from the cucumbers—juices which contain fermentable sugars. The sugars are changed to lactic acid by natural bacterial action, the process we call fermentation. It is the lactic acid thus produced that gradually changes cucumbers into pickles and gives them their desirable flavor. (In quick-process pickles, vinegar is the acid that does the pickling.) During this fermentation or curing period, which takes from 4 to 6 weeks, the salt also helps prevent spoilage.

The amount of salt used to start this process and keep it going is criti-

cal. If the brine is too salty, it draws juices from the cucumbers too fast— pickles shrivel. Also, too much salt inhibits the lactic acid bacteria. On the other hand, a brine that is too weak permits yeasts to crowd out the lactic acid bacteria—pickles may be bloated, hollow and poorly flavored. Like Goldilocks' porridge, the brine must be *just right*.

The directions that follow tell you how to mix a 10% brine solution and how to maintain it. As the cucumbers cure, they release moisture which dilutes the brine; small amounts of salt must be added at regular intervals to maintain brine strength. Weigh your cucumbers before you put them in cure and write it down. The amount of salt you add from time to time is based on original weight of cucumbers.

While the cucumbers are curing, yeasts and molds will grow on the surface of the brine—you'll see this as a film or scum. If these organisms take over, the lactic acid bacteria lose out and cucumbers will spoil. To prevent this, skim the brine *every day*. The directions that follow are detailed, but the daily procedure will quickly become routine. And your pickles will be crisp and well-flavored.

BRINE-CURED PICKLES

Cucumbers should be a pickling variety, freshly picked, slightly immature. Make brine with soft or softened water (see directions for softening water at beginning of this chapter), and use pure fine-granulated canning/pickling salt.

The ideal room temperature for fermentation is 80° to 85° F.

1. Wash cucumbers carefully, taking care not to break the skin. Cut $\frac{1}{16}$" slice off blossom end.

2. Weigh cucumbers; write down their weight. Put them in a clean, *scalded* crock or glass container. Cover with brine (1 c. salt to 2 qts. water).

3. To keep cucumbers submerged in brine, cover them with a *scalded* heavy plate that fits inside the crock. Put a weight on top of the plate; a glass jar filled with water, capped and *scalded*, makes a good weight. Brine should cover plate by 2" for easy skimming. Cover crock with a clean terry cloth towel; tie with twine. Keep crock at room temperature of 80° to 85° F.

4. Next day, add 1 c. pickling salt for each 5 pounds of cucumbers. Add salt on top of the plate to prevent its going to the bottom and forming too strong a brine there.

5. At the end of the week (7th day) and once a week for 4 or 5 succeeding weeks, add ¼ c. pickling salt for each 5 pounds of cucumbers. Add salt as before, on top of plate.

6. EACH DAY, uncover crock and remove any yeast or mold film which has formed on top of brine—skim with a *scalded* stainless steel spoon. If mold forms on plate and weight, wash and *scald* them. Re-cover with towel; change towel for a clean one each week.

7. Fermentation (bubbling) will continue for 4 weeks or so. Test for bubbles by tapping container on the side with your hand. When bubbling ceases, the cucumbers are cured. Add no more salt. (For another check on whether fermentation is complete, cut a cucumber in half. If it looks translucent, if color is even with no rings or white spots, cucumber is cured.)

8. Cucumbers may be stored in this brine solution until they're made into pickles. Pour a ¼" layer of melted paraffin on top of cucumbers; cover with a lid and store in a cool, dark place (60° F. or lower).

Note: Fresh cucumbers may be added to the crock during the first day or two of the curing process. Weigh them; write down weight. Add enough brine to cover, making brine as directed in Step 2. The next day, add 1 c. pickling salt for each 5 lbs. of additional cucumbers, pouring salt on top of plate as directed in Step 4. Thereafter, consider total weight of cucumbers in crock and follow directions in Step 5 for adding salt each week to maintain a 10% brine solution.

DESALTING CURED CUCUMBERS

Before you make them into pickles, brine-cured cucumbers must be desalted.

• Drain cucumbers. Put them in a crock and cover with hot water (180° F.) Let stand about 4 hours. Stir occasionally. Do not weight down.

• Lift cucumbers out of water. Empty and rinse out crock. Return cucumbers to crock; cover with hot water. Let stand 4 hours, stirring occasionally.

• Repeat above procedure again; let stand 4 hours. Empty and rinse out crock. Prick cucumbers in several places to prevent shriveling, using a table fork. Cover with a weak vinegar solution (1 part water and 3 parts cider vinegar). Let stand 12 hours.

• Test to see if sufficient salt has been removed; if not, let stand 12 hours longer. Drain. Immediately place the desalted cucumbers into pickling solution; recipes for Sour and Sweet Pickles follow. Both call for Spiced Vinegar, which you can get ready while your cucumbers are still curing.

SPICED VINEGAR

1 gal. 5% acid strength vinegar
3 tblsp. whole allspice
3 tblsp. whole cloves
1 stick cinnamon
1 piece mace

• Pour vinegar into a 5-qt. kettle. Tie spices in cheesecloth bag, add to vinegar. Bring to a boil; simmer 15 minutes. Cool, cover (or rebottle) and let stand 3 weeks before removing spice bag.

SWEET PICKLES

• Place desalted brined cucumbers into a clean, scalded crock or glass container.

• Heat Spiced Vinegar to boiling (1 pt. vinegar for each quart of pickles). Pour boiling liquid over cucumbers. Let stand overnight.

• Drain liquid and measure; place in kettle. Add ¾ c. sugar for each 2 c. liquid. Bring to boiling; pour over cucumbers. Let stand overnight.

• Drain cucumbers; reserve liquid. Add ½ c. sugar for each 2 c. liquid. Heat to boiling.

• Pack cucumbers into sterilized hot quart jars. Pour boiling liquid over cucumbers, filling to within ¼ " of jar top. Wipe jar rim; adjust lids.

• Process in boiling water bath 10 minutes. Start to count processing time when water in canner returns to boiling. Remove jars and complete seals unless closures are self-sealing type.

SOUR PICKLES

• For each 6 qts. desalted brined cucumbers, use 1 gal. Spiced Vinegar. Put it in a large kettle; stir in 2 c. sugar. (For less astringent pickle, use 4 c. sugar to 1 gal. vinegar.) Bring to boiling. Add ¼ of cucumbers at a time; boil 2 minutes; remove cucumbers. Repeat with remaining cucumbers.

· Pack cucumbers into hot quart jars. Bring Spiced Vinegar to boiling; pour over cucumbers, filling to within ¼″ of jar top. Wipe jar rim; adjust lids.

· Process in boiling water bath 10 minutes. Start to count processing time when water in canner returns to boiling. Remove jars and complete seals unless closures are self-sealing type.

Homemade Sauerkraut

Country cooks often make sauerkraut when there's a surplus of cabbage. They find its sharp flavor adds a pleasing change of pace in winter meals. And it's a real country convenience food—ready to serve cold in salads. Or try it as an appetizer, adding a dash of celery or caraway seeds and a few chunks of chilled pineapple. Many families prefer it hot alongside meats. They know that the sharpness of its flavor depends on how long it is cooked. For the most tang and greatest crispness, they only heat it. They cook it longer for a milder flavor.

CROCK SAUERKRAUT

Use pure fine-granulated canning/pickling salt to make sauerkraut and do measure the salt accurately—use a knife to level the tablespoon. The cabbage will not ferment properly if you add too much salt. To put down 50 lbs. of cabbage, you'll need a 10-gal. stoneware crock or glass container (or two 4-gal. crocks) and about 1½ c. salt.

Follow each step explicitly. You'll see the word *scalded* repeated several times. Absolute sanitation is as important for good sauerkraut as it is for crisp pickles.

1. Remove and discard outer leaves from firm, matured heads of cabbage (late cabbage is best as it is higher in sugar). Wash, drain, cut in halves or quarters. Remove and discard cores.

2. Shred 5 lbs. cabbage at a time with shredder or sharp knife. Shreds should be no thicker than a dime. Place in large mixing bowl.

3. Sprinkle 3 tblsp. canning/pickling salt over each 5 lbs. cabbage. Mix thoroughly with clean hands or stainless steel spoon.

4. Pack firmly and evenly into crock that has been washed with soapy water, rinsed, and *scalded*. Juices will form. Keep cabbage covered with juice as you pack by packing slowly and pressing cabbage down.

5. Repeat shredding and salting of cabbage in 5 lb. lots until crock is filled not more than 5″ from top.

6. Make sure juice covers cabbage. If not, making additional brine by mixing 1½ tblsp. salt in 1 qt. boiling water. Cool to room temperature before adding to crock.

7. Now the cabbage needs to be covered and weighted down, to keep it submerged in brine. Fit 1 large food-grade plastic bag inside another to make a double bag. Fill with brine solution (1½ tblsp. salt to 1 qt.

water) and lay over cabbage. Bag should fit snugly against inside of crock to seal surface from exposure to air; this will prevent growth of yeast film or molds. It also serves as a weight; the amount of brine in the bag can be adjusted to give just enough pressure to keep cabbage submerged, that is, covered with brine. Twist and tie to seal bag. (Bag is filled with brine as protection in case of leak or accidental spillage. If bag were filled with plain water, spillage would dilute brine in crock. Maintaining the proper brine solution is essential for fermentation.)

8. Cover crock with plastic food wrap and then with a heavy terry towel. Tie twine around crock to hold plastic wrap and towel in place. Do not open until fermentation time is completed.

9. Fermentation will begin the day following packing. How long it takes depends on room temperature. For best quality sauerkraut, a room temperature of 75° F. is ideal, and it will take about 3 weeks. At 70° F., allow about 4 weeks; at 65° F., about 5 weeks; and at 60° F., about 6 weeks. Temperature above 75° F. will result in earlier fermentation or possible spoilage.

10. Keep track of temperature so you know when to check kraut. Remove cover. Fermentation is complete if bubbling has stopped and no bubbles rise when crock is tapped gently.

The Old-fashioned Way: Instead of weighting the cabbage with brine-filled plastic bag (step 7), you can give sauerkraut *daily care* as follows:

Cover cabbage with a clean white cloth, tucking edges down between cabbage and inside of crock. Cover with a *scalded* heavy plate that fits snugly inside crock. For a weight fill clean glass jars with water; cap with lids and screw bands; scald jars before setting on plate. Use enough weight to bring brine 2" above plate —this makes daily skimming easier. Make additional brine (step 6) if necessary. Cover crock with clean heavy terry towel and top with plastic food wrap to help prevent evaporation. Tie with twine. EACH DAY, uncover crock, remove yeast film or mold with scalded stainless steel spoon. Have a second set of jar weights ready and scalded to replace weights when you've finished skimming. Cover again with clean terry towel and plastic food wrap.

KEEPING SAUERKRAUT

Sauerkraut can be stored in the refrigerator. Or you can keep it in a cold room with temperature of 55° F. or lower, if you'll be using it before winter ends. Or, it can be canned.

To Can Sauerkraut: Pour sauerkraut into large kettle; bring slowly to a boil, stirring constantly. Remove from heat. Pack sauerkraut into hot jars, pressing down as you pack to remove air bubbles. Cover kraut with juice to within ½" of jar top. (If more brine is needed, follow step 6—but do not cool.) Wipe rim of jar; adjust lids. Process in boiling water bath.

Pint jars   15 minutes
Quart jars   20 minutes
    Remove jars from canner and com

plete seals unless closures are self-sealing type.

Each quart jar holds about 2½ lbs. sauerkraut; 50 lbs. cabbage makes about 18 quarts.

Note: Double-check all canning jar lids used to can sauerkraut; make sure there is no scratch in the enamel lining.

GLASS JAR SAUERKRAUT

If you want to make small quantities of sauerkraut, mix shredded cabbage with salt following proportions in recipe for Crock Sauerkraut. Pack cabbage into glass jars and weight with brine-filled plastic bags, as in Step 7. When fermentation is complete, store covered jars in refrigerator.

SAUERKRAUT ANSWERS

Q. What makes sauerkraut turn pink?
A. The color is caused by certain kinds of yeast. They often grow when the salt is not evenly distributed, or when too much salt is used, and when the cabbage is too loosely packed in a crock or stone jar.

Q. What causes soft sauerkraut?
A. The salt content may be too low. High temperatures during curing also have a softening effect. Use of containers not cleaned thoroughly sometimes causes this, as does improper packing of cabbage which traps air.

Q. Why is sauerkraut sometimes slimy?
A. Certain bacteria are responsible and usually they flourish when not enough salt is used and the temperatures are too high.

Q. What makes sauerkraut darken at times?
A. Failure to clean the cabbage properly, uneven salting and a high curing temperature may be at fault.

Q. Is use of too much salt harmful?
A. Yes, it prevents fermentation.

Q. Can you prevent the formation of a white film?
A. The trick is to cover the sauerkraut and adjust the weights so very little kraut juice or kraut is exposed to the air. This treatment discourages the growth of this yeast.

Mixed Pickles Galore

Mixed pickles, like crazy quilts and hit-and-miss rag rugs, use up scraps—often the tag ends of garden vegetables. These pickles have a tremendous following in the country. Farm women praise the marvelous blended flavors of many vegetables and seasonings—they consider them a potpourri of summer's best.

It's a rare country fruit closet that does not have rows of jars filled with these sweet-sour-spicy pickles. We give our readers credit for the recipes.

CHOW-CHOW

Old-fashioned special that today's children like on hot dogs

1 medium head cabbage, cored
6 medium onions, peeled
   (1½ lbs.)
6 green peppers, seeded and
   stemmed
6 sweet red peppers, seeded
   and stemmed
4 c. green tomatoes,
   cored (2 lbs.)
¼ c. canning/pickling salt
2 tblsp. prepared mustard
6 c. 5% acid strength cider
   vinegar
2½ c. sugar
1½ tsp. ground turmeric
1 tsp. ground ginger
2 tblsp. mustard seeds
1 tblsp. mixed pickling spices

• Put vegetables through food chopper. Place in 1-gal. crock. Add salt and mix well. Cover; let stand 12–18 hours at room temperature. Drain.
• In 6-qt. kettle blend prepared mustard and vinegar with wire whisk; add sugar, turmeric, ginger and mustard seeds. Tie pickling spices in cheesecloth bag and add to kettle. Bring to boil; simmer, uncovered, for 20 minutes. Add vegetables; simmer, uncovered, 10 minutes. Remove spice bag.
• Ladle into 8 hot pint jars, filling to within ¼" of jar top. Wipe jar rim; adjust lids.
• Process in boiling water bath 15 minutes. Start to count processing time when water in canner returns to boiling. Remove jars and complete seals unless closures are self-sealing type. Makes 8 pints.

ORANGE/CUCUMBER PICKLES

An exceptional recipe

2 qts. chopped, peeled pickling
   cucumbers (5 lbs.)
⅔ c. chopped green peppers
   (3 large)
⅔ c. chopped sweet red peppers
   (1½ large)
2 tblsp. canning/pickling salt
2 oranges
1 c. 5% acid strength cider
   vinegar
1 c. brown sugar, firmly packed
½ tsp. mustard seeds
½ tsp. celery seeds

• Combine vegetables in 1-gal. crock. Add salt; mix well. Cover. Let stand 12–18 hours at room temperature. Drain.
• Squeeze juice from oranges. Remove seeds. Grind pulp and peel.
• Combine vegetables, orange juice and pulp, vinegar, sugar, mustard seeds and celery seeds in 4-qt. kettle. Bring to boil; simmer, uncovered, for 15 minutes, stirring often.
• Ladle into 4 hot pint jars, filling to within ¼" of jar top. Wipe jar rim; adjust lids.
• Process in boiling water bath 15 minutes. Start to count processing time when water in canner returns to boiling. Remove jars and complete seals unless closures are self-sealing type. Makes 4 pints.

Note: For most accurate measure of chopped ingredients, use Water Displacement Method—see Index.

HOMEMADE CUCUMBER PICKLES

Recipe used every year by an expert pickle maker

- 6 lbs. (3–5") pickling cucumbers
- ⅔ c. chopped green pepper (3 peppers)
- 1½ c. chopped celery
- 6 medium (2") onions
- ¼ c. prepared mustard
- 4⅔ c. 5% acid strength cider vinegar
- ½ c. canning/pickling salt
- 3½ c. sugar
- 2 tblsp. mustard seeds
- 3 tblsp. celery seeds
- ½ tsp. ground turmeric
- ½ tsp. whole cloves

· Wash cucumbers; cut ⅛" off each end. Cut crosswise into ⅛ to ¼" slices.

· Measure and chop peppers and celery using Water Displacement Method. Peel and slice onions.

· In 8-qt. kettle, combine mustard and vinegar. Blend well with wire whisk. Add salt, sugar, mustard seeds, celery seeds and spices. Bring to boil. Add all vegetables. Cover and slowly bring to boil. *Do not overcook.*

· Pack vegetables into 10 sterilized hot pint jars; cover with syrup, filling to within ¼" of jar top. Wipe jar rim; adjust lids.

· Process in boiling water bath 5 minutes. Start to count processing time when water in canner returns to boiling. Remove jars and complete seals unless closures are self-sealing type. Makes 10 pints.

HOMEMADE MIXED PICKLES

For lighter color pack, use white vinegar instead of cider vinegar

- 2 lbs. (3–4") onions
- 2 lbs. (3–5") pickling cucumbers, sliced
- 2 c. carrot slices, ½" lengths
- 2 c. small cauliflower flowerets
- 2 c. chopped sweet red peppers
- 4 c. celery slices, 1½" lengths
- ¼ c. prepared mustard
- 4⅔ c. 5% acid strength cider vinegar
- ½ c. canning/pickling salt
- 3½ c. sugar
- 2 tblsp. mustard seeds
- 3 tblsp. celery seeds
- ½ tsp. ground turmeric
- ½ tsp. whole cloves

· Peel and quarter onions; chop or slice remaining vegetables.

· In 8-qt. kettle, combine mustard and vinegar. Blend well with wire whisk. Add salt, sugar, mustard seeds, celery seeds and spices. Bring to boil. Add all vegetables. Cover and slowly bring to boil. *Do not overcook.*

· Pack vegetables into 12 sterilized hot pint jars; cover with syrup, filling to within ¼" of jar top. Wipe jar rim; adjust lids.

· Process in boiling water bath 5 minutes. Start to count processing time when water in canner returns to boiling. Remove jars and complete seals unless closures are self-sealing type. Makes 12 pints.

MIXED PICKLE RELISH

You'll be glad to have this medley of vegetables when winter comes

2   c. peeled, sliced carrots
2   c. green beans (or 1 c. green
        and 1 c. yellow beans),
        1" lengths
2   c. green lima beans
2   c. cauliflower flowerets
2   c. sweet corn
1   branch celery, sliced
3   green peppers, coarsely
        chopped
3   sweet red peppers, coarsely
        chopped
Water
1½ tsp. canning/pickling salt
4   c. small pickling cucumbers
2   c. small onions, peeled
2½ c. sugar
¼   c. mustard seeds (4 tblsp.)
2½ tsp. canning/pickling salt
½   tsp. ground turmeric
2   tblsp. celery seeds
8   c. 5% acid strength cider
        vinegar

• Place carrots, green beans, lima beans, cauliflower, corn, celery and peppers into 8-qt. kettle. Cover with water and add 1½ tsp. salt. Bring to boil; simmer, covered, 20 minutes. Drain. Add cucumbers, onions, sugar, mustard seeds, 2½ tsp. salt, turmeric, celery seeds and vinegar. Bring to boil; simmer 5 minutes, uncovered.
• Ladle into 10 sterilized hot pint jars, filling to within ¼" of jar top. Wipe jar rim; adjust lids.
• Process in boiling water bath 5 minutes. Start to count processing time when water in canner returns to boiling. Remove jars and complete seals unless closures are self-sealing type. Makes 10 pints.

1000-ISLAND PICKLES

Sunday-dinner pickle from a Wyoming ranch home. Good with game

4   lbs. pickling cucumbers (8
        large or 12 c. chopped)
2   medium onions (1 c. chopped)
2   green peppers
1   sweet red pepper
1   large head cauliflower
        (1¾ lbs. or 8 c. chopped)
½   c. canning/pickling salt
1   tblsp. mustard seeds
1   tblsp. celery seeds
6   tblsp. dry mustard
1   tblsp. ground turmeric
6   c. sugar
¾   c. flour
2⅔ c. water
5   c. 5% acid strength cider
        vinegar

• Peel cucumbers and onions. Remove seeds and stems from peppers. Chop all vegetables. Layer salt and vegetables in 2-gal. crock. Mix well. Let stand 1 hour. Drain. Rinse.
• Combine spices, sugar and flour in 8-qt. kettle. Add water and blend well with wire whisk. Add vinegar and gradually bring to boil, stirring constantly. Add drained vegetables. Cook 20 minutes, uncovered, stirring often.
• Ladle into 12 hot pint jars; cover with syrup, filling to within ¼" of jar top. Wipe jar rim; adjust lids.
• Process in boiling water bath 15 minutes. Start to count processing time when water in canner returns to boil. Remove jars and complete seals unless closures are self-sealing type. Makes 12 pints.

CRISP MUSTARD PICKLES

Secret of these pickles is the sauce

6   (5") pickling cucumbers
        (1¼ lbs.)
2   c. chopped onion
3   sweet red peppers, chopped
1   medium head cauliflower,
        cut in small pieces
2   c. small pickling onions
1   c. canning/pickling salt
4   c. sugar
4   c. 5% acid strength cider
        vinegar
¾   c. flour
¼   c. dry mustard
1½ tsp. ground turmeric
1   tblsp. celery salt

· Wash cucumbers; cut ¹⁄₁₆" off blossom ends; chop. Layer cucumbers, onion, peppers, cauliflower, onions and salt in 1-gal. crock; cover with cold water. Let stand 12–18 hours in cool place. Drain; rinse.
· Combine sugar, vinegar, flour and spices in 4-qt. kettle. Blend well with wire whisk; gradually bring to boil, stirring constantly. Add vegetables; bring to boil; simmer, uncovered, 15 minutes.
· Ladle into 6 hot pint jars, filling to within ¼" of jar top. Wipe jar rim; adjust lids.
· Process in boiling water bath 15 minutes. Start to count processing time when water in canner returns to boiling. Remove jars and complete seals unless closures are self-sealing type. Makes 6 pints.

Note: If tiny pearl onions (cured or dried) are not available, substitute canned pearl cocktail onions, drained.

MIXED PICKLES

An excellent way to put the garden in jars for use in winter meals

1   qt. chopped pickling
        cucumbers (2 lbs.)
1   qt. (2") pickling cucumbers,
        pierced (1½ lbs.)
1   qt. chopped onions (2 lbs.)
1   qt. small pickling onions
        peeled
6   green peppers, seeded and
        chopped
1   sweet red pepper, seeded
        and chopped
1   head cauliflower, cut in
        small pieces
¾   c. canning/pickling salt
8   c. brown sugar, firmly packed
⅔   c. flour
2   qts. 5% acid strength cider
        vinegar
1   tblsp. mustard seeds
2¼ tsp. celery seeds
2   tblsp. ground turmeric

· Combine chopped and whole cucumbers, chopped and whole onions, peppers and cauliflower with salt in 2-gal. crock. Cover. Let stand 12–18 hours at room temperature. Drain.
· Combine sugar, flour, vinegar, mustard seeds, celery seeds and turmeric in 8-qt. kettle. Mix well with wire whisk. Bring to boil, stirring constantly. Add vegetables; bring to boil; simmer, uncovered, 15 minutes.
· Ladle into 14 sterilized hot pint jars, filling to within ¼" of jar top. Wipe jar rim; adjust lids.
· Process in boiling water bath 5 minutes. Start to count processing time when water in canner returns to boiling. Remove jars and complete seals unless closures are self-sealing type. Makes 14 pints.

END-OF-GARDEN RELISH

No one ever discovered a better relish

1 c. sliced pickling cucumbers
1 c. chopped green peppers (4)
1 c. chopped onion (1 large)
1 c. chopped cabbage
½ c. canning/pickling salt
2 qts. water
1 c. green beans, 1″ lengths
1 c. yellow beans, 1″ lengths
1 c. lima beans
1 c. carrots, pared, ½″ lengths
1 c. chopped celery
1 tsp. salt
1 tblsp. celery seeds
2 tblsp. mustard seeds
2 c. 5% acid strength cider
    vinegar
2 c. sugar
2 tblsp. ground turmeric

• Place cucumbers, peppers, onions and cabbage in 1-gal. crock. Dissolve ½ c. salt in 2 qts. water; pour over cucumbers. Cover. Let stand 12–18 hours in cool place. Drain.
• Place beans, carrots, celery and 1 tsp. salt in 4-qt. kettle; cover with water; boil 15 minutes. Drain. Add uncooked vegetables, celery seeds, mustard seeds, vinegar, sugar and turmeric. Bring to boil; simmer 15 minutes.
• Ladle into hot pint jars, filling to within ¼″ of jar top. Wipe jar rim; adjust lids.
• Process in boiling water bath 15 minutes. Start to count processing time when water in canner returns to boiling. Remove jars and complete seals unless closures are self-sealing type. Makes 3½ pints.

GARDEN WALK PICKLES

A delicious way to prepare a variety of pickled vegetables

2 lbs. (4–5″) pickling cucumbers
    (12 medium)
2 lbs. green tomatoes, cored and
    coarsely chopped (1 qt.)
1½ lbs. onions, peeled and sliced
    (6 medium)
2 sweet red peppers, seeded and
    cut in strips
2 green peppers, seeded and cut
    in strips
1 c. canning/pickling salt
1 gal. cold water
2 c. tender green beans, 1″
    lengths
2 c. lima beans
1 lb. carrots, pared, ½″ lengths
2 c. celery, 1″ lengths
1 medium cauliflower, cut in
    flowerets
1 tblsp. canning/pickling salt
8 c. 5% acid strength vinegar
7 c. sugar
¼ c. mixed pickling spices
4 tblsp. mustard seeds
2 tblsp. celery seeds

• Wash cucumbers; cut ⅛″ slice off each end. Cut cucumbers crosswise in ⅛ to ¼″ slices.
• Combine cucumbers, tomatoes, onions and peppers in 2-gal. crock. Dissolve 1 c. salt in 1 gal. cold water and pour over vegetables. Cover. Let stand 12–18 hours at room temperature. Drain; pour boiling water over vegetables; drain immediately. Set aside.
• Place green beans, lima beans, carrots, celery and cauliflower into 8-qt. kettle. Add 1 tblsp. salt and water to

cover. Cook, covered, for 20 minutes. Drain.

• Combine vinegar, sugar, pickling spices (tied in cheesecloth bag), mustard seeds and celery seeds with all vegetables in kettle. Bring to boil; simmer 5 minutes. Remove spice bag.

• Ladle into 7 sterilized hot quart jars, filling to within ¼" of jar top. Wipe jar rim; adjust lids.

• Process in boiling water bath 5 minutes. Start to count processing time when water in canner returns to boiling. Remove jars and complete seals unless closures are self-sealing type. Makes 7 quarts.

Garden Special Pickles

More cucumbers end up in pickle jars than all the other vegetables put together. But some of the other garden vegetables, like corn, beets and green beans, when pickled, also contribute tasty morsels to meals. Do try the recipes that follow. You're bound to like them.

CRISP DILLED BEANS

Chill thoroughly and serve for supper

2 lbs. tender, mature green or yellow beans
1 tsp. red pepper
4 cloves garlic
4 large heads fresh dill
2 c. water
¼ c. canning/pickling salt
2 c. 5% acid strength cider vinegar

Wash and stem beans; pack uniformly into 4 sterilized hot pint jars, blossom ends down.

To each pint, add ¼ tsp. red pepper, 1 clove garlic and 1 head dill.

Combine water, salt and vinegar in 2-qt. saucepan. Bring to rolling boil, pour over beans, filling to within ¼" of jar top. Wipe jar rim; adjust lids.

Process in boiling water bath 5 minutes. Start to count processing time when water in canner returns to boiling. Remove jars and complete seals unless closures are self-sealing type. Makes 4 pints.

Most pickles need time in the jar to develop flavors, but these Pickled Beets are ready to serve. Children called them "yummers" and ate them like candy. The entire taste panel rated them tops, "best we ever tasted."

PICKLED BEETS

Mild in flavor—perfect in spices

24 small beets (2–2½") or 40 baby beets (1–1½")
Water
3 medium onions (2–2½")
2 c. 5% acid strength cider vinegar
1¼ c. sugar
2 tblsp. canning/pickling salt
1 c. water
6 whole cloves
1 stick cinnamon

• Remove beet tops, leaving 1" stems and the roots. Wash beets well. Place

in 6-qt. kettle, cover with boiling water. Cook, covered, until beets are tender when pricked with fork. Drain.
• Cool beets; remove skins, roots and stems. Cut small beets crosswise in ¼" slices or leave baby beets whole. Peel and slice onions crosswise in ¼" slices, keeping each slice intact.
• Combine vinegar, sugar, salt, 1 cup water, and spices (tied in cheesecloth bag) in 4-qt. kettle. Bring to boil; add beets and onions; simmer 5 minutes. Remove spice bag.
• Ladle beets and onions into 4 hot pint jars, cover with liquid, filling to within ¼" of jar top. Wipe jar rim; adjust lids.
• Process in boiling water bath 30 minutes. Start to count processing time when water in canner returns to boiling. Remove jars and complete seals unless closures are self-sealing type. Makes 4 pints.

BEET RELISH

Adds a spot of color to the plate

3   c. chopped or ground beets
      (about 9 beets, 2½")
Water
1   c. 5% acid strength cider
      vinegar
6½ c. sugar
2   tsp. prepared horseradish
¼   tsp. ground cinnamon
¼   tsp. ground cloves
¼   tsp. ground allspice
1   bottle liquid pectin

• Wash beets. Remove tops, leaving 1" stems and roots. Place beets in 4-qt. kettle, cover with boiling water. Cook, covered, until beets are tender when pricked with fork. Drain and

cool. Remove skins, stems and roots. Chop or grind beets.
• Combine beets, vinegar, sugar, horseradish and spices in 4-qt. kettle. Bring to full rolling boil over high heat; boil rapidly for 1 minute, stirring constantly. Remove from heat and stir in liquid pectin at once. Skim and discard foam.
• Ladle into 8 hot half-pint jars, filling to within ⅛" of jar top. Wipe jar rim; adjust lids.
• Process in boiling water bath 1 minutes. Start to count processing time when water in canner returns to boiling. Remove jars and complete seals unless closures are self-sealing type. Makes 8 half pints.

Note: Wait 6 weeks before opening until relish jells nicely and flavor blend. Make this on a wintry day using canned beets.

IOWA CORN RELISH

Young, tender corn at its flavor peak makes the best-tasting relish

20 ears sweet corn (2½ qts.
      whole kernels)
1   c. chopped green pepper (3
      large)
1   c. chopped sweet red pepper
      (3 large)
1   c. chopped onion (2 medium)
1   c. chopped celery
1½ c. sugar
1   tblsp. canning/pickling salt
1   tblsp. mustard seeds
1   tsp. celery seeds
½   tsp. ground turmeric
2⅔ c. 5% acid strength cider
      vinegar
2   c. water

• Boil husked, silked corn ears 5 minutes. Cool in cold water. Cut whole kernels from cobs; measure.

• Combine all ingredients in 4-qt. kettle. Simmer 20 minutes.

• Ladle relish into 6 hot pint jars, covering with liquid, and filling to within ¼" of jar top. Wipe jar rim; adjust lids.

• Process in boiling water bath 15 minutes. Start to count processing time when water in canner returns to boiling. Remove jars and complete seals unless closures are self-sealing type. Makes 6 pints.

DILLED OKRA

Unusually crisp and amazingly good

3 lbs. young okra (4" long)
4 c. water
2 c. 5% acid strength cider vinegar
½ c. canning/pickling salt
3 tsp. celery seeds
6 cloves garlic
6 heads dill

• Pierce each okra with fork; do not stem or slice.

• Combine water, vinegar and salt in 3-qt. saucepan; bring to boil.

• Place ½ tsp. celery seeds, 1 clove garlic and 1 head dill into each of 6 sterilized hot pint jars. Pack okra lengthwise into jars. Pour boiling brine over okra, filling to within ¼" of jar top. Wipe jar rim; adjust lids.

• Process in boiling water bath 5 minutes. Start to count processing time when water in canner returns to boiling. Remove jars and complete seals

unless closures are self-sealing type. Makes 6 pints.

Note: Let stand 3 to 4 weeks before using to let flavors blend.

PICKLED ONIONS

Some like it hot—spices give these onions a moderately hot taste

4 qts. small pickling onions (3 lbs.) (See Note)
1 c. canning/pickling salt
8 c. 5% acid strength white vinegar
2 c. sugar
¼ c. pickling spices

• Peel onions; place in 1-gal. crock. Add salt and mix well. Let stand 12–18 hours at room temperature.

• Place onions in colander. Rinse thoroughly with cold water and drain well.

• Combine vinegar and sugar in 3-qt. saucepan; add pickling spices (tied in cheesecloth bag). Bring to boil; boil 10 minutes.

• Pack onions into 8 sterilized hot pint jars. Remove spice bag. Pour boiling syrup over onions, filling to within ¼" of jar top. Wipe jar rim; adjust lids.

• Process in boiling water bath 5 minutes. Start to count processing time when water in canner returns to boiling. Remove jars and complete seals unless closures are self-sealing type. Makes 8 pints.

Note: Pickling onions means cured onions. Onions are cured by drying. Onions pulled fresh from the garden do not pickle well.

CRISP SQUASH PICKLES

Good—even if you don't like squash

5 lbs. zucchini
3 medium onions, thinly sliced (2¼" diameter)
½ c. canning/pickling salt
Ice cubes
3 c. 5% acid strength cider vinegar
3 c. sugar
2 tsp. celery seed
2 tsp. mustard seed
1½ tsp. turmeric
1 tsp. ginger
½ tsp. pepper

• Wash zucchini; cut ⅛" off each end. Cut crosswise into ⅛ to ¼" slices.

• Combine zucchini, onions and salt in bowl. Top with layer of ice cubes. Cover and let stand 3 hours. Drain and rinse in cold water.

• Combine zucchini mixture with vinegar, sugar, celery seed, mustard seed, turmeric, ginger and pepper in a large kettle. Heat to boiling; reduce heat and simmer for 2 minutes.

• Ladle into 6 sterilized hot pint jars, filling to within ¼" of jar top. Wipe jar rim; adjust lids.

• Process in boiling water bath 5 minutes. Start to count processing time when water in canner returns to boiling. Remove jars and complete seals unless closures are self-sealing type. Makes 6 pints.

Green Tomato Pickles

When shorter days are noticeable and the sumac leaves start to look like red flames, farm women know the last of the year's pickle making is at hand. Baskets filled with green tomatoes come to the kitchen for the cook's magic that turns them into pickles.

GREEN TOMATO PICKLES

Pickles are spicy and sharp with a sweet tang—brown sugar adds flavor

4 lbs. green tomatoes (2 qts. thinly sliced)
3 tblsp. canning/pickling salt
2 c. 5% acid strength cider vinegar
⅔ c. dark brown sugar, firmly packed
1 c. sugar
3 tblsp. mustard seeds
½ tsp. celery seeds

1 tsp. ground turmeric
3 c. thinly sliced onions
2 sweet red peppers, seeded, chopped
1 hot green or red pepper, seeded, chopped

• Layer tomatoes and salt in 2-qt. bowl or crock. Cover. Let stand at room temperature 12 hours. Drain well.

• Combine vinegar, sugars and spices in 4-qt. kettle; bring to boil; add sliced onions and boil gently, uncovered, 5 minutes. Add drained tomatoes and peppers; bring slowly to boil. Simmer 5 minutes, stirring occasionally.

Pack vegetables into 9 sterilized hot half-pint jars; cover with syrup, filling to within ¼" of jar top. Wipe jar rim; adjust lids.

• Process in boiling water bath 5 minutes. Start to count processing time when water in canner returns to boiling. Remove jars and complete seals unless closures are self-sealing type. Makes 9 half pints.

INDIAN PICKLE

A chopped pickle that's been made in one family for 3 generations

4   lbs. green tomatoes
      (8 c. chopped)
4   lbs. ripe tomatoes, peeled
      (7 c. chopped)
3   medium (2" diameter) onions,
      peeled (1½ c. chopped)
3   sweet red peppers, seeded
      (1½ c. chopped)
3   green peppers, seeded
      (1 c. chopped)
1   large pickling cucumber
      (1 c. chopped)
7   c. chopped celery (4½ lbs.)
⅔ c. canning/pickling salt
6   c. 5% acid strength cider
      vinegar
3   lbs. brown sugar (6¾ c., firmly
      packed)
1   tsp. dry mustard
1   tsp. ground black pepper

• Coarsely chop all vegetables. Combine with salt in 2-gal. crock. Cover. Let stand 12–18 hours at room temperature. Drain.
• Pour into 6-qt. kettle. Add vinegar, sugar, mustard and pepper. Bring to boil; simmer, uncovered, for 30 minutes, stirring occasionally.
• Ladle into 8 hot pint jars; cover with liquid, filling to within ¼" of jar top. Wipe jar rim; adjust lids.
• Process in boiling water bath 15 minutes. Start to count processing time when water in canner returns to

boiling. Remove jars and complete seals unless closures are self-sealing type. Makes 8 pints.

Note: For most accurate measure of chopped ingredients, use Water Displacement Method—see Index.

BEST-EVER PICCALILLI

Perfect relish to enhance flavor of meat sandwiches

22   medium-size green tomatoes
        (6 lbs.)
2   large (3") onions
        (2 c. chopped)
6   green peppers
6   sweet red peppers
6   c. 5% acid strength cider
        vinegar
3½   c. sugar
¼   c. canning/pickling salt
1½   tsp. ground allspice
1½   tsp. ground cinnamon
4   tsp. celery seeds
½   c. mustard seeds

• Wash vegetables; core tomatoes; peel onions; remove seeds, membranes and stem from peppers. Put vegetables through food chopper, or chop with water in electric blender; drain.
• Place vegetables in 8-qt. kettle; add 4 c. vinegar. Boil 30 minutes, uncovered, stirring frequently. Drain and discard liquid. Return vegetables to kettle; add remaining 2 c. vinegar, sugar, salt and spices. Bring to boil; simmer 3 minutes.
• Ladle into 8 hot pint jars, filling to within ¼" of jar top. Wipe jar rim; adjust lids.
• Process in boiling water bath 15 minutes. Start to count processing time when water in canner returns to

boiling. Remove jars and complete seals unless closures are self-sealing type. Makes 8 pints.

RUMMAGE PICKLE

An old-time pickle to make of garden remnants when Jack Frost is due

2½ lbs. ripe tomatoes, peeled
    (1 qt. diced)
4 lbs. green tomatoes
    (2 qts. chopped)
1 small head cabbage
3 green peppers, seeded
3 sweet red peppers, seeded
1 large yellow or green cucumber
3 large (3" diameter) onions
    (1½ lbs.)
6 c. celery (4 lbs.)
½ c. canning/pickling salt
1½ c. 5% acid strength cider
    vinegar
1½ c. water
4 c. sugar
1 tsp. dry mustard

· Dice ripe tomatoes. Put remaining vegetables through food chopper. Pour into 2-gal. crock. Add salt; mix well. Cover. Let stand overnight at room temperature. Drain.
· Combine vinegar, water, sugar and mustard in 10-qt. kettle. Bring to boil; simmer, covered, stirring often, for 1 hour.
· Ladle into 10 hot pint jars, pressing down so liquid covers pickles as you pack. Fill to within ¼" of jar top. Wipe jar rim; adjust lids.
· Process in boiling water bath 15 minutes. Start to count processing time when water in canner returns to boiling. Remove jars and complete

seals unless closures are self-sealing type. Makes 10 pints.

GREEN TOMATO/PEPPER RELISH

This pretty relish is great with pork— looks like confetti

4 qts. green tomatoes, cored and
    chopped (8 lbs.)
1⅓ c. chopped onions (2 medium)
2⅔ c. chopped green pepper
    (4 large)
1⅓ c. chopped sweet red pepper
    (2 large)
¼ c. canning/pickling salt
3 c. 5% acid strength cider
    vinegar
1 c. water
2 c. sugar
1 tsp. mixed pickling spices

· Combine tomatoes, onion, peppers and salt in large bowl. Cover and let stand overnight at room temperature. Drain well.
· Combine vinegar, water, sugar and pickling spices (tied in cheesecloth bag) in large 5-qt. kettle. Bring to boiling; add vegetables; return to boiling. Simmer 30 minutes, stirring occasionally. Remove spice bag.
· Immediately ladle into 8 hot pint jars, filling to within ¼" of jar top. Wipe jar rim; adjust lids.
· Process in boiling water bath 15 minutes. Start to count processing time when water in canner returns to boiling. Remove jars and complete seals unless closures are self-sealing type. Makes 8 pints.

Note: For most accurate measure of chopped ingredients, use Water Displacement Method—see Index.

OLD-FASHIONED RAG PICKLES

Add to mayonnaise—serve with fish

8 c. pickling cucumbers, peeled and chopped (4 lbs.)
8 c. chopped green tomatoes (4 lbs.)
8 c. cored, chopped cabbage (4 lbs.)
10 medium onions, peeled and chopped (3 lbs.)
2 tsp. ground turmeric
2 tsp. celery seeds
2 tsp. dry mustard
3 tsp. canning/pickling salt
2 c. sugar
4 c. 5% acid strength cider vinegar

• Combine all ingredients in 8-qt. kettle. Bring to boil; boil 20 minutes, uncovered, stirring occasionally.
• Ladle into 12 hot pint jars, filling to within ¼" of jar top. Wipe jar rims; adjust lids.
• Process in boiling water bath 15 minutes. Start to count processing time when water in canner returns to boiling. Remove jars and complete seals unless closures are self-sealing type. Makes 12 pints.

DANISH GREEN RELISH

Excellent with peanut butter sandwiches, baked beans and meats

30 green tomatoes, cored (8 lbs. or 4 qts. chopped)
5 large onions (2½ lbs. or 5 c. chopped)
3 sweet red peppers, stemmed and seeded
3 green peppers, stemmed and seeded
¼ c. canning/pickling salt
¼ c. dry mustard
1 tblsp. celery seeds
4 c. sugar
5 c. 5% acid strength cider vinegar

• Put vegetables through food chopper. Place in 1-gal. crock, add salt, mix well. Cover. Let stand 12–18 hours at room temperature. Drain.
• Pour into 8-qt. kettle; add mustard, celery seeds, sugar and vinegar. Bring to boil; simmer, uncovered, 15 minutes.
• Ladle into 8 hot pint jars, filling to within ¼" of jar top. Wipe jar rim; adjust lids.
• Process in boiling water bath 15 minutes. Start to count processing time when water in canner returns to boiling. Remove jars and complete seals unless closures are self-sealing type. Makes 8 pints.

Colorful Pickled Peppers

You may have trouble saying the Simple Simon tongue-twister about pickled peppers, but you won't have difficulty making relishes with the recipes that follow. Plump, meaty, sweet peppers, both red and green, are favorites for pickling. Perhaps their cheerful colors have something to do with their popularity, but their superior flavors have more to do with it. Many country cooks like to fix pepper relishes to serve during the holidays. In fact, they often call them Christmas relishes. Certainly their sharp flavor and gala colors fit beautifully into turkey and chicken dinners.

Hot peppers, prized so highly in the Southwest, sometimes wind up in peppy relishes. The example that follows, Hot Pepper Relish, won blue ribbons for a FARM JOURNAL reader at the New Mexico State Fair. And Pepper Slaw captured the same honors for an Eastern farm woman at the Connecticut State Fair.

Use fresh, tender, young peppers to make relishes with the best taste.

HOT PEPPER RELISH

Made with hot chili peppers; won first prize at New Mexico State Fair

18   red chili peppers, seeded and stemmed
18   green chili peppers, seeded and stemmed
4   lbs. onions, peeled
1   tblsp. canning/pickling salt
Boiling water
2½ c. 5% acid strength cider vinegar
2½ c. sugar

· Put peppers and onions through food chopper, or chop in water in electric blender and drain. Place in 6-qt. kettle.
· Add salt; cover with boiling water. Let stand 10 minutes. Drain and discard liquid. Add vinegar and sugar to vegetables. Bring to boil; simmer 20 minutes.
· Ladle into 7 hot pint jars, pressing down as you pack so liquid covers vegetables, filling to within ¼" of jar top. Wipe jar rim; adjust lids.
· Process in boiling water bath 15

minutes. Start to count processing time when water in canner returns to boiling. Remove jars and complete seals unless closures are self-sealing type. Makes 7 pints.

Note: If hand-chopping peppers, wear rubber gloves to protect hands.

PEPPER RELISH

All pepper relishes are best served within six months after canning

4   c. chopped green peppers (12 large)
6   c. chopped sweet red peppers (12 large)
3   c. chopped onions (1 lb.)
Boiling water
3   c. 5% acid strength cider vinegar
1½ c. sugar
4   tsp. canning/pickling salt
2   tsp. celery seeds

· Remove stems, membranes, and seeds from peppers and peel onions before chopping.
· Place vegetables in 4-qt. kettle, cover with boiling water; let stand 10 minutes. Drain off water. Add vinegar, sugar, salt and celery seeds. Boil over low heat 15 minutes.
· Ladle into 8 hot pint jars, filling to within ¼" of jar top. Wipe jar rim; adjust lids.
· Process in boiling water bath 15 minutes. Start to count processing time when water in canner returns to boiling. Remove jars and complete seals unless closures are self-sealing type. Makes 8 pints.

Note: For most accurate measure of chopped ingredients, use Water Displacement Method—see Index.

CABBAGE/PEPPER RELISH

One of the relishes women in the Ozarks serve with justified pride

2 medium heads cabbage (4½ lbs.)
6 green peppers
6 sweet red peppers
8 carrots (1 lb.)
6 medium onions (1½ lbs.)
½ c. canning/pickling salt
8 c. 5% acid strength cider vinegar
6 c. brown sugar, firmly packed
1 tblsp. celery seeds
1 tblsp. mustard seeds

• Core cabbage. Remove stems, seeds and membranes from peppers. Peel carrots and onions. Put vegetables through food chopper. Combine with salt in 2-gal. crock. Cover. Let stand 3 hours at room temperature. Drain.

• Combine vinegar, sugar and seeds in 8-qt. kettle; bring to boil; add chopped vegetables. Bring to boil; simmer, uncovered, for 15 minutes.

• Ladle into 10 hot pint jars, filling to within ¼" of jar top. Wipe jar rim; adjust lids.

• Process in boiling water bath 15 minutes. Start to count processing time when water in canner returns to boiling. Remove jars and complete seals unless closures are self-sealing type. Makes 10 pints.

CANADIAN PEPPER RELISH

A Northern neighbor shares her favorite relish recipe with us

1 c. chopped green peppers (3 large)
1 c. chopped sweet red peppers (2 large)
7 c. sugar
1½ c. 5% acid strength cider vinegar
1 bottle liquid pectin

• Combine peppers, sugar and vinegar in 4-qt. kettle. Bring to a *full rolling boil*, stirring constantly; boil 2 minutes. Remove from heat; add pectin; skim to remove foam.

• Ladle into 7 sterilized hot half-pint jars, filling to within ⅛" of jar top. Wipe jar rim; adjust lids.

• Process in boiling water bath 5 minutes. Start to count processing time when water in canner returns to boiling. Remove jars and complete seals unless closures are self-sealing type. Makes 7 half pints.

SPICY PEPPER RELISH

From a North Carolina home where it rates tops among canned relishes

1 gal. green tomatoes, cored (16 lbs.)
6 large green peppers, stemmed and seeded
3 large sweet red peppers, stemmed and seeded
1 medium head cabbage, cored
2 large (3") onions, peeled
4 c. 5% acid strength cider vinegar
1 c. sugar
1 c. brown sugar, firmly packed
1 tsp. ground cinnamon
1 tsp. ground allspice
½ c. canning/pickling salt

• Put tomatoes, peppers, cabbage and onions through food chopper. Or chop in water in electric blender; drain thoroughly.

• Combine vegetables, vinegar, sugars, spices and salt in 12-qt. kettle. Bring

to boil; simmer, uncovered, for 30 minutes.
· Ladle into 12 hot pint jars, filling to within ¼″ of jar top. Wipe jar rim; adjust lids.
· Process in boiling water bath 15 minutes. Start to count processing time when water in canner returns to boiling. Remove jars and complete seals unless closures are self-sealing type. Makes 12 pints.

FRENCH DRESDEN PICKLES

Chill a jar in the refrigerator to add a cool, spicy note to a meal

12 green peppers, stemmed and seeded (3 c. chopped)
12 sweet red peppers, stemmed and seeded (4 c. chopped)
3 small heads cabbage, cored (5 lbs.)
8 large (3″ diameter) onions, peeled (4 lbs.; 7 c. chopped)
¾ c. canning/pickling salt
8 c. 5% acid strength cider vinegar
3½ c. sugar
2 tblsp. mustard seeds
2 tblsp. celery salt

· Prepare vegetables and put through food chopper. Pour into 2-gal. crock; add salt; mix well. Cover. Let stand 12–18 hours in cool place. Drain.
· Add vinegar, sugar, mustard seeds and celery salt to vegetables in 8-qt. kettle. Bring to boil; simmer, uncovered, 15 minutes, stirring often.
· Ladle into 12 hot pint jars, filling to within ¼″ of jar top. Wipe jar rim; adjust lids.
· Process in boiling water bath 15 minutes. Start to count processing time when water in canner returns to

boiling. Remove jars and complete seals unless closures are self-sealing type. Makes 12 pints.

FINE CHOPPING MADE EASY

If you have an electric blender, do use it to chop the vegetables to make this slaw. After seeding, peeling and removing cores as directed in the recipe, put vegetable pieces in the blender with water to cover. Chop and drain well. (Or chop dry in a food processor, if you have one.)

The recipe gives amounts of vegetables whole; in parentheses, you'll find the yield when vegetables are chopped. If you use the recommended Water Displacement Method of measuring, you can do it in the blender. For example, to get 1 c. chopped pepper, put 1 c. water in blender; add pepper pieces until water line reaches 2 c. Chop and drain.

People who tasted this slaw thought it had been marinated instead of canned. It adds good taste and good looks to any meal, and would be an excellent addition to a salad bar. Serve cold.

PEPPER SLAW

Takes the place of salad in a hurry-up sandwich-and-soup lunch

12 green peppers (3 c. chopped)
12 sweet red peppers (4 c. chopped)
12 medium onions (3 lbs.; 5 c. chopped)
2 large heads cabbage (5 lbs.; about 14 c. chopped)
¼ c. canning/pickling salt

6 c. sugar
2½ tblsp. celery seeds
2½ tblsp. mustard seeds
4 c. 5% acid strength cider
    vinegar

• Remove membranes and seeds from peppers. Peel onions. Core cabbage. Coarsely chop all vegetables; pour into 2-gal. crock. Add salt; mix well; cover. Let stand at room temperature 12–18 hours. Drain.
• Combine sugar, spices and vinegar in 8-qt. kettle. Bring to boil; add chopped vegetables; simmer gently, uncovered, for 25 minutes, stirring often.
• Ladle into 10 hot pint jars, pressing down as you pack so syrup covers slaw. Fill to within ¼″ of jar top. Wipe jar rim; adjust lids.
• Process in boiling water bath 15 minutes. Start to count processing time when water in canner returns to boiling. Remove jars and complete seals unless closures are self-sealing type. Makes 10 pints. Serve cold.

"What Went Wrong with My Pickles?"

Q. What makes my pickles soft and slippery?
A. If pickles are both slippery and soft, this is evidence of spoilage caused by growth of bacteria and yeasts. The most likely reasons are: failing to remove scum from top of brine; not heating pickles long enough to destroy spoilage microorganisms; jars not sealing; using brine or vinegar that's too weak; failing to keep cucumbers covered with liquid; failing to remove blossom ends from cucumbers; storing jars of pickles at too high a temperature—above 80° F.

If pickles are soft but not slippery, with no signs of spoilage, you may simply have processed them too long.

Q. Why are my cucumber pickles hollow?
A. (1) Often this is due to growing conditions or the wrong variety of cucumbers (use pickling varieties). (2) Or the cucumbers may have waited too long between gathering time and pickling. (3) A gaseous type of spoilage may have occurred.

Note: When making whole pickles, you can check the cucumbers that are hollow—usually they float in water.

Q. Why are my pickles shriveled and sometimes tough?
A. Shriveling is usually the result of "shocking" the cucumber with too much salt or sugar too quickly, using too heavy a syrup at the beginning of cooking, or too strong vinegar. Another reason is letting cucumbers wait too long after picking, before pickling.

Q. Why are my pickles dark?
A. The darkening may be due to corrosion of some metal utensils or jar lids, iron in water, use of ground rather than whole spices, or leaving spice bag in jars. Dark spots are caused by holding cucumbers too long after picking before pickling them.

Q. Why does the garlic clove in my pickles turn green?

A. The garlic was not dry enough. Recently, some home canners have been puzzled by garlic turning purple —it's a new variety which reacts that way in the presence of acid, and is harmless.

Q. My pickles have a bitter flavor. What went wrong?

A. Among the possible causes: using cucumbers that had a very dry growing season; too much spice; packing whole spices with the pickles.

CHAPTER 14

Fruit Pickles, Relishes, Ketchups and Chutneys

Recipes for Spiced Apples, Crab Apples, Apricots, Blueberries, Cantaloupe, Cranberries · Pickled Figs, Prunes, Peaches, Pears, Quinces, Pineapple, Watermelon · Mixed Fruit and Vegetable Relishes · Ketchups · Chili Sauces · Chutneys

Halfway through summer, country kitchens begin to fill with the most tantalizing of all food fragrances—fruits and tomatoes, spiced just right, cooking in sugar-vinegar syrup. Country cooks depend greatly on spiced and pickled fruits to give meals a flavor lift—and a complementary spot of color.

In this chapter, you will find old friends like crisp watermelon pickles and golden peach halves that glorify chicken dinners . . . not to mention whole crab apple pickles and quartered slices of golden spicy pineapple.

Try to decide which recipe for red relish sauces to try first; if you have tomatoes ripening, you have the makings for superior chili sauce and ketchup. Don't overlook other intriguing fruit ketchups. As for chutneys, we have four delicious variations made with apples, pineapples, peaches and tomatoes. They're so good with rice and poultry dishes.

In this chapter, you will find choice recipes from Western ranches, Southern, New England and Midwestern kitchens. Among them are the combination fruit/vegetable relishes that are conversation pieces at any party, and delicious products to donate to food sales or take along to community suppers. Recipes for Victory Relish (the name still sticks) came to us from all parts of the country—a reminder of the Forties when so many townspeople patriotically planted gardens. The version we print is a zippy combination of tomatoes, pears, peaches, onions and peppers which has a praise-winning record.

When we retested the pickle and relish recipes, we enrolled a panel of adults to serve as taste-testers—but the children wanted to participate, too. The comments of our tasters—young and old—will help you pick out recipes to please your family.

Wonderful Sweet-Sours

Fruits are acid foods and the vinegar added to fruit pickles is for flavor —to give them a sweet-sour taste.

However, this chapter also includes recipes for relishes and ketchups made with a combination of fruits and vegetables. Vegetables are low-acid foods and the amount of vinegar in such recipes is necessary to acidify the mixture and make it safe for canning in the boiling water bath. Review the information about ingredients and equipment needed for pickles and relishes at the beginning of Chapter 13. Do not reduce the amount of vinegar or change the proportion of vinegar to food solids.

The way you measure ingredients for relishes will make a difference in yields. In testing recipes, we measured *before* chopping, using the Water Displacement Method described in Chapter 13 (see Index).

Farm women prefer slightly underripe fruits for pickling—peaches and pears not quite ripe enough for freezing or canning. Wash fruit thoroughly and prepare as recipe directs.

Prepare jars and lids. Jars should be sterilized if recipe calls for processing less than 15 minutes. Pack pickles into hot jars; cover with boiling syrup, leaving recommended head space. Remove air bubbles, wipe sealing edge of jar and adjust lids. Process pickles and relishes in the boiling water canner as directed. For complete directions on all canning procedures, see "Steps to Successful Canning" in Chapter 9.

ALTITUDE CORRECTIONS

All processing times given for the boiling water bath are for altitudes less than 1,000 feet above sea level. See Index for Altitude Corrections if you live above 1,000 feet.

SPICED APPLE STICKS

A relish that's spiced just right— makes other foods taste better

10	medium apples (5 lbs. or 2 qts. sliced)
3	qts. water
3	tblsp. vinegar
1	c. sugar
½	c. light corn syrup
1	c. 5% acid strength cider or white vinegar
⅔	c. water
2	tsp. whole cloves
1½	sticks cinnamon

• Wash, peel, core and slice apples into eighths lengthwise. To prevent discoloration, drop slices directly into 5-qt. bowl containing solution of 3 qts. water and 3 tblsp. vinegar.

• Combine sugar, corn syrup, 1 c. vinegar, ⅔ c. water and spices (tied in cheesecloth bag) in 4-qt. kettle; bring to boil. Add drained apples, cover. Boil until apples are translucent, about 3 minutes.

• Pack apple sticks into 3 hot pint jars, cover with syrup, filling to within ¼" of jar top. Wipe jar rim; adjust lids.

• Process in boiling water bath 15 minutes. Start to count processing time when water in canner returns to boiling. Remove jars and complete seals unless closures are self-sealing type. Makes 3 pints.

OLD-FASHIONED GINGER APPLES

Grandmother's rule will work for you —use early tart apples

4 c. chopped tart apples (15 medium)
1 lemon
2 c. light brown sugar, firmly packed
2 c. water
5 pieces ginger root (½" squares)

· Wash, peel, quarter and core apples. Chop apples or dice into ½" cubes. Place in 2-qt. mixing bowl. Grate lemon peel; extract juice; combine with apples, mixing well.
· Combine sugar, water and ginger root (tied in a cheesecloth bag) in 4-qt. kettle. Bring to boil; boil 15 minutes.
· Add apple mixture. Cook over low heat 90 minutes, or until slightly thickened and dark, stirring often. Remove spice bag.
· Ladle into 3 hot pint jars, filling to within ⅛" of jar top. Wipe jar rim; adjust lids.
· Process in boiling water bath 15 minutes. Start to count processing time when water in canner returns to boiling. Remove jars and complete seals unless closures are self-sealing type. Makes 3 pints.

Note: To measure apples, pour 1 qt. water in 2-qt. measuring bowl. Add apple quarters until water reaches 2-qt. line. Drain and chop.

CINNAMON APPLE RINGS

Candy does the spicing and tinting— glamorous relish to serve with pork

5 lbs. tart apples (about 18 medium)
6 c. sugar
3 c. water
1 (9 oz.) pkg. red cinnamon candies
3 drops red food color (optional)

· Wash and peel apples; cut crosswise into ¼ slices; cut out center core with melon baller. Place apple rings in salt water solution (3 tblsp. salt to 4 qts. water) to prevent darkening.
· Combine sugar, water, cinnamon candies and food color in a large 5-qt. kettle. Bring to a boil; boil 3 minutes, stirring constantly.
· Drain apples; rinse well; add to syrup. Cook until apples are tender and translucent, about 5 minutes.
· Pack apple rings into 4 hot pint jars. Pour syrup over apples, filling to within ½" of jar top. Wipe jar rim; adjust lids.
· Process in boiling water bath for 25 minutes. Start to count processing time when water in canner returns to boiling. Remove jars and complete seals unless closures are self-sealing type. Makes 4 pints.

Note: Use York, Northern Spy, Golden Delicious or Gravenstein apples.

CRAB APPLE PICKLES

No one has found a more colorful or tastier garnish for meat platters

4   lbs. crab apples
3½ c. 5% acid strength cider
        vinegar
3   c. water
6   c. sugar
1   tblsp. whole cloves
3   sticks cinnamon
1   piece fresh or dried ginger root
        (1″ cube)

• Wash crab apples; remove blossom ends; do not remove stems. Prick each apple in several places (fork works well) to prevent fruit from bursting while cooking.
• Bring vinegar, water and sugar to boil in 4-qt. kettle. Add spices tied in cheesecloth bag.
• Cook half of crab apples in syrup for 2 minutes; remove crab apples with slotted spoon, place in 1-gal. crock or glass bowl. Add remainer uncooked crab apples to kettle and cook for 2 minutes. Pour apples, syrup and spice bag into crock. Cover. Let stand overnight (12–18 hours) in cool place.
• Remove spice bag. Drain syrup into 2-qt. saucepan. Pack apples into 7 hot pint jars to within ½″ of jar top. Heat syrup to boiling; pour over apples; leave ½″ head space. Remove air bubbles. Wipe jar rim; adjust lids.
• Process in boiling water bath 20 minutes. Start to count processing time when water in canner returns to boiling. Remove jars and complete seals unless closures are self-sealing type. Makes 7 pints.

CHOPPED SPICED
CRAB APPLES

Excellent with ham—you'll like these

6   lbs. large ripe crab apples
2   oranges
2   (15 oz.) pkgs. seedless raisins
2   c. 5% acid strength cider vinegar
5   c. brown sugar, firmly packed
2   tsp. ground cinnamon
½   tsp. ground cloves

• Wash crab apples, remove stems, blossoms and cores; dice. Peel oranges; discard peel and seeds; dice.
• Combine all ingredients in 6-qt. kettle. Bring to boil. Simmer, covered, until apples are tender, about 30 minutes.
• Ladle into 8 hot pint jars; cover with syrup, filling to within ¼″ of jar top. Wipe jar rim; adjust lids.
• Process in boiling water bath 15 minutes. Start to count processing time when water in canner returns to boiling. Remove jars and complete seals unless closures are self-sealing type. Makes 8 pints.

CALIFORNIA APRICOT
PICKLES

A most pleasant blend of sugar and spices—doesn't taste pickled

4   qts. medium apricots (6 lbs.)
    Whole cloves
4   c. sugar
4   c. brown sugar, firmly packed
4   c. 5% acid strength cider vinegar
6   sticks cinnamon

• Wash apricots, do not peel. Stud each apricot with 3 cloves.
• Combine sugars, vinegar and cinnamon sticks in 5-qt. kettle. Bring to boil. Add half of the apricots; simmer only until skins begin to shrivel, about 5 minutes. *Do not overcook*. Remove with slotted spoon. Simmer remaining

apricots; remove with slotted spoon. (Apricots should not stand in hot syrup.)
• Pack apricots into 8 hot pint jars. Heat syrup to boiling. Remove cinnamon sticks. Pour boiling syrup over apricots, filling to within ¼" of jar top. Wipe jar rim; adjust lids.
• Process in boiling water bath 20 minutes. Start to count processing time when water in canner returns to boiling. Remove jars and complete seals unless closures are self-sealing type. Makes 8 pints.

PICKLING BLUEBERRIES

We received the following recipe for Spiced Blueberries from one of our readers with these comments: "My mother often served this sharp relish with game or with what she called 'heavy meats' like pork, beef and mutton. And when she did, the compliments went round the table."

SPICED BLUEBERRIES

New, delicious way to use blueberries

3 qts. fresh blueberries
1 c. 5% acid strength cider vinegar
1 c. sugar
2 tblsp. whole cloves

• Wash blueberries. Combine with vinegar and sugar in 5-qt. kettle. Add cloves (in cheesecloth bag). Bring to boil; simmer 25 minutes, uncovered, or until liquid begins to jell (see Index for Jelly Test). Remove spice bag. Skim and discard foam.
• Ladle into 4 hot pint jars, filling to within ⅛" of jar top. Wipe jar rim; adjust lids.

• Process in boiling water bath 15 minutes. Start to count processing time when water in canner returns to boiling. Remove jars and complete seals unless closures are self-sealing type. Makes 4 pints.

Note: Wait 3 weeks before opening, until jell develops fully and flavors blend.

CANTALOUPE PICKLES

Prove you're a gourmet; serve this

1 large slightly underripe cantaloupe
4 c. 5% acid strength cider vinegar
2 c. water
1 tsp. ground mace
2 sticks cinnamon
2 tblsp. whole cloves
4 c. sugar (or 2 c. white and 2 c. brown, firmly packed)

• Peel cantaloupe and cut in 1" cubes. Place in 1 gal. crock or glass bowl.
• Combine vinegar, water and mace in 3-qt. saucepan. Add cinnamon and cloves (tied in cheesecloth bag). Bring to boil; pour over cantaloupe. Let stand, covered, 12–18 hours in cool place.
• Drain cantaloupe, reserving liquid in 4-qt. kettle. Bring liquid to boil; stir in sugar; add cantaloupe. Simmer, uncovered, until cantaloupe is translucent, about 1 hour. Remove spice bag. Remove cantaloupe with slotted spoon. Return liquid to boiling; boil uncovered 10 more minutes.
• Pack cantaloupe into 2 sterilized hot pint jars; cover with syrup to within ¼" of jar top. Wipe jar rim; adjust lids.
• Process in boiling water bath 5 minutes. Start to count processing time

when water in canner returns to boiling. Remove jars and complete seals unless closures are self-sealing type. Makes 2 pints.

PICKLED CHERRIES

If you like really tart relishes, pickled cherries are for you. Some cooks trim the stems slightly and leave these "handles" on the fruit. Wrap the jars of pickles in heavy brown paper when storing them in the fruit closet. It will protect their color.

PICKLED CHERRIES

Good with roast pork. Festive addition to a relish tray

4 lbs. large sweet cherries,
    stemmed
2 c. 5% acid strength cider vinegar
1 c. sugar
½ tsp. ground cinnamon
½ tsp. ground mace
⅛ tsp. ground cloves

• Wash cherries; do not pit. Combine sugar, vinegar and spices in 5-qt. kettle. Bring to boil; add cherries. Bring to boil; simmer 1 to 2 minutes or until skins begin to crack. Pour into 4-qt. glass bowl or crock. Cover, let stand at room temperature for 4 hours.
• Drain cherries, reserving liquid in 1-qt. saucepan. Pack cherries into 4 hot pint jars. Bring liquid to a boil; pour over cherries, filling to within ¼" of jar top. Wipe jar rim; adjust lids.
• Process in boiling water bath 15 minutes. Start to count processing time when water in canner returns to boiling. Remove jars and complete seals unless closures are self-sealing type. Makes 4 pints.

PICKLED FIGS

Brown sugar and spices give figs an unusual, delicious new taste

6 qts. figs (8 lbs.)
Salt water solution (see below)
8 c. brown sugar, firmly packed
4 c. 5% acid strength cider vinegar
7 sticks cinnamon
2 tsp. whole cloves

• Wash figs; cover with salt water solution (1 tblsp. salt to 1 gal. water) in 10-qt. kettle; bring to boil; simmer 15 minutes, uncovered. Drain.
• Combine sugar, vinegar and spices (tied in cheesecloth bag) in 5-qt. kettle. Bring to boil. Pour over figs and simmer, uncovered, for 1 hour.
• Pack figs loosely into 8 hot pint jars; cover with syrup, filling to within ¼" of jar top. Wipe jar rim; adjust lids.
• Process in boiling water bath 20 minutes. Start to count processing time when water in canner returns to boiling. Remove jars and complete seals unless closures are self-sealing type. Makes 8 pints.

PRUNE PICKLES

These plump, spicy prunes are a real flavor boost with goose and duck

1 lb. dried prunes
2 c. water
1 c. 5% acid strength cider vinegar
1 c. brown sugar, firmly packed
½ tsp. whole cloves
4 sticks cinnamon

• Wash prunes; place in 1-qt. mixing bowl; cover with water. Soak for 1 hour; drain, reserving ½ c. liquid.

• Combine vinegar, reserved liquid, brown sugar and spices (tied in cheesecloth bag) in 2-qt. saucepan. Simmer 12 minutes. Remove spice bag.

• Pack prunes into 2 hot pint jars; pour hot syrup over prunes, filling to within ¼" of jar top. Wipe jar rim; adjust lids.

• Process in boiling water bath 20 minutes. Start to count processing time when water in canner returns to boiling. Remove jars and complete seals unless closures are self-sealing type. Makes 2 pints.

Note: Let prunes stand at least 3 weeks before opening to give flavors time to blend.

PEACH PICKLES

A pickle to be used as a sidedish of fruit—a delicious snack for children

4   lbs. ripe peaches
        (22, 2" diameter)
1   tblsp. whole cloves
4   c. sugar
1½ c. 5% acid strength cider
        vinegar
¾   c. water
1   piece fresh ginger root
        (1" cube)
2   sticks cinnamon

• Dip peaches into boiling water for ½ minute to loosen skins. Cool in cold water; drain. Remove skins. Stud each peach with a clove. To prevent darkening, place in 4-qt. bowl containing 1½ tblsp. salt and 2 qts. water.

• Combine sugar, vinegar and ¾ c. water in 4-qt. kettle. Tie ginger root, remaining cloves and cinnamon in cheesecloth bag and add to pot. Bring to boil. Cook half of the peaches in syrup for 10 minutes; remove with slotted spoon; cook remaining peaches 10 minutes. Place peaches into 1-gal. crock or glass bowl; pour syrup over, cover and let stand 12–18 hours in a cool place.

• Drain peaches, reserving syrup in 2-qt. saucepan. Pack peaches into 4 hot pint jars. Heat syrup to boiling. Remove spice bag. Pour syrup over peaches, filling to within ¼" of jar top. Wipe jar rim; adjust lids.

• Process in boiling water bath 20 minutes. Start to count processing time when water in canner returns to boiling. Remove jars and complete seals unless closures are self-sealing type. Makes 4 pints.

PEAR PICKLES

Use Bartlett pears for a special gourmet treat you'll never forget

3½ lbs. ripe pears (10 large, 14 to
        16 medium)
2½ c. sugar
1¼ c. 5% acid strength vinegar
1   c. water
1   (2") piece fresh ginger root
2   tblsp. whole cloves
7   (3") sticks cinnamon

• Wash, pare, halve and core pears. To prevent darkening, place in 4-qt. bowl containing 1 qt. water and 1 tblsp. vinegar.

• Combine sugar, 1¼ c. vinegar and 1 c. water in 1-qt. saucepan. Tie ginger, cloves, and cinnamon in cheesecloth bag; add to saucepan. Bring to boil;

cover; boil 5 minutes. Remove spice bag.

• Drain pears. Pack pears into 3 hot pint jars; cover with boiling syrup, filling to within ¼" of jar top. Wipe jar rim; adjust lids.

• Process in boiling water bath 15 minutes. Start to count processing time when water in canner returns to boiling. Remove jars and complete seals unless closures are self-sealing type. Makes 3 pints.

GOURMET GINGER PEARS

A relish to serve with meats; spread for hot biscuits and toast

4 lbs. ripe pears (14 medium)
5 c. sugar
¼ c. fresh lemon juice
⅓ c. finely chopped candied ginger
2 tsp. grated lemon peel (peel from 1 lemon)
¼ c. 5% acid strength cider or white vinegar

• Wash, pare, quarter and core pears. To prevent darkening, place in 4-qt. bowl containing 1 qt. water and 1 tblsp. vinegar. Drain.

• Pour thin layer of sugar over bottom of 4-qt. bowl or crock; begin layering pears, cup side down, covering generously with sugar as you pack. Cover. Let stand 6 to 8 hours at room temperature.

• Pour into 4-qt kettle; add lemon juice and candied ginger. Bring to boil; simmer, uncovered, over low heat, stirring frequently, until pears are tender and translucent, about 1 hour. Add lemon peel and vinegar 5 minutes before cooking time is completed.

• Ladle pears into 5 hot half-pint jars;

cover with syrup, filling to within ¼" of jar top. Wipe jar rim; adjust lids.

• Process in boiling water bath 20 minutes. Start to count processing time when water in canner returns to boiling. Remove jars and complete seals unless closures are self-sealing type. Makes 5 half pints.

Note: Candied ginger is expensive; you can substitute ⅓ c. chopped ginger root. Tie it in a cheesecloth bag and remove it before packing pears.

PEAR RELISH

"Deep South" specialty

4 lbs. Kieffer pears (4 c. chopped)
3 c. 5% acid strength cider vinegar
3 c. sugar
2 tblsp. canning/pickling salt
1½ tsp. celery seeds
1 tblsp. allspice (optional)
6 medium onions, peeled (4 c. chopped)
15 large green peppers, seeded and stemmed (5 c. chopped)
4 hot red peppers, seeded and stemmed

• Wash pears; peel, core, and slice directly into salt water solution (1 tblsp. salt to 1 qt. water) to prevent darkening of fruit.

• Combine vinegar, sugar, salt and celery seeds in 5-qt. kettle. If desired, add allspice tied in cheesecloth bag. Bring to boil.

• Put pears, onions and peppers through food chopper; add to kettle. Simmer, uncovered, until onions and pears are translucent, about 40 minutes. Remove spice bag.

• Ladle into 6 hot pint jars, pressing down as you pack so syrup covers relish. Fill to within ¼" of jar top. Wipe jar rim; adjust lids.

· Process in boiling water bath 15 minutes. Start to count processing time when water in canner returns to boiling. Remove jars and complete seals unless closures are self-sealing type. Makes 6 pints.

PINEAPPLE PICKLES

Put on relish tray when you barbecue chicken—guests gobble it up!

2   large pineapples
1½ c. sugar
¾   c. water
⅓   c. 5% acid strength vinegar
10 whole cloves
1   stick cinnamon

· Cut pineapple in ¼" crosswise slices; peel, remove eyes. Leave slices whole or cut into quarters or eighths; remove cores.

· Combine pineapple, sugar, water and vinegar in 4-qt. kettle; add cloves and cinnamon (tied in cheesecloth bag). Bring to boil; simmer uncovered 30 minutes. Remove spice bag.

· Ladle pineapple into 3 hot pint jars. Cover with syrup, filling to within ¼" of jar top. Wipe jar rim; adjust lids.

· Process in boiling water bath 20 minutes. Start to count processing time when water in canner returns to boiling. Remove jars and complete seals unless closures are self-sealing type. Makes 3 pints.

QUINCE PICKLE

Lovely apricot-peach color

8 ripe yellow quinces
2 c. water
Whole cloves
2 oranges
6 c. sugar
2 c. 5% acid strength vinegar
1 (3") stick cinnamon
1 apple, peeled, cored, diced

· Wash, peel, core and quarter quinces. Place in 4-qt. kettle, add water and cook until tender. Drain, reserving the liquid.

· Stick 2 cloves in each quince piece.

· Slice unpeeled oranges, discarding seeds; combine in kettle with sugar, vinegar, 1½ c. reserved liquid from quinces (add water if measure is short), cinnamon and apple. Simmer over low heat 10 minutes. Add quince slices and simmer gently 30 minutes more.

· Ladle fruit into 6 hot pint jars to within ½" of jar top. Cover with hot syrup, leaving ¼" head space. Wipe jar rim; adjust lids.

· Process in boiling water bath 20 minutes. Start to count processing time when water in canner returns to boiling. Remove jars and complete seals unless closures are self-sealing type. Makes 6 pints.

Watermelon Pickles

"No matter how busy I am in summer, I always make time to put up watermelon and peach pickles," says a Georgia farm woman. "No relish adds more to chicken dinners than one of these favorites."

Through the years this superior cook has made watermelon pickles, she has accumulated knowledge that pickle makers can use. Here are her tips:

The rind of watermelon is better for pickling if the melon is not overripe and if not grown late in the season.

The pink flesh of the melon will not become crisp—leave only a very thin line of pink.

A 16-pound melon yields 5 to 6 pounds of rind. And 1 pound of rind measures 1 quart. One quart of rind makes 1 pint of pickles.

WATERMELON PICKLES

Not too spicy, delicate in flavor

4–5 qts. watermelon rind
　　(about 4 lbs.)
Salt water solution (see below)
2 c. 5% acid strength cider
　　vinegar
7 c. sugar
1 tblsp. whole cloves
3 sticks cinnamon
1 piece dried or fresh ginger root
　　(1″ cube)
Red or green food coloring
　　(optional)

• Choose melon with thick, firm rind. Trim off outer green skin and pink flesh, leaving very thin line of pink. Cut into 1″ squares. Place in 2-gal. crock.
• Cover rind with salt water solution (2 tblsp. salt to 4 qts. water). Cover. Let stand 24 hours at room temperature. Drain; rinse with cold water. Cover with ice water. Let stand 1 hour. Drain.
• Place rind in 4-qt. kettle. Cover with boiling water. Bring to boil; re-

duce heat and simmer until tender, about 10 minutes. Drain.
• Combine vinegar, sugar and spices (tied in cheesecloth bag) in 4-qt. kettle. Bring to boil. Add rind. Cook gently until rind is translucent. Remove spice bag. Add coloring if desired.
• Turn rind and syrup into 1-gal. crock. Cover. Let stand 24 hours at room temperature.
• Drain off syrup; heat to boiling. Pack rind into 6 sterilized hot pint jars; cover with syrup, filling to within ¼″ of jar top. Wipe jar rim; adjust lids.
• Process in boiling water bath 5 minutes. Start to count processing time when water in canner returns to boiling. Remove jars and complete seals unless closures are self-sealing type. Makes 6 pints.

3-DAY WATERMELON PICKLES

They're worth the effort

8 qts. watermelon rind (7 lbs.)
Salt water solution (see below)
7 c. sugar
2 c. 5% acid strength vinegar
¼ tsp. oil of cloves (or 1 tblsp. whole
　　cloves)
¼ tsp. oil of cinnamon (or 3 sticks
　　cinnamon)
Red food coloring (optional)

• Choose a melon with thick, firm rind. Trim off outer green skin and pink flesh, leaving very thin line of pink. Stamp out rind with small cookie cutter or cut into neat 1″ squares. Place in 2-gal. crock.
• Cover rind with salt water solution

(¼ c. salt to 1 qt. water) and soak for 2 hours. Drain; rinse.

· Place rind in 4-qt. kettle and cover with cold water. Bring to boil; cook until tender, about 10 minutes; drain. Place in 2-gal. crock.

· Combine sugar, vinegar and spices (if using whole spices, tie in cheesecloth bag); heat to boiling. Add coloring if desired. Pour over rind. Let stand overnight at room temperature.

· Drain off syrup; heat syrup to boiling and pour over rind. Let stand overnight.

· Heat rind in syrup. Remove spice bag. Pack rind into 8 sterilized hot pint jars; cover with syrup, filling to within ¼" of jar top. Wipe jar rim; adjust lids.

· Process in boiling water bath 5 minutes. Start to count the processing time when water in canner returns to boiling. Remove jars and complete seals unless closures are self-sealing type. Makes 8 pints.

Note: If you like a strongly spiced pickle, increase oil of cloves and oil of cinnamon to ½ tsp. each.

ROSY WATERMELON PICKLES

A small bottle of maraschino cherries gives pickles a pretty color

4–5 qts. watermelon rind (about 4 lbs.)
Salt water solution (see below)
2 c. 5% acid strength cider vinegar
1⅓ c. water

6⅔ c. sugar
2 tsp. whole cloves
2 sticks cinnamon
1 tsp. peppercorns
1 piece fresh or dried ginger root (1" cube)
½ c. (4 oz. jar) maraschino cherries

· Choose melon with thick, firm rind. Trim off outer green skin and pink flesh, leaving very thin line of pink. Cut into 1" squares. Place in 2-gal. crock. Cover with salt water solution (3 tblsp. salt to 4 qts. cold water). Let stand 24 hours at room temperature. Drain.

· Cover rind with boiling water and boil gently 1½ hours. Drain. Place rind in 1-gal. crock or glass bowl, cover with ice water until thoroughly chilled (about 1 hour). Drain.

· Combine vinegar, 1⅓ c. water, sugar and spices (tied in cheesecloth bag) in 4-qt. kettle. Bring to boil. Add rind; boil gently 30 minutes; remove spice bag. Place rind and syrup in 1-gal. crock. Let stand in cool place for 24 hours.

· Pour rind and syrup in 4-qt. kettle, add maraschino cherries. Bring to boil.

· Pack into 5 sterilized hot pint jars; cover rind with syrup, filling to within ¼" of jar top. Wipe jar rim; adjust lids.

· Process in boiling water bath 5 minutes. Start to count processing time when water in canner returns to boiling. Remove jars and complete seals unless closures are self-sealing type. Makes 5 pints.

Fruit/Vegetable Relishes

Fruits and vegetables team together in many delightful relishes. None of them is more popular than Victory Relish. It goes by many names in various parts of the country. Some women call it a chutney, others a sauce. Its name is not important—taste is what counts. And just about everyone who samples, praises it and takes a second helping.

VICTORY RELISH

Relish has zip—cut down on spices if you prefer a milder taste

20 large (3½" diameter) ripe
    tomatoes (6 lbs.)
8 pears, peeled and cored (2 lbs.)
8 peaches, peeled and pitted (2 lbs.)
6 large (3" diameter) onions,
    peeled (1½ lbs.)
2 large sweet red peppers,
    stemmed and seeded (½ lb.)
3 c. 5% acid strength cider vinegar
4 c. sugar
2 tblsp. canning/pickling salt
4 oz. whole pickling spices

• Peel and coarsely chop tomatoes into 8-qt. kettle.
• Put pears, peaches, onions, and peppers through coarse blade of food chopper. Combine with tomatoes.
• Add vinegar, sugar, salt and spices (tied in cheesecloth bag). Cook slowly, uncovered, stirring occasionally, until mixture is thick and rounds up on spoon, about 2 hours.
• Ladle into 7 hot pint jars, filling to within ¼" of jar top. Wipe jar rim; adjust lids.
• Process in boiling water bath 15 minutes. Start to count processing time when water in canner returns to

boiling. Remove jars and complete seals unless closures are self-sealing type. Makes 7 pints.

FRUITED VEGETABLE RELISH

Garden and orchard share honors in this prized relish—extra-good

6 lbs. peeled ripe tomatoes
    (12 very large)
½ c. chopped celery
1 c. chopped onion
1 sweet red pepper, cored and
    chopped
1 green pepper, cored and
    chopped
5 peaches, peeled, pitted, chopped
3 pears, peeled, cored, chopped
1 hot pepper, whole
1 tblsp. mixed pickling spices
2½ c. sugar
1 tblsp. canning/pickling salt
1 c. 5% acid strength cider
    vinegar

• Dice tomatoes into 8-qt. kettle. Add chopped celery, onion, peppers, peaches and pears. Add hot pepper and pickling spices (tied in cheese-cloth bag); stir in sugar, salt and vinegar. Bring to a boil; cook slowly, uncovered, for 2 hours or until mixture is thick and rounds up on spoon. Remove hot pepper and spice bag.
• Ladle into 5 hot pint jars, filling to within ¼" of jar top. Wipe jar rim; adjust lids.
• Process in boiling water bath 15 minutes. Start to count processing time when water in canner returns to boiling. Remove jars and complete seals unless closures are self-sealing type. Makes 5 pints.

INDIA RELISH

Not too spicy, not too sweet—just right! Good with beef

7 lbs. peeled ripe tomatoes
    (3 qts. chopped)
3 lbs. tart apples, peeled
    (6 c. chopped)
1 lb. onions, peeled (3 c. chopped)
1 sweet red pepper, stemmed and
    seeded
1 c. raisins
2 tsp. canning/pickling salt
1 c. sugar
2 c. 5% acid strength cider vinegar

• Chop coarsely tomatoes, apples, onions and pepper.
• Combine all ingredients in 5-qt. kettle. Cook slowly, uncovered, stirring occasionally, 1½ hours or until mixture is thick and rounds up on spoon.
• Ladle into 7 hot pint jars, filling to within ¼" of jar top. Wipe jar rim; adjust lids.
• Process in boiling water bath 15 minutes. Start to count processing time when water in canner returns to boiling. Remove jars and complete seals unless closures are self-sealing type. Makes 7 pints.

1½ c. 5% acid strength cider
    vinegar
2¾ c. sugar
1 tblsp. canning/pickling salt
2 tblsp. mustard seeds
½ tblsp. whole cloves
6 (3") sticks cinnamon

• Dice tomatoes into 4-qt. kettle. Add chopped onion, celery, apples, peppers, vinegar, sugar, salt and spices (tied in cheesecloth bag). Bring to boil; simmer, uncovered, stirring often, for 1½ hours or until mixture is thick and rounds up on spoon. Remove spice bag.
• Ladle into 3 hot pint jars, filling to within ¼" of jar top. Wipe jar rim; adjust lids.
• Process in boiling water bath 15 minutes. Start to count processing time when water in canner returns to boiling. Remove jars and complete seals unless closures are self-sealing type. Makes 3 pints.

Note: For most accurate measure of chopped ingredients, use Water Displacement Method—see Index.

TOMATO/APPLE RELISH

Apples and tomatoes are flavorful companions in this spicy treat

2 qts. peeled, diced ripe tomatoes
    (about 5 lbs.)
1 c. chopped onion
1 c. chopped celery
2 c. chopped unpeeled tart apples
2 green peppers, cored and
    chopped
2 sweet red peppers, cored and
    chopped

GOLDEN RELISH

Appealing color, extremely good taste

4 c. chopped carrots (1½ lbs.)
4 c. chopped celery (2⅓ lbs.)
2 green peppers, chopped
2 sweet red peppers, chopped
8 unpeeled cooking apples, cored
    and chopped
4 c. sugar
2 c. 5% acid strength cider vinegar
2 c. water
2 tsp. celery seeds
2 tblsp. canning/pickling salt

• Combine all ingredients in 8-qt. kettle; bring to boil; simmer until carrots are tender when pricked with a fork, about 45 minutes.

• Ladle into 7 hot pint jars, filling to within ¼" of jar top. Wipe jar rim; adjust lids.

• Process in boiling water bath minutes. Start to count processi time when water in canner returns boiling. Remove jars and comple seals unless closures are self-sealir type. Makes 7 pints.

Red Relish Sauces

Of all the relish sauces, chili sauce and ketchup have the widest circle of friends. Country cooks like to make them with ripe tomatoes so heavy that they are about to drop from the vines. Their full flavors produce marvelous sauces. A pair of recipes for each follows.

It's a good idea to wrap the filled jars with brown paper before putting them away. Excluding the light helps to retain their cheerful color.

EASTERN CHILI SAUCE

New Englanders like to serve this at their famed baked bean suppers

8 lbs. ripe tomatoes (25–
      30 medium)
1½ c. 5% acid strength cider
      vinegar
2 tsp. whole cloves
1 stick cinnamon, broken
1 tsp. celery seeds
1 c. sugar
1 tblsp. chopped onion
½ tsp. ground red pepper
      (cayenne)
1 tblsp. canning/pickling salt

• Dip tomatoes into boiling water f ½ minute to loosen skins. Cool cold water. Remove skins, cores a quarter tomatoes into 4-qt. bowl.

• Combine vinegar, cloves, cinnam and celery seeds in 1-qt. saucepa Bring to boil; remove from heat a let stand to blend flavors.

• Combine half of tomatoes with ½ sugar, onion and pepper in 4-qt. k tle. Simmer 45 minutes, uncovere stirring often.

• Add remaining tomatoes and sug to boiling tomato mixture. Simme uncovered, 45 minutes more, stirri frequently.

• Strain vinegar and discard spice Add spiced vinegar and salt to toma mixture and cook, stirring constantl about 15 minutes or until sauce thick and begins to round up spoon.

• Ladle into 2 hot pint jars, filling within ⅛" of jar top. Wipe jar rir adjust lids.

• Process in boiling water bath minutes. Start to count processi time when water in canner returns boiling. Remove jars and comple seals unless closures are self-sealir type. Makes 2 pints.

TOMATO KETCHUP

*You can double recipe but cooking
time will be longer, ketchup darker*

8   lbs. ripe tomatoes (25–
    30 medium)
1   c. chopped onions
¼   tsp. ground red pepper
    (cayenne)
1   c. 5% acid strength cider
    vinegar
1½ tsp. whole cloves
1   stick cinnamon, broken into
    pieces
½   tsp. whole allspice
1   tsp. celery seeds
½   c. sugar
4   tsp. canning/pickling salt

• Dip tomatoes into boiling water for
½ minute to loosen skins. Place into
cold water, remove skins and cores,
cut in quarters into 5-qt. kettle.
• Combine tomatoes, onions and red
pepper; cook, uncovered, 20 minutes,
stirring occasionally.
• Combine vinegar and spices (tied in
a cheesecloth bag) in 1-qt. saucepan.
Bring to boil; remove from heat. Let
stand to mingle flavors while ketchup
cooks down.
• Put tomato mixture through food
mill or sieve. Return to kettle; stir in
sugar and salt. Cook, uncovered, stirring frequently, until volume is reduced one-half, about 75 minutes.
• Strain vinegar to remove spices; stir
into tomato mixture. Continue boiling, stirring constantly, until mixture
rounds up on spoon with no separation of liquid and sauce, about 30
minutes.
• Ladle into 2 hot pint jars, filling to
within ⅛" of jar top. Wipe jar rim;
adjust lids.

• Process in boiling water bath for 15
minutes. Start to count processing
time when water in canner returns to
boiling. Remove jars and complete
seals unless closures are self-sealing
type. Makes 2 pints.

GRANDMOTHER'S SHIRLEY SAUCE

*Unspiced, tangy sweet-sour—perfect
teammate for beef*

6   lbs. ripe tomatoes
    (10 c. chopped)
⅔   c. chopped green peppers (2
    large)
2   c. chopped onions (2 large)
2   c. sugar
2   c. 5% acid strength cider vinegar
2   tblsp. canning/pickling salt

• Dip tomatoes into boiling water ½
minute to loosen skins. Cool in cold
water. Remove skins and cores. Blend
or put through food chopper. Place in
8-qt. kettle.
• Remove stems, membranes and
seeds from peppers and peel onions
before chopping. Add to tomatoes;
stir in sugar, vinegar and salt. Simmer, uncovered, stirring frequently,
for 2 hours or until thick and sauce
begins to round up on spoon.
• Ladle into 3 hot pint jars, filling to
within ⅛" of jar top. Wipe jar rims;
adjust lids.
• Process in boiling water bath 15
minutes. Start to count processing
time when water in canner returns to
boiling. Remove jars and complete
seals unless closures are self-sealing
type. Makes 3 pints.

WESTERN GOURMET KETCHUP

Hot peppers add a hint of chili flavor. Great on beef

18 lbs. ripe tomatoes (about 72, 2½")
3 tblsp. canning/pickling salt
1 c. sugar
1 tblsp. paprika
¼ tsp. ground red pepper (cayenne)
1 tblsp. dry mustard
1 tblsp. whole peppercorns
1 tblsp. whole allspice
1 tblsp. mustard seeds
4 bay leaves
1 tblsp. basil leaves
4 hot peppers, sliced and seeded
2 c. 5% acid strength cider vinegar

• Dip tomatoes into boiling water ½ minute to loosen skins. Cool in cold water. Remove skins and cores. Place in 12-qt. kettle, cook until soft, about 20 minutes. Press through sieve or food mill. (Makes 7 quarts purée.) Return purée to kettle.

• Add salt, sugar, paprika, red pepper (cayenne) and mustard. Tie peppercorns, allspice, mustard seeds, bay and basil leaves and hot peppers in cheesecloth bag; add to kettle. Bring to boil; reduce heat and simmer until ketchup is very thick, about 3 hours. Add vinegar and simmer about 10 minutes longer, or until ketchup rounds up on spoon and there is no separation of liquids and sauce. Remove spice bag.

• Ladle into 4 hot pint jars, filling to within ⅛" of jar top. Wipe jar rim; adjust lids.

• Process in boiling water bath 15 minutes. Start to count processing time when water in canner returns to boiling. Remove jars and complete seals unless closures are self-sealing type. Makes 4 pints.

BLENDER KETCHUP

Texture is chunky—more like sauce— but it has a good ketchup taste

8 lbs. ripe tomatoes (about 32, 2½")
2 medium-size sweet red peppers
2 medium-size green peppers
4 onions (2" diameter), peeled and quartered
3 c. 5% acid strength cider vinegar
3 c. sugar
3 tblsp. canning/pickling salt
3 tsp. dry mustard
½ tsp. ground red pepper (cayenne)
½ tsp. whole allspice
1½ tsp. whole cloves
1 stick cinnamon

• Dip tomatoes into boiling water for ½ minute to loosen skins. Place in cold water. Remove skins and cores. Quarter.

• Remove seeds and membranes from peppers; cut into strips.

• Put tomatoes, peppers and onion into blender container, filling jar ¾ full. Blend at high speed 4 seconds; pour into 5-qt. kettle. Repeat until all vegetables are blended.

• Cook, stirring constantly, about 5 minutes. Add vinegar, sugar, salt, dry mustard, red pepper and spices (tied in cheesecloth bag). Cook, uncovered, over medium heat about 2 hours; stir frequently to prevent scorching. When volume is reduced to one half and mixture rounds up on spoon with

no separation of liquid and sauce, remove from heat; remove spice bag.
• Ladle into 4 hot pint jars, filling to within ⅛" of jar top. Wipe jar rim; adjust lids.
• Process in boiling water bath 15 minutes. Start to count processing time when water in canner returns to boiling. Remove jars and complete seals unless closures are self-sealing type. Makes 4 pints.

VARIATION

BLENDER APPLE/TOMATO KETCHUP: Add 2 c. thick applesauce to cooked ketchup and mix thoroughly. Makes 6 pints.

SMOOTH BLENDER KETCHUP: Follow recipe for Blender Ketchup, but put tomato mixture through a food mill after 20 minutes cooking; return puréed mixture to kettle and cook until volume is reduced by one-half. Makes 5 half pints.

HOMEMADE FRUIT KETCHUP

Makes a good sandwich spread, but also use it like tomato ketchup

12 tart apples (4 lbs.), peeled, cored, sliced
2 medium onions (2"), finely chopped
½ c. water
1 c. sugar
1½ c. 5% acid strength cider vinegar
¼ tsp. ground black pepper
1 tsp. canning/pickling salt
1 tsp. dry mustard
1 tsp. ground cloves
1 tsp. ground cinnamon
¼ tsp. ground allspice
Red food coloring (optional)

• Place apples, onions and water in 4-qt. kettle. Cook until apples are tender. Put through sieve or food mill.
• Combine apple mixture, sugar, vinegar and seasonings in 4-qt. kettle. Cook, uncovered, over medium heat, stirring often, until thick and mixture rounds up on spoon with no separation of liquid and sauce, about 1 hour. Add coloring if desired.
• Pour into 3 hot pint jars, filling to within ⅛" of jar top. Wipe jar rim; adjust lids.
• Process in boiling water bath for 15 minutes. Start to count processing time when water in canner returns to boiling. Remove jars and complete seals unless closures are self-sealing type. Makes 3 pints.

GOVERNOR'S SAUCE

Makes a pot roast meal special

14 lbs. slightly ripe tomatoes (42 medium or 26 large)
¼ c. canning/pickling salt
6 large green peppers
6 large sweet red peppers
6 large unpeeled red apples
1 medium head cabbage
6 c. sugar
¼ c. canning/pickling salt
4 c. 5% acid strength cider vinegar
6 tblsp. mixed pickling spices (1¼ oz. pkg.)

• Dip tomatoes in boiling water until skins crack. Cool in cold water. Drain. Remove cores and skins. Dice into 2-gal. crock or glass bowl. Add

¼ c. salt; mix well. Cover. Let stand 3 hours at room temperature. Drain. Rinse. Place in 6-qt. kettle.
• Remove seeds and membranes from peppers; remove cores from apples and cabbage. Put through food chopper (coarse blade); combine with tomatoes. Add sugar, ¼ c. salt, vinegar and spices (tied in cheesecloth bag). Bring to boil; boil, uncovered, 30 minutes, stirring often. Remove spice bag. Pour into 2-gal. crock or glass bowl. Cover. Let stand overnight (12–18 hours) in a cool place.
• Return ingredients to 6-qt. kettle. Cover. Reheat to boiling, stirring often.
• Ladle into 12 hot pint jars, pressing solids down as you pack until sauce covers solids. Fill to within ¼" of jar top. Wipe jar rim; adjust lids.
• Process in boiling water bath 15 minutes. Start to count processing time when water in canner returns to boiling. Remove jars and complete seals unless closures are self-sealing type. Makes 12 pints.

CRANBERRY KETCHUP

Quick to cook, beautiful color. Great spread for chicken sandwiches

4 lbs. cranberries (4 qts.)
1 lb. onions, peeled and chopped
2 c. water
5½ c. sugar
2 c. 5% acid strength cider vinegar
1 tblsp. ground cloves
1 tblsp. ground cinnamon
1 tblsp. ground allspice
1 tblsp. canning/pickling salt
1 tsp. black pepper

• Place cranberries, onions and water in 5-qt. kettle. Cook, uncovered, until cranberries burst and onions are translucent. Put through sieve or food mill.
• Return purée to pot, add sugar, vinegar and spices; boil and stir constantly until thick, about 8 minutes.
• Pour into 6 hot pint jars, filling to within ⅛" of jar top. Wipe jar rim; adjust lids.
• Process in boiling water bath 15 minutes. Start to count processing time when water in canner returns to boiling. Remove jars from canner and complete seals unless closures are self-sealing type. Makes 6 pints.

CRANBERRY PICKLE SAUCE

Thick relish with fine flavor—gets along famously with white meats

4 c. (1 lb.) cranberries
1½ c. sugar
½ c. 5% acid strength cider vinegar
½ tsp. ground cinnamon
½ tsp. ground nutmeg
½ tsp. ground cloves

• Combine cranberries, sugar and vinegar in 3-qt. saucepan. Bring to boil; simmer, uncovered, until all cranberries burst. Add spices; cook, uncovered, over low heat 15 minutes.
• Ladle into 3 hot half-pint jars, filling to within ⅛" of jar top. Wipe jar rim; adjust lids.
• Process in boiling water bath 15 minutes. Start to count processing time when water in canner returns to boiling. Remove jars and complete seals unless closures are self-sealing type. Makes 3 half pints.

Country-Style Chutneys

You don't have to travel to India to enjoy chutneys. Country cooks make American versions of these highly spiced relishes from fruits they have —cranberries, for instance. Traditionally, you serve them with curried dishes—a splendid combination. But they're excellent with almost all chicken and rice combinations.

"When I hear the children at play rattling in autumn's fallen leaves," says a FARM JOURNAL reader, "I know it's time to make Apple/Tomato Chutney. We think it's tops with poultry and pork and I like to take it to covered dish suppers." Here is the recipe for it and several others.

TOMATO-APPLE CHUTNEY

A sweet, spicy sauce to serve as an accompaniment for meats

- 6   lbs. ripe tomatoes, peeled and chopped (18 medium)
- 4   lbs. apples, peeled and finely chopped
- 2   c. chopped onions
- 1   c. chopped green peppers
- 2   lbs. light brown sugar (5⅓ c., firmly packed)
- 2   c. dark, seedless raisins
- 4   c. 5% acid strength cider vinegar
- 4   tsp. salt
- 4   tsp. ground ginger
- ½   tsp. ground allspice
- 4   large garlic cloves, minced

• Combine all ingredients in 12-qt. kettle. Bring to boil; cook, over low heat, stirring frequently, for 2 hours or until mixture thickens and rounds up on spoon.

• Ladle into 8 hot pint jars, filling to within ⅛" of jar top. Wipe jar rim; adjust lids.

• Process in boiling water bath 15 minutes. Start to count processing time when water in canner returns to boiling. Remove jars and complete seals unless closures are self-sealing type. Makes 8 pints.

APPLE CHUTNEY

Just right for chicken or game dinner

- 8   c. chopped, peeled apples (10 medium)
- 2   c. seedless light raisins
- Peel of 2 oranges, finely chopped
- 4½ c. sugar
- ½   c. 5% acid strength cider vinegar
- ⅓   tsp. ground cloves
- 1   c. chopped pecans (optional)

• Combine all ingredients in 4-qt. kettle. Bring to rolling boil; reduce to simmer. Cook until apples are tender and translucent, about 30 minutes.

• Ladle into 7 hot half-pint jars, filling to within ⅛" of jar top. Wipe jar rim; adjust lids.

• Process in boiling water bath 15 minutes. Start to count processing time when water in canner returns to boiling. Remove jars and complete seals unless closures are self-sealing type. Makes 7 half pints.

HAWAIIAN CHUTNEY

Spicy-hot with sweet, fruity taste

4   c. chopped, cored, peeled apples
1   (1 lb. 3 oz.) can crushed,
      unsweetened pineapple
½   c. 5% acid strength cider
      vinegar
½   c. light brown sugar, firmly
      packed
1   c. seedless light raisins
1½ tsp. canning/pickling salt
2   cloves garlic, finely chopped
½   tsp. ground cloves
½   tsp. ground cinnamon
½   tsp. red pepper (optional)
2   tblsp. candied ginger, finely
      chopped (or ½ ground
      ginger)
½   c. slivered almonds (optional)

• Combine all ingredients in 3-qt. saucepan. Bring to rolling boil; simmer uncovered for 30 minutes, stirring often.
• Ladle into 6 hot half-pint jars, cover with syrup, filling to within ⅛" of jar top. Wipe jar rim; adjust lids.
• Process in boiling water bath 15 minutes. Start to count processing time when water in canner returns to boiling. Remove jars and complete seals unless closures are self-sealing type. Makes 6 half pints.

PEACH/LIME CHUTNEY

Lime juice lends a distinctive tang. Good with cold meats, chicken, turkey, pork, also with rice casseroles

5   lbs. peaches
2   c. 5% acid strength cider vinegar
¼   c. fresh lime juice
1   lemon, quartered, seeded and
      diced
½   lb. seedless raisins
½   lb. dates, chopped
½   c. chopped nuts (optional)
3   c. sugar
½   c. candied ginger, coarsely
      chopped (or 2 tsp. ground
      ginger)

• Dip peaches into boiling water for 30 seconds to loosen skins; place in cold water; remove skins and pits. Dice peaches into vinegar and lime juice in 5-qt. kettle. Add lemon, raisins, dates and nuts. Cook mixture over low heat until peaches are tender and translucent, stirring often.
• Add sugar; cook over low heat until chutney has thickened, about 1½ hours. Stir in ginger.
• Ladle into 5 hot pint jars, filling to within ⅛" of jar top. Wipe jar rim; adjust lids.
• Process in boiling water bath 15 minutes. Start to count processing time when water in canner returns to boiling. Remove jars and complete seals unless closures are self-sealing type. Makes 5 pints.

CHAPTER 15

Jams, Jellies and
Other Spreads

Ingredients and Equipment • The Paraffin Question • How to Make
Jellies and Jams • Recipes for Short-cook, Longer-cook and Uncooked (Frozen) Jellies and Jams • Fruit Preserves • Special Occasion Conserves • Old-fashioned Fruit Butters • Cheerful Marmalades • The Gourmet Corner • Wild Fruit Specialties

With promises of jam on the table, a farm mother has always managed to
get the berry patch weeded. Today's farm children still search the fence-
rows and hillsides for wild berries. Now, families also come together from
towns and cities to buy sun-ripened fruit at farm stands or on "pick-your-
own" farms. The tradition of summer preserving continues strong. But
with some notable improvements!

Grandmother boiled her preserves in the heat of summer, canning in
rhythm with the ripening fruit. Today most farm women utilize both
freezer and canning kettle—partly for convenience, but also for taste.
They praise the short-cook methods and the easier-to-fix frozen spreads
that retain so much fresh fruit flavor and aroma. You will find recipes in
this chapter for the best canned and frozen sweet spreads that FARM JOUR-
NAL's readers make.

Many fruit growers' wives have developed specialties which they find
profitable to sell at roadside stands, by mail and in gourmet shops. Don't
miss this chapter's Gourmet Corner, filled with potential sellers and winners in fair exhibits. And be sure to note the delicacies made from wild
fruits, often to be had for the picking—bright red chokecherries, clusters
of wild grapes on tangled vines, sweet strawberries ripening in the pasture.

These direct-from-the-farm-to-you recipes will add sparkle and flavor
to your meals and boost your fame as a good cook. And they may inspire
you to earn money in your own kitchen!

Preserving Time in the Country

When gentle spring breezes brush the purple iris, country cooks know it's time to make rhubarb jelly and strawberry jam. But if your schedule doesn't mesh with Nature's timing, do what busy farm women do: Quickly freeze the fresh fruit so you can make jam whenever the spirit moves you. Or extract and can juice now; boil jelly in January. Certainly there is something satisfying about ladling luscious fruity sweets into canning jars when blizzards whip the trees. "The kitchen is so comfortable and cheery in stormy winter weather," a farm woman says. "I feel I've conquered nature on the warpath if I make jelly and jam then."

You can also use commercially frozen or dried fruit or canned juices to make sweet spreads as you need them. Some of our recipes specify these products.

Cook jellied fruits in small batches. The small-size recipes in this chapter are easy to work with and your chances of success will be much greater. Double batches take longer to cook and this affects flavor and texture. We do not recommend doubling these recipes.

KNOW YOUR FRUIT

Grandmother knew that tart apples, tart blackberries, crab apples, cranberries, red currants, gooseberries, Concord Grapes, lemons, loganberries, May haws, muscadines, quinces and many varieties of tart plums were good jelly makers. She used the poor jelly makers for jams, marmalades and preserves—apricots, figs, peaches, pears, pomegranates, prunes, raspberries, strawberries, ripe apples, tart cherries, elderberries, grapefruit, loquats and oranges.

If she was in an adventuresome mood, she mixed equal parts of juices from a poor jelly-making fruit and a good one—raspberries with plums, perhaps. Sometimes her efforts paid off with delicate tasty jelly. Often she was disappointed. Her rule for both jellies and jams was to use slightly underripe fruits or at least ¼ *underripe* and ¾ *ripe*.

Today we know that jellies jell and jams set when the proper amounts of fruit, pectin, acid and sugar are present. The greatest concentration of pectin and acid occurs in slightly underripe fruit; the best flavor comes from ripe fruit. If they're not too ripe, some fruits (see the "good jelly makers" listed above) have enough natural pectin and acid to jell when cooked in the traditional long-boil method.

But fruit that's fully ripe, and particularly the second list of fruits—the "poor jelly makers"—will need some added pectin and added acid to make a jelly firm enough to hold its shape.

You can find out how much pectin and acid your fruit contains with simple tests. Or you can do what most women do today—skip the pectin test and simply choose a recipe calling for ripe fruit, commercial pectin and—usually—a little lemon juice.

PECTIN

Pectin is the jelling agent—a natural substance found in all fruits and berries. The amount varies according to the kind of fruit and degree of ripeness; even the variety of the fruit or growing conditions can affect the concentration of pectin.

Recipes in this chapter for the longer-cook jellies and jams call for fruits that are dependably rich in pectin, or the recipe may direct you to make a pectin test, to insure success.

Pectin Test: Measure 1 tblsp. ordinary rubbing alcohol into a cup. Gently stir in 1 tsp. cooked, cooled juice. *Do not taste—rubbing alcohol is poisonous.* If the juice is rich in pectin, it will form a gelatinous mass you can pick up with a fork. Juice low in pectin will remain liquid or form only a few bits of jelly-like material.

Jelmeter Test for Pectin: A jelmeter is a calibrated glass tube open at both ends, available in some hardware stores or housewares departments. You fill the tube with fruit juice and let it flow through. According to directions that come with the jelmeter, the time it takes indicates the amount of pectin in the juice.

If your test shows that your juice is low in natural pectin, you can use a "short-cook" recipe that calls for the addition of commercial pectin. Or follow directions for longer cook jellies (see Index).

Commercial Pectin. The powdered or liquid pectins which you buy in grocery stores are made by extracting natural pectin from apples or the skins of citrus fruits. Either form will give you good results if you follow recipes exactly. However, you cannot use powdered pectin in a recipe that calls for liquid pectin—the two forms are not interchangeable. Powdered fruit pectin won't dissolve in high sugar concentration so it must be stirred into the fruit before sugar is added. Liquid fruit pectin is added after fruit and sugar have come to a boil.

Buy only as much pectin each season as you need. If stored too long, it may lose its jelling strength. Store pectin in a cool, dry place. If powdered pectin becomes brown or caked, don't use it. If liquid pectin looks very thin and watery, don't use it. An opened bottle of liquid pectin should be recapped, the remainder stored in the refrigerator and used within a month. When you take it out of the refrigerator, let it stand at room temperature for an hour; if it is still congealed or set in the bottle, don't use it.

With commercial pectins, you can make jellied preserves from any fruit at its peak of ripeness and flavor. The boiling time is short, which saves cooking fuel as well as the fresh-fruit flavor.

Yields are higher, because you don't boil away the juice. Uncooked or frozen jellies and jams are also jelled with commercial pectin.

ACID

A certain concentration of acid is necessary in the fruit/pectin/sugar mix; without it, the product won't jell. Underripe fruits contain more acid than ripe, just as they contain more pectin. You may need to add more acid than is in the fruit. Most women add lemon juice, but you can also buy crystalline citric acid at the drugstore and substitute it for lemon juice, ⅛ tsp. for 1 tblsp. lemon juice. Commercial pectins also contain some acid.

Acid Test: You do not need to make an acid test when you follow short-cook recipes using commercial pectin. If you are making longer-cook jellies, it's a good idea to see if your fruit juice is tart enough. Make this taste comparison. Mix 1 tsp. lemon juice with 3 tblsp. water and ½ tsp. sugar. If fruit juice is not as tart as the diluted lemon juice, add 1 tblsp. strained lemon juice to every cup of fruit juice. A good recipe will often call for lemon juice if needed.

SUGAR

There's no denying it, jams and jellies are about half fruit, half sugar. A specific ratio of sugar to pectin is necessary to achieve a jelly set; sugar also adds flavor and serves as a preserv-

ative. There's no way you can try to make a less-sweet jelly by reducing sugar. Either you add the correct amount of sugar at the start, or the jelly will have to boil down until evaporation brings the proportion of sugar into balance with the acid and pectin.

Recipes using commercial pectin may seem to call for more sugar in proportion to the fruit or juice than longer-cook recipes, but after boiling, the concentration of sugar to fruit will be about equal.

Use either cane or beet sugar, but do not use brown sugar unless the recipe specifically calls for it. The flavor and color of brown sugar affect the taste and looks of jellied fruits, and the sweetening power is not the same, which could affect the jelly set.

Equipment for Making Jellies and Jams

Although jelly batches are relatively small, you need a large kettle to cook them in—from 8 to 12 quarts capacity. The kettle should be enameled, or stainless steel, with a broad, flat bottom. This will allow the rapidly boiling fruit-sugar mixture to expand without boiling over.

For extracting juice, you can sew a jelly bag from firm muslin or cotton flannel (napped side inside), or you can use a colander lined with cheesecloth, or a fruit press.

A good size for a jelly bag is about 10×15″. Stitch it down both sides and across the bottom; leave top open. When it's filled with crushed or cooked fruit, knot cord around the top and hang the bag so juice can drip

into a bowl. A stand for hanging the jelly bag is convenient.

If you're making short-cook and uncooked jellies and jams, you'll need a clock/timer that will signal 1 minute. For the long-cook method, a jelly thermometer will tell you when the fruit has reached the jellying point.

Kitchen scales will come in handy; so will the following kitchen tools: 1 quart and 1 cup measures, measuring spoons, paring and utility knives, food chopper, food mill, grater, masher, reamer, wire basket, sieve or strainer, long-handled spoon and bowls.

Finally, you'll need canning jars and lids, a boiling water bath canner, jar lifter, jar funnel and ladle, as described in Chapter 9.

The Paraffin Question

If you have any spoilage problems with fruit spreads, it is more likely to be caused by molds than anything else, and it's usually the result of an imperfect or damaged paraffin seal as used in open kettle canning. When you discovered such a jar, it used to be standard practice simply to scrape the mold off and use the remainder. While you regretted the spoilage, mold was regarded more as a nuisance and a waste than a danger.

Now there is evidence that compounds called mycotoxins can form in some foods as a result of mold growth. Some of these toxins have caused cancer in animals; whether they affect humans in the same way is yet unknown. Certainly the safest thing to do, when you find moldy jelly or jam, is throw it out.

Better yet, follow new recommendations for sealing and processing fruit spreads, so you won't have any more problems with mold. Or with other kinds of spoilage, such as fermentation.

As of this writing, a few authorities are still permissive about sealing *jelly* with paraffin, but all agree that jams, marmalades, conserves and preserves should be filled into half-pint or pint canning jars designed with regular Mason screw threads and capped with the self-sealing metal lids and screw bands.

Furthermore, to effect a good seal, and to kill any spoilage organisms that might have gotten into the jar while it was being filled, preserves should then be processed in a boiling water bath. Jellies, too, can be (and probably should be) sealed and steri-

lized the same way. Because it sometimes shrinks or cracks, paraffin is not a sure seal against airborne molds and other spoilage organisms. Moreover, paraffin is highly flammable and if spilled can cause serious burns. (Also see "Unsafe Canning Practices" in Chapter 9.)

How long should jellies and jams be processed? Recommended times range from 5 minutes to 15 minutes (or even longer for certain fruit mixtures). Current research at several state universities suggests that the longer processing time—10 or 15 minutes—will provide better protection against spoilage. On the other hand, longer processing seems to aggravate the problem of floating fruit and soft textures in some recipes. Research on this is continuing, but as of this writing, the USDA has not yet revised its 1975 recommendation that jams and preserves be filled into presterilized jars and processed for 5 minutes in the boiling water canner.

If you are careful to sterilize jars before filling; if you are sure jars have sealed; if you have the recommended "cool, dry place" to store your jams, and if you will be using them within 6 months or so, the 5-minute processing time will probably give you adequate protection against spoilage from molds. However, if temperatures in your storage area approach 80° F. or if they fluctuate, and especially if you've had trouble in the past with mold forming on your fruit spreads, you may want to try the longer processing times.

You can skip the step of sterilizing jars if you fill clean hot jars with hot jelly or jam; add lids according to manufacturer's instructions and process for 15 minutes in the boiling water canner to sterilize both jar and contents. If the texture of some recipes is too soft, use presterilized jars and 5 minutes in a boiling water canner next time.

How to Make Jellies and Jams

Even though new ways of making fruity sweets are popular, the yardstick for measuring their quality is the same as Grandmother's. The first requirement is that they carry the flavor and bright color of the fruit.

Jelly at its best is sparkling and clear. It quivers slightly when you turn it out in a dish, but it holds its shape. When you cut it with a spoon, it retains the angle of the cut. Topnotch jelly is tender and spreads easily.

Excellent jams are jelly-like in consistency and they are fairly smooth because fruits are crushed before cooking. Soft textures of berries make them ideal for making jams. Our readers sent us many more jam than jelly recipes. They say jams are easier to make.

You can take your pick of three methods of making jellies and jams—the *uncooked, short-cook and longer-cook* ways. The taste and consistency of the products obtained varies according to the method followed. You will want to choose the one your family and friends like best.

STERILIZING JARS

Before you start to make jam or jelly, get the jars and lids ready. Sterilize jars before filling if contents will not be processed for 15 minutes. After washing and rinsing them, stand them upright in a large container on a rack, add water to fill and cover jars by at least 1", bring to a boil and boil for 15 minutes. Leave jars in hot water until you need them. When ready to fill, drain jars upside down on a clean towel. Jars should be hot and dry inside when you fill them.

Follow manufacturer's directions for preparing lids. If you will use a jar funnel and/or a ladle, both of these should be boiled with the jars.

PREPARING FRUIT

No matter what method you use to make jelly, the first steps—preparing the fruit and extracting the juice—are the same. Wash sound, firm fruit. Discard blemished and spoiled spots and remove stem and blossom parts. Core, but do not peel hard fruits like apples. Cut fruit in small pieces or put it through the coarse blade of food chopper to insure extraction of the pectin. Crush soft fruits like berries to start flow of juices.

You may not need to heat juicy berries, but most fruits will require heating to extract the juice. Add a minimum of water to fruit or you'll dilute the juice. Firm fruits, like apples, need about 1 c. water to 1 lb. fruit. Firm plums, like the Damson, require ½ to ¾ c. water to 1 lb. fruit, but juicy plums need only ¼ c. water to 1 lb. fruit. Soft fruits and berries require no water unless they are not juicy—then you add ¼ c. to 1 lb. fruit.

Bring the fruit (with water added if needed) quickly to a boil, cover and boil gently until soft, 5 to 20 minutes, depending on fruit. Too much boiling reduces jelling strength and destroys fruit flavors. Stir, if necessary, to prevent scorching.

To extract juice, spread 4 layers of cheesecloth into a colander, set it over a bowl and pour in softened fruit. Or pour fruit into a damp jelly bag, suspend it over a bowl, and *let juice drip out* for the clearest jelly. Or gently twist the bag or cheesecloth to press out the juice. The juice then is ready for jelly making, although you may want to strain it again through layers of cheesecloth if you have forced it through the bag or press.

To make jam, follow recipe directions for chopping, grinding or crushing fruit.

SHORT-COOK JELLIES AND JAMS

Another way to describe short-cook recipes is to say "with added pectin." Use ripe fruit with fully developed flavor. Choose the pectin your recipe calls for—powdered or liquid—and follow the recipe exactly. (Or choose a recipe that matches the pectin you have.) Both pectins do the same job, but they work differently. Powdered pectin will not dissolve in high sugar concentrations, so it goes into the fruit or fruit juice before heating, and before adding sugar. Liquid pectin is added after fruit and sugar have come to a boil.

The boiling time is brief—usually 1 minute—and you won't be successful if you change it. Use a timer!

Directions with both powdered and liquid pectins also describe what the boil looks like. You time the boil when it reaches a *full rolling boil*— one which cannot be stirred down.

Measure fruit or juice and sugar accurately with standard level measures. Short-cook recipes are standardized, so you don't have to make acid or pectin tests, or determine the jellying point. But the averages depend on your measuring carefully. If you don't have enough juice for your jelly recipe, mix some water with the pulp in the jelly bag and squeeze it again. To measure for jam, pack fruit and juice solidly in the measuring cup; add water if you're only a little bit short; otherwise, prepare more fruit.

After you've made one batch, you may decide you want jam or jelly that's softer, or firmer. For softer sets, add a little more fruit or juice; for firmer sets, use a little less (see directions with the packaged pectins for these changes). The amount of natural pectin in the fruit also affects the jelly set. Since this varies with the variety, growing conditions and ripeness of the fruit, you'll always notice slight variations in firmness.

LONGER-COOK JELLIES AND JAMS

Fruits with enough natural pectin and acid can be jelled with longer boiling. They are listed at the beginning of this chapter, along with Grandmother's rule to use slightly underripe fruit or to mix ¼ underripe with ¾ ripe fruit. It is not necessary to make pectin tests for jams, but you'll avoid jelly failure if you find out how much pectin and acid the fruit juice contains—see Index for pectin and acid tests. The amount of sugar you use in longer-cook recipes depends on the amount of pectin and acid in the juice.

If your pectin test indicates that juice is low in pectin, it is best to add pectin before you try to make jelly with it. Here's how to determine the right amount to add:

Measure 1 c. fruit juice and stir in 1 tblsp. liquid or liquefied powdered pectin (directions follow). Use 1 tsp. of this juice to make the pectin test again. Repeat the additions and pectin tests until you get a strong pectin test. Keep track of the number of tablespoons of pectin added; add the same amount to each cup of juice.

To Liquefy Powdered Fruit Pectin: Empty contents of box into a small saucepan. Add ½ c. water. Bring mixture to a boil and boil 1 minute, stirring constantly. Pour into a measuring cup and add cold water to make 1 c. Stir to mix thoroughly.

Amount of Sugar: If the pectin test gives you a jelled mass you can pick up with a fork, measure 1 c. sugar for each cup juice. If the test is moderately strong (large jelly flakes), use ¾ c. sugar for each cup juice. When juice contains only a moderate amount of pectin, use ⅔ to ¾ c. sugar to 1 c. juice. (Too much sugar results in syrup, not jelly.)

To Cook Jelly: Cook only 4 to 6 c. juice at a time, using your large jelly kettle. Measure juice (and pectin, if you use it) accurately and set the kettle over high heat. When juice comes to a boil, add the correct amount of sugar and stir until it dissolves. Boil rapidly, but stand ready to check for doneness (see tests below). Do not overcook because you will lose fruit flavor and color and destroy the jellying power of the pectin.

To Cook Jam: Cook jams in small batches; do not double recipes. Measure fruit and sugar into a large kettle and stir over low heat until sugar dissolves. Then bring to a boil and cook rapidly until it thickens. Stir frequently to prevent sticking and scorching. When it starts to thicken, test for doneness (see below).

Testing for the Jellying Point: There are three tests you can use to determine when longer-cook jelly is done. The most dependable is the *temperature test,* using a jelly-candy or deep fat thermometer. The *jellying point for your altitude* may be determined in this way: Boil some water before you start to make jelly. Note the temperature at which it boils; then add 8° F. to get temperature of finished jelly. If water boils at 212° F. jelly is done when boiling juice reaches 220° F.

Grandma used the *spoon test.* Many women find it a good idea to test the jelly this way in addition to using the thermometer. Dip a metal spoon into boiling juice mixture and hold it at least 12 inches above the kettle, out of the steam, and let the syrup run off the sides. As it is approaching the jellying point, the syrup will run off in two drops. When it reaches the jellying point, the drops slide together and drop off the spoon in a sheet (that's why some people call this the *sheet test*).

Or give jelly the *refrigerator test.*

When you think it's done by the spoon test, remove jelly from heat. Pour a spoonful on a cold plate and set it in the freezing compartment of your refrigerator. If it jells in a minute or two, it is cooked enough.

Testing Jam for Doneness. The most reliable test for jam cooked without added pectin is the same as for jelly—the *temperature test*. With this change: Cook jam mixtures to a temperature 9° F. higher than the boiling point of water at your altitude. Stir the jam mixture thoroughly just before reading its temperature.

If you don't have a thermometer, use the *refrigerator test* when jam begins to thicken or hold its shape in the spoon. (It's almost impossible to use a sheet test on jam.)

FILLING AND SEALING JARS

As soon as jellies or jams test done, remove kettle from heat, skim, ladle into sterilized hot jars and seal. Here are detailed directions:

Skimming: Let jelly stand for a minute after it is removed from heat; this permits a film to form. Then skim—circle the top of the hot jelly with a spoon gathering up the foam and film.

When you remove jam from heat, skim as necessary to remove foam. The skimming is mainly for appearance—the foam is harmless. A half-teaspoonful of butter added to fruit or juice before boiling helps reduce foaming.

Sealing with Lids: Ladle hot jelly or jam into hot sterilized jars. Fill one jar at a time and adjust lid before filling next jar. Fill jars to within

⅛″ of top. Wipe jar rim clean with a damp paper towel. Put lid on jar with sealing compound next to the glass; screw the screw band on tight. Process half pints and pints in a boiling water bath, following directions in Chapter 9. (If you live above 1,000 feet, see Index for Altitude Corrections.)

Sealing with Paraffin: Use the paraffin seal only for jellies—not for jams or other fruit preserves. If you've had problems with mold forming on jelly, use jars and lids and process jelly same as jam. (See discussion on paraffin earlier in this chapter.)

To melt paraffin, put it in a container (a coffee can will do; an old small coffee pot with a pouring lip is better) and set the container in boiling water. (Paraffin is highly flammable. Never melt it directly over heat—always use some kind of double boiler arrangement.)

Ladle jelly boiling hot into sterilized glass, filling to within ½″ of top. Fill and seal one glass at a time. Wipe the inside lip with a damp paper towel. Cover jelly immediately with ⅛″ melted paraffin. Be sure paraffin flows to all edges. Prick any bubbles that form; otherwise, in cooling, they may form holes. It takes about 1 tblsp. paraffin to cover a glass approximately 2½″ in diameter. The single thin layer is better than a double layer or thick layer, because it adjusts for expansion or contraction if temperatures change in the storage area.

COOLING AND STORING

When paraffin seal turns white, examine it closely to be sure it's secure. If not, remove it and seal again with melted paraffin. Don't try to cover a faulty seal with another layer of wax. Cover glass with a metal or paper lid, label, date and store in a cool, dry place. Enjoy jelly while it's still fresh; if stored too long, it loses flavor.

If you processed jam or jelly, remove jars from the boiling water as processing time is up. Place jars on a rack or folded towel, a few inches apart, and let cool for 12 hours. Test for seals. Remove screw bands, clean jars, label, date and store following directions in Chapter 9.

Short-cook Jellies and Jams

One of FARM JOURNAL's readers makes especially beautiful and interesting jellies—she calls them Mystery Jellies. Her friends always praise them. She gathers most of the berries and fruits in the woods and on their Wisconsin farm.

All throughout the busy summer, she collects remnants of fruit juices, the leftovers from her regular jelly making. She pours them into containers and freezes them. When one container is full of mixed juices, she starts all over again. From them she makes jelly throughout the winter, "confident," she says, "that when I give a friend a glass of it for Christmas or at any other time, she will not have another one like it."

We asked her to freeze the different kinds of juices separately and experiment with measured proportions of them in making jellies. Each of her five boys—her husband and four sons —picked his favorite. Here is the quintet of recipes she sent us.

MYSTERY JELLIES

No. 1

2 c. frozen ripe gooseberry juice
1¾ c. frozen red raspberry juice

1¾ c. frozen red currant juice
1 pkg. powdered fruit pectin
6 c. sugar

No. 2

2 c. frozen red raspberry juice
2 c. frozen black raspberry juice
1½ c. frozen red currant juice
1 pkg. powdered fruit pectin
7 c. sugar

No. 3

1 c. frozen red currant juice
2¼ c. frozen tart cherry juice
1¾ c. frozen blueberry juice
1 pkg. powdered fruit pectin
7 c. sugar

No. 4

1½ c. frozen black raspberry juice
1 c. frozen plum juice
1 c. frozen ripe gooseberry juice
¾ c. frozen red currant juice
1 pkg. powdered fruit pectin
7 c. sugar

No. 5

3 c. frozen plum juice
2½ c. frozen ripe gooseberry juice
1 pkg. powdered fruit pectin
7 c. sugar

• Thaw fruit juices, combine with pectin in large kettle and bring to a full rolling boil, stirring constantly.

• Add sugar and bring to a full boil again over high heat; boil for 1 minute, stirring constantly. Remove from heat; skim.

• Pour into sterilized hot jars to within ⅛″ of jar top. Wipe jar rim; adjust lids. Process in boiling water bath 5 minutes. Remove from canner and complete seals unless closures are self-sealing type. Makes 8 to 9 half pints.

• Or pour into sterilized jelly glasses to within ½″ of jar top. Cover immediately with ⅛″ melted paraffin.

WINTER APPLE JELLY

Makes wonderful glaze when spread over baking ham for last half hour

1 qt. bottled or canned apple
    juice
5 drops (about) red food color
1 pkg. powdered fruit pectin
5½ c. sugar

• Combine juice, color and pectin in large saucepan; bring to a full boil.

• Stir in sugar; return to boil; boil 2 minutes, stirring constantly. Remove from heat; skim.

• Pour into sterilized hot jars to within ⅛″ of jar top. Wipe jar rim; adjust lids. Process in boiling water bath 5 minutes. Remove from canner and complete seals unless closures are self-sealing type. Makes about 7 half pints.

• Or pour into sterilized jelly glasses to within ½″ of jar top. Cover immediately with ⅛″ melted paraffin.

VARIATIONS

CRANBERRY/APPLE: Use equal parts bottled cranberry juice cocktail and apple juice, omitting food color.

SPICE MINTED APPLE: Substitute green for red food color; tie 1 tblsp. whole cloves in muslin bag; add to juice with pectin. Remove bag after cooking; add ½ tsp. peppermint extract.

GERANIUM: Lay a washed rose-geranium leaf in each jar before pouring in the hot jelly mixture.

GRAPE: Use bottled grape juice for apple juice.

GRAPEFRUIT/LEMON: Use unsweetened canned grapefruit juice for apple juice; add juice of 2 lemons and about 10 drops yellow food color.

CRANBERRY: Use bottled cranberry juice cocktail for apple juice.

ORANGE JELLY

For extra glamor, wrap in festive paper and tie with holiday ribbons

2 c. water
1 pkg. powdered fruit pectin
3½ c. sugar
1 (6 oz.) can frozen orange juice
    concentrate, thawed

• Mix water and pectin in saucepan; bring quickly to full rolling boil; boil hard 1 minute; add sugar and concentrate; stir until dissolved (don't boil). Remove from heat; skim.

• Pour into sterilized hot jars to

within ⅛″ of jar top. Wipe jar rim; adjust lids. Process in boiling water bath 5 minutes. Remove from canner and complete seals unless closures are self-sealing type. Makes 5 half pints.
• Or pour into sterilized jelly glasses to within ½″ of jar top. Cover immediately with ⅛″ melted paraffin.

PINEAPPLE JELLY: Substitute frozen pineapple juice concentrate for orange.

CINNAMON JELLY

A red beauty—right for Christmas, but good for breakfast any day

1 qt. bottled or canned apple juice
1 pkg. powdered fruit pectin
4½ c. sugar
1 or 2 tblsp. red cinnamon candies

• Combine juice and pectin in large saucepan. Bring to full rolling boil.
• Add sugar and candies; stirring constantly, return to boil. Boil 2 minutes. Remove from heat. Let boiling subside and skim.
• Pour into sterilized hot jars to within ⅛″ of jar top. Wipe jar rim; adjust lids. Process in boiling water bath 5 minutes. Remove from canner and complete seals unless closures are self-sealing type. Makes 7 half pints.
• Or pour into sterilized jelly glasses to within ½″ of jar top. Cover immediately with ⅛″ melted paraffin.

MARJORAM JELLY

Gourmet special: Serve with meats— it's tangy, crystal clear, tasty

2 tblsp. marjoram leaves
1 c. boiling water
⅓ c. lemon juice
3 c. sugar
½ c. liquid fruit pectin

• Combine marjoram and water. Let stand 15 minutes. Strain through fine mesh cheesecloth. Measure liquid and add water to make 1 c.
• Strain lemon juice through cheesecloth. Combine lemon juice, sugar and herb liquid in a saucepan. Place over high heat. Bring to a boil; stir in pectin. Stirring constantly, bring to a full boil. Boil ½ minute. Remove from heat. Skim.
• Pour into sterilized hot jars to within ⅛″ of jar top. Wipe jar rim; adjust lids. Process in boiling water bath 5 minutes. Remove from canner and complete seals unless closures are self-sealing type. Makes 3 half pints.
• Or pour into sterilized jelly glasses to within ½″ of jar top. Cover immediately with ⅛″ melted paraffin.

PLUM JELLY

Melt a little jelly, thin with hot water and use to baste the ham loaf

4 lbs. red plums
1 c. water
6½ c. sugar
½ bottle liquid fruit pectin

• Wash plums. Add water and cook until soft. Strain through jelly bag; you should have 4 c. juice.
• Measure juice into large kettle. Add sugar and mix well; bring to a boil, stirring constantly. Stir in pectin and bring to a full rolling boil. Boil hard for 1 minute, stirring constantly. Remove from heat; skim off foam.

• Pour into sterilized hot jars to within ⅛" of jar top. Wipe jar rim; adjust lids. Process in boiling water bath 5 minutes. Remove from canner and complete seals unless closures are self-sealing type. Makes 4 pints.
• Or pour into sterilized jelly glasses to within ½" of jar top. Cover immediately with ⅛" melted paraffin.

RHUBARB JELLY

Pink-red and just right with veal. Use it often to garnish desserts

3½ c. rhubarb juice (about 3 lbs. fresh rhubarb)
7   c. sugar
1   bottle liquid fruit pectin

• Cut unpeeled red rhubarb stalks into 1" lengths. Grind. Place in jelly bag and squeeze out juice.
• Measure juice into large kettle. Add sugar, mix well. Bring to a boil, stirring constantly. Stir in pectin. Bring to a rolling boil; boil hard 1 minute, stirring constantly. Remove from heat, skim off foam.
• Pour into sterilized hot jars to within ⅛" of jar top. Wipe jar rim; adjust lids. Process in boiling water bath 5 minutes. Remove from canner and complete seals unless closures are self-sealing type. Makes about 10 half pints.
• Or pour into sterilized jelly glasses to within ½" of jar top. Cover immediately with ⅛" melted paraffin.

VARIATION

RHUBARB/STRAWBERRY JELLY: Follow recipe for Rhubarb Jelly but use 1½ lbs. rhubarb and 1½ qts. ripe strawberries, 6 c. sugar. Grind rhubarb, crush berries and mix. Put in jelly bag and squeeze out juice.

RUBY JELLY

Reflects the glow of jewels—a top favorite of all taste-testers

1   qt. loganberries
1½ qts. red raspberries
1   pkg. powdered fruit pectin
5½ c. sugar

• Crush fully ripe berries and place in jelly bag. Squeeze out juice.
• Measure 4 c. juice into large saucepan. Stir in pectin; bring to boil.
• Add sugar and return to a full rolling boil, stirring constantly. Boil 1 minute. Remove from heat and skim.
• Pour into sterilized hot jars to within ⅛" of jar top. Wipe jar rim; adjust lids. Process in boiling water bath 5 minutes. Remove from canner and complete seals unless closures are self-sealing type. Makes 3½ pints.
• Or pour into sterilized jelly glasses to within ½" of jar top. Cover immediately with ⅛" melted paraffin.

APRICOT JAM

Like sunshine on the breakfast table —has a rich, tangy sweet taste

3   c. diced fresh apricots
¼   c. lemon juice
7   c. sugar
½   bottle liquid fruit pectin

• Mix apricots, lemon juice and sugar. Boil until apricots are soft. Remove from heat; stir in pectin; skim.
• Ladle into sterilized hot jars to within ⅛" of jar top. Wipe jar rim;

adjust lids. Process in boiling water bath 5 minutes. Remove from canner and complete seals unless closures are self-sealing type. Makes 3½ pints.

BLUEBERRY/RASPBERRY JAM

"My mother gave me this recipe—her mother used it," a reader said

1 pt. blueberries
1 qt. red raspberries
7 c. sugar
1 bottle liquid fruit pectin

• Crush berries; measure 4 c. (if necessary, add water to make 4 c.)
• Add sugar, mix well. Heat to full rolling boil; boil hard 1 minute, stirring constantly. Remove from heat; stir in pectin; skim.
• Ladle into sterilized hot jars to within ⅛" of jar top. Wipe jar rim; adjust lids. Process in boiling water bath 5 minutes. Remove from canner and complete seals unless closures are self-sealing type. Makes 10 half pints.

BLUEBARB JAM

The union of blueberries and rhubarb is a miracle flavor-blend

3 c. finely cut rhubarb
3 c. crushed blueberries
7 c. sugar
1 bottle liquid fruit pectin

• Combine rhubarb and blueberries in large saucepan, add sugar; mix.

• Place over high heat; bring to full, rolling boil and boil hard 1 minute, stirring constantly. Remove from heat; stir in pectin; skim.
• Ladle into sterilized hot jars to within ⅛" of jar top. Wipe jar rim; adjust lids. Process in boiling water bath 5 minutes. Remove from canner and complete seals unless closures are self-sealing type. Makes about 9 half pints.

Note: This recipe was tested with fresh rhubarb and unsweetened, frozen blueberries.

PEACHY PEAR JAM

Guaranteed to add pleasing variety to your collection of sweet spreads

3½ c. mashed peaches and pears
6½ c. sugar
Juice of 2 lemons
½ tsp. ground cinnamon
½ tsp. ground nutmeg
½ bottle liquid fruit pectin

• Combine fruit, sugar, lemon juice and spices in a large kettle. Place over high heat and bring to a full rolling boil. Boil hard for 1 minute, stirring constantly. Remove from heat; stir in pectin; skim.
• Ladle into sterilized hot jars to within ⅛" of jar top. Wipe jar rim; adjust lids. Process in boiling water bath 5 minutes. Remove from canner and complete seals unless closures are self-sealing type. Makes 4 pints.

Longer-cook Jellies

APPLE/GERANIUM JELLY

Gourmet shops report splendid sales of this old-fashioned delicacy

6 c. apple juice (about 5 lbs. apples)
4 c. sugar
12 small rose geranium leaves

• Prepare juice by removing stem and blossom ends of tart, red apples. Slice and put in kettle with water, barely to cover. Cook until very tender. Turn into jelly bag, let juice drip into bowl.
• Measure 6 c. juice into large kettle; bring quickly to a boil. Add sugar, stirring until dissolved. Boil rapidly until jellying point is reached.
• Quickly place 2 small (or 1 large) rose geranium leaves in each hot jar. Skim jelly.
• Pour into sterilized hot jars to within ⅛" of jar top. Wipe jar rim; adjust lids. Process in boiling water bath 5 minutes. Remove from canner and complete seals unless closures are self-sealing type. Makes 6 half pints.
• *Or* pour into sterilized jelly glasses to within ½" of jar top. Cover immediately with ⅛" melted paraffin.

Note: If apples are not tart, test juice for pectin before making jelly.

VARIATION

MINT JELLY: Omit geranium leaves. Tint jelly delicately with green food color; add ½ tsp. peppermint extract.

CRAB APPLE JELLY

Spread between folds of a puffed-up omelet, or use to glaze baked ham

5 lbs. crab apples
8 c. water
Sugar
1 tsp. vanilla (optional)

• Remove stem and blossom ends from washed crab apples, cut in halves and place in large kettle. (Red fruit makes the most colorful jelly.) Add water and cook until fruit is very soft, about 10 minutes.
• Strain mixture through jelly bag, but do not squeeze or force juice through bag.
• Measure juice; you should have about 7 c. Pour into large kettle. Stir in ¾ c. sugar for every cup of juice. Bring to a boil quickly and cook rapidly until jellying point is reached.
• Skim off foam, stir in vanilla and pour into sterilized hot jars to within ⅛" of jar top. Wipe jar rim; adjust lids. Process in boiling water bath 5 minutes. Remove from canner and complete seals unless closures are self-sealing type. Makes about 4 half pints.
• *Or* pour into sterilized jelly glasses to within ½" of jar top. Cover immediately with ⅛" melted paraffin.

RED CURRANT JELLY

Bright red gem on shelf or table—it's tart enough to accompany meats

2½ qts. red currants
1 c. water
4 c. sugar

• Sort, wash and crush currants without removing stems. Place in kettle, add water, bring quickly to a boil over high heat. Lower heat, simmer

10 minutes. Let drip through jelly bag.

• Measure juice. There should be 4 c. Stir in sugar. Boil until jellying point is reached. Remove from heat; skim off foam quickly.

• Pour into sterilized hot jars to within ⅛" of jar top. Wipe jar rim; adjust lids. Process in boiling water bath 5 minutes. Remove from canner and complete seals unless closures are self-sealing type. Makes about 4 half pints.

• Or pour into sterilized jelly glasses to within ½" of jar top. Cover immediately with ⅛" melted paraffin.

TWO-STEP GRAPE JELLY

Make it this way and you'll not find crystals in the amethyst

3½ lbs. Concord grapes
½ c. water

• Select ¼ underripe and ¾ ripe grapes. Wash and remove stems. Place in kettle, crush, add water, cover and bring quickly to a boil.

• Lower heat and simmer 10 minutes. Let juice drip through jelly bag.

• Cover juice and let stand in a *cool place overnight.* Strain through two thicknesses of damp cheesecloth to remove crystals.

Second Step

4 c. grape juice
3 c. sugar

• Pour juice into large kettle; stir in sugar. Boil over high heat until jellying point is reached. Remove from heat; skim off foam quickly.

• Pour into sterilized hot jars to within ⅛" of jar top. Wipe jar rim;

adjust lids. Process in boiling water bath 5 minutes. Remove from canner and complete seals unless closures are self-sealing type. Makes about 4 half pints.

• Or pour into sterilized jelly glasses to within ½" of jar top. Cover immediately with ⅛" melted paraffin.

PLUM AND ORANGE JELLY

Beat jelly and add to mayonnaise for fruit salads—¼ jelly to ¾ dressing.

5 lbs. red plums
6 large oranges, peeled and sliced
1 lemon, sliced
Sugar

• Wash plums and cover with water. Add oranges and lemon. Cook until skin and pits of plums separate from pulp. Strain through jelly bag. Boil juice for 20 minutes.

• Add 3½ c. sugar for each 4 c. juice. Boil rapidly to jellying point. Remove from heat; skim.

• Pour into sterilized hot jars to within ⅛" of jar top. Wipe jar rim; adjust lids. Process in boiling water bath 5 minutes. Remove from canner and complete seals unless closures are self-sealing type. Makes 7 pints.

• Or pour into sterilized jelly glasses to within ½" of jar top. Cover immediately with ⅛" melted paraffin.

CURRANT JELLY SAUCE

Try this with venison

1½ c. currant jelly
¾ tsp. dry mustard

• Melt jelly over low heat. Stir until smooth. Add mustard, mixing well. Makes 8 servings.

N IOWA FARMER'S JELLY SAUCE:
.dd 3 tblsp. prepared horseradish to
urrant Jelly Sauce and serve with
ork. Makes 8 to 10 servings.

APPLE JELLY SAUCE

1akes baked ham something special

½ c. apple or crab apple jelly
2 whole cloves

Melt jelly with cloves over low heat.
tir until smooth. Remove cloves.
erve warm. Makes 8 servings.

MINTED ORANGE SAUCE

Perfect with lamb roast or chops

1½ c. mint jelly
1½ tsp. grated orange peel

• Melt jelly over low heat. Stir until
smooth. Mix in orange peel. Serve
warm. Makes 8 servings.

"What Went Wrong with My Jelly?"

Many questions about jelly troubles
ave come to the Food Editors of
'ARM JOURNAL. Here are questions
sked frequently, and our answers:
Q. My jelly did not set; it is thick and
ticky. What caused the trouble?
A. You used too much sugar or
ooked the mixture too slowly and too
ong. Or there was not enough acid or
•ectin in the fruit juice.
Q. Why is my jelly soft?
A. Chances are it was not cooked
nough. You may have used too
nuch juice and too little sugar. Or
here was not enough acid in the
uice. Soft jelly also occurs when you
nake too big a batch of jelly at a
ime.
Q. Why does jelly "weep"?
A. The causes are too much acid in
he juice, or storage of jelly in a place
oo warm or with widely fluctuating
emperatures. Very rapid jellying also
auses "weeping." This occurs when
he fruit is too green.
Q. Why is jelly cloudy?

A. Cloudiness results when jelly is
poured too slowly into jars or glasses,
when the mixture stands too long before pouring and when the juice is not
strained properly and some of the
fruit pulp remains.
Q. Some of the jelly I made is tough.
Why?
A. The mixture was cooked too long
before it reached the jellying point, a
result of using too little sugar. Too
much pectin will also make jelly
tough.
Q. Why do crystals form in jelly?
A. Too much sugar, lack of acid or
pectin, or overcooking often result in
crystals. In the case of grape jelly, the
difficulty may be caused by using fruit
that is too ripe, failure to let the juice
stand overnight before making the
jelly or stirring up the sediment when
pouring the juice into the kettle for
cooking. Crystals forming at top of
opened glass are due to evaporation.
Q. One batch of jelly I made is
gummy. What mistake was made?

A. The jelly mixture was overcooked.

Q. A few jars and glasses of jelly are dark at the top. Why?

A. It was stored in a place too warm or the seal was imperfect. Oxidation causes darkening; you may have left too much head space when filling jars with self-sealing lids. Or jelly may not have been hot enough when you filled jars.

Q. Some of my prized red jellies have turned brown. What did I do?

A. The jelly was stored in a place with too much light. Or it may be an of the reasons for jelly turning dark (see above).

Q. Why does jelly ferment?

A. The containers were not sterilize properly or the storage place was to warm or damp. Perhaps the seal wa not airtight.

Q. What causes mold on jelly?

A. The storage place was damp, th seal was imperfect or the containe were not properly sterilized.

Longer-cook Jams

STRAWBERRY JAM

This old-time sweet never loses its appeal—an All-American favorite

8 c. strawberries
Lemon juice (optional)
6 c. sugar

• Wash, drain, hull and crush berries. If they are very ripe and sweet, add 1 tblsp. lemon juice to every cup of crushed berries. Place in large saucepan.

• Add sugar and cook slowly, stirring constantly until sugar dissolves. Bring quickly to a boil and boil rapidly until the jellying point is reached.

• Ladle into sterilized hot jars to within ⅛" of jar top. Wipe jar rim; adjust lids. Process in boiling water bath 5 minutes. Remove from canner and complete seals unless closures are self-sealing type. Makes 6 half pints.

APPLE/RASPBERRY JAM

The woman who shares this recipe says: "A delight to make and eat"

9 c. sugar
2 c. water
6 c. finely chopped tart apples
    (about 2 lbs.)
3 c. red raspberries, washed and
    drained

• Combine sugar and water in larg kettle. Boil until mixture spins thread (230° F.).

• Stir in apples and boil 2 minute Add raspberries and boil 20 minute longer, stirring often.

• Ladle into sterilized hot jars t within ⅛" of jar top. Wipe jar rim adjust lids. Process in boiling wate bath 5 minutes. Remove from cann and complete seals unless closures a self-sealing type. Makes 5 pints.

APRICOT/BLACK
RASPBERRY JAM

One of the tastiest ways to preserv summer's bounty in glasses

Homemade pickles and relishes (clockwise from top): Rosy Watermelon Pickles, Sweet Cucumber Pickles, Pepper Slaw, Cinnamon Cucumber Rings, Crisp Mustard Pickles. Recipes in Chapters 13 and 14.

We share a bountiful collection of choice pickle recipes from country kitchens —sweets and sours, whole and sliced, quick-process and brine-cured. Recipes for pickling cucumbers and other vegetables in Chapter 13.

Tangy Blender-made Ketchup is a snap to make. It's only one of several ketchup and chili sauce recipes in this book. Can some of these red relishes when tomatoes are ripe, to spice up winter meals. Recipes in Chapter 14.

Before frost visits your garden, gather up the last tomatoes and make Green Tomato Mincemeat (recipe in Chapter 10). It cans and freezes beautifully, makes excellent pies. So does our Venison Mincemeat recipe in Chapter 33.

Sugar
2 qts. blackcaps (black raspberries,
  about 7 c.)
4 qts. diced fresh apricots

• Add sugar equal in weight to fruit. Mix in large kettle. Boil to consistency desired.
• Ladle into sterilized hot jars to within ⅛" of jar top. Wipe jar rim; adjust lids. Process in boiling water bath 5 minutes. Remove from canner and complete seals unless closures are self-sealing type. Makes 10 pints.

APRICOT/PINEAPPLE JAM

Pure gold—extend your pleasure by giving a glass of it to a friend

4 c. diced fresh apricots
4 c. sugar
1 (1 lb. 4 oz.) can crushed
  pineapple

• Mix apricots and sugar in large saucepan. Boil for 20 minutes. Add drained pineapple. Let come to a boil again.
• Ladle into sterilized hot jars to within ⅛" of jar top. Wipe jar rim; adjust lids. Process in boiling water bath 5 minutes. Remove from canner and complete seals unless closures are self-sealing type. Makes 3 pints.

FRENCH APRICOT JAM

An old French recipe—crack a few pits and cook the kernels with fruit

5 c. sugar
3 lbs. (24 to 30) apricots, peeled,
  pitted and chopped

• Pour 2 c. sugar over apricots; let stand 8 hours.
• Add remaining sugar; boil gently in large saucepan until thick (mash with spoon while mixture cooks). Remove from heat; skim.
• Ladle into sterilized hot jars to within ⅛" of jar top. Wipe jar rim; adjust lids. Process in boiling water bath 5 minutes. Remove from canner and complete seals unless closures are self-sealing type. Makes 6 half pints.

GREEN BLUFF STRAWBERRIES

Perfect for topping vanilla ice cream

2½ c. sugar
½ c. water
1 qt. fresh strawberries (measure
  after washing and hulling)

• Mix 2 c. sugar and ½ c. water in 3-qt. saucepan. Boil until the soft crack stage (290° F.).
• Quickly add all strawberries at once. They will sear and sizzle, but lift edges with a fork gently and juice and melted sugar will soon appear. When boiling begins again, cook steadily and gently 10 minutes.
• Add the remaining ½ c. sugar and boil 5 minutes longer.
• Pour into a shallow pan and let stand in cool place overnight.
• In the morning, bring to a boil, skim and ladle into sterilized hot jars to within ⅛" of jar top. Wipe jar rim; adjust lids. Process in boiling water bath 5 minutes. Remove from canner and complete seals unless closures are self-sealing type. Makes about 4 half pints.

APRICOT/ORANGE JAM

Tasty winter-made specialty with nice blend of flavor and color

1 lb. dried apricots
2 oranges
2 (1 lb. 4 oz.) cans crushed
    unsweetened pineapple
8 c. sugar

· Soak apricots overnight; cook in soaking water until soft. Cut oranges in quarters and discard seeds; do not peel. Put apricots and oranges through food chopper.
· Mix all fruit and sugar in a large saucepan; cook 20 to 25 minutes.
· Ladle into sterilized hot jars to within ⅛" of jar top. Wipe jar rim; adjust lids. Process in boiling water bath 5 minutes. Remove from canner and complete seals unless closures are self-sealing type. Makes 6 pints.

CHERRY/LOGANBERRY JAM

Full-fruit flavors—favorite recipe of Michigan cherry grower's family

2 lbs. tart red or light cherries
    (5½ c. stemmed and pitted)
7 c. sugar
2 lbs. loganberries (7 c.)

· Pit cherries and cover with 2 c. sugar. Add berries and cover with 2 c. sugar. Allow to stand 1 hour.
· Add remaining sugar to mixed fruits. Cook to desired consistency in large saucepan.
· Ladle into sterilized hot jars to within ⅛" of jar top. Wipe jar rim; adjust lids. Process in boiling water bath 5 minutes. Remove from canner and complete seals unless closures are self-sealing type. Makes 4½ pints.

TRI-CHERRY JAM

Recipe from a Michigan farm woman. Delicious served on ice cream

2 c light sweet cherries
2 c. dark sweet cherries
2 c. tart red cherries
6 c. sugar
Juice of 1 lemon

· Pit and measure cherries. Grind coarsely. Place in large saucepan; simmer about 5 minutes to soften skins. Add sugar and lemon juice.
· Cook until thickened, 15 to 20 minutes (218° F.). Do not cook more than 20 minutes.
· Ladle into sterilized hot jars to within ⅛" of jar top. Wipe jar rim; adjust lids. Process in boiling water bath 5 minutes. Remove from canner and complete seals unless closures are self-sealing type. Makes 8 to 9 half pints.

SWEET CHERRY/BERRY JAM

Gladdens the top of baked custard, French toast, waffles, ice cream

4 c. pitted, sweet red cherries
4 c. red or black raspberries
8 c. sugar

· Cut cherries in halves or grind very coarsely. Place in large saucepan; cover, heat slowly. Simmer 5 to 7 minutes or until skins are slightly softened. Add raspberries and sugar; stir gently. Cook briskly to 218° F., about 15 minutes. Do not cook more than 20 minutes.
· Ladle into sterilized hot jars to within ⅛" of jar top. Wipe jar rim; adjust lids. Process in boiling water bath 5 minutes. Remove from canner and complete seals unless closures are

self-sealing type. Makes 11 to 12 half pints.

Note: If using black raspberries, use plump large berries—not small seedy ones.

CURRANT JAM

If you don't mind the seeds, you'll like this; if you do mind, make jelly

5¼ c. stemmed currants
4½ c. sugar
3 tblsp. lemon juice

• Crush currants in large saucepan; heat until some juice forms. Add sugar and stir until dissolved. Add lemon juice. Bring to a boil, shaking pan in circular motion for a time. Boil until consistency is like jelly, stirring constantly.
• Ladle into sterilized hot jars to within ⅛" of jar top. Wipe jar rim; adjust lids. Process in boiling water bath 5 minutes. Remove from canner and complete seals unless closures are self-sealing type. Makes 5 half pints.

ORANGE/FIG JAM

You can use either white or dark figs

1 large orange
¼ c. water
2 c. sugar
½ c. peeled, chopped figs
3 tblsp. lemon juice

With vegetable peeler, remove the orange part of orange skin in thin strips. Cut into slivers (there should be about ⅓ c.) and cook in water until tender.

• Chop orange pulp, discarding seeds; combine pulp, peel, sugar and figs in a large saucepan. Cook slowly until thick, about 30 minutes, stirring frequently.
• Add lemon juice. Remove from heat and ladle into sterilized hot jars to within ⅛" of jar top. Wipe jar rim; adjust lids. Process in boiling water bath 5 minutes. Remove from canner and complete seals unless closures are self-sealing type. Makes 3 to 4 half pints.

Note: Increase lemon juice to ¼ c. if using dark figs.

GOOSEBERRY JAM

Excellent with meats. Use ripe berries for the most attractive color

2 qts. gooseberries
4 c. sugar
Juice of 1 lemon

• Remove stems and tails from gooseberries. Wash. Grind with a medium coarse blade of food chopper. Place in large kettle; stir in sugar and lemon juice.
• Cook until mixture clings to the wooden spoon, showing it has begun to thicken. Stir constantly.
• Ladle into sterilized hot jars to within ⅛" of jar top. Wipe jar rim; adjust lids. Process in boiling water bath 5 minutes. Remove from canner and complete seals unless closures are self-sealing type. Makes 4 half pints.

Note: Half-ripe or ripe berries may be used. With half-ripe or green berries, omit lemon juice.

PEACH/APPLE JAM

A good cook's company special

2   c. chopped peeled peaches
2   c. chopped peeled apples
3   tblsp. lemon juice
2   tsp. grated lemon peel
3½ c. sugar

• Combine ingredients in saucepan. Cook slowly until thick and transparent, about 1 hour, stirring often.
• Ladle into sterilized hot jars to within ⅛" of jar top. Wipe jar rim; adjust lids. Process in boiling water bath 5 minutes. Remove from canner and complete seals unless closures are self-sealing type. Makes 2 pints.

PEACH/APRICOT JAM

Thick, tangy and richly golden

1 lb. peaches
2 lbs. apricots
Sugar

• Peel and slice fruits; combine and measure—there should be 6 c. Depending on tartness of fruit, add ¾ to 1 c. sugar for each cup fruit. Let stand 2 hours to draw out juices.
• Simmer gently until thick, about 20 minutes, stirring often.
• Ladle into sterilized hot jars to within ⅛" of jar top. Wipe jar rim; adjust lids. Process in boiling water bath 5 minutes. Remove from canner and complete seals unless closures are self-sealing type. Makes 7 half pints.

PEACH/PLUM JAM

The blending of two fruits, pointed up with lemon, is the flavor secret

4 c. peeled, chopped peaches (about 3 lbs.)
5 c. chopped red plums
8 c. sugar
1 lemon, very thinly sliced

• Combine peaches, plums, sugar and lemon slices (seeds discarded) in large kettle. Mix well. Boil rapidly, stirring, until jellying point is reached, or until thick. Remove from heat; skim.
• Ladle into sterilized hot jars to within ⅛" of jar top. Wipe jar rim; adjust lids. Process in boiling water bath 5 minutes. Remove from canner and complete seals unless closures are self-sealing type. Makes 12 half pints.

PEACH/PINEAPPLE JAM

Three fruits mingle their flavors—this is excellent with ham

12 large peaches
Grated peel and juice from 1 orange
1   (8¼ oz.) can crushed
        unsweetened pineapple (1 c.)
Sugar

• Peel and mash peaches; combine with orange peel, juice and pineapple. Measure. Depending on tartness of peaches, add ¾ to 1 c. sugar for each cup fruit. Mix in large saucepan and cook rapidly to desired consistency, stirring frequently.
• Ladle into sterilized hot jars to within ⅛" of jar top. Wipe jar rim; adjust lids. Process in boiling water bath 5 minutes. Remove from canner and complete seals unless closures are self-sealing type. Makes 4 pints.

ORIENTAL PEAR JAM

*You'll like what ginger does to pears
—serve with chicken or pork*

**8 medium pears
1 (8¼ oz.) can crushed unsweetened
   pineapple (1 c.)
4 c. sugar
1 (4 oz.) pkg. crystallized ginger,
   diced**

· Peel and core pears; put through food chopper. Combine pears, pineapple, sugar and ginger in a large saucepan. Cook approximately 30 minutes, stirring frequently, until of desired consistency.
· Ladle into sterilized hot jars to within ⅛" of jar top. Wipe jar rim; adjust lids. Process in boiling water bath 5 minutes. Remove from canner and complete seals unless closures are self-sealing type. Makes 2½ pints.

Note: For less pronounced spicing, reduce amount of ginger.

RASPBERRY/PEAR JAM

*Invented by a farm woman who says:
"Perfect filling for jelly rolls"*

**2½ lbs. pears (10 medium)
5 c. sugar
Juice of 1 lemon
1 pt. frozen red raspberries**

· Peel and core pears and put through food chopper.
· Cook pears, sugar and lemon juice for 20 minutes. Add raspberries and cook until thick and transparent.
· Ladle into sterilized hot jars to within ⅛" of jar top. Wipe jar rim; adjust lids. Process in boiling water bath 5 minutes. Remove from canner

and complete seals unless closures are self-sealing type. Makes 3 pints.

RED PLUM JAM

A deep red jam with a slightly tart taste. Excellent with meat

**5 lbs. red plums
½ c. water
Sugar**

· Chop plums; you should have about 11¼ c. Cook in water until soft. Put through food mill; measure. For each cup plum purée, add 1½ c. sugar. Combine in large kettle. Boil rapidly, stirring frequently, to jellying point, approximately 10 minutes.
· Ladle into sterilized hot jars to within ⅛" of jar top. Wipe jar rim; adjust lids. Process in boiling water bath 5 minutes. Remove from canner and complete seals unless closures are self-sealing type. Makes 6 pints.

CRUSHED RASPBERRY JAM

For variety, make with black or wild raspberries—extra-delicious

**1 c. crushed red raspberries
4 c. diced rhubarb, scalded
5 c. sugar**

· Combine raspberries, rhubarb and 2½ c. sugar in saucepan.
· Bring to a boil and cook for 3 minutes, stirring constantly. Add remaining sugar and boil 5 minutes.
· Ladle into sterilized hot jars to within ⅛" of jar top. Wipe jar rim; adjust lids. Process in boiling water bath 5 minutes. Remove from canner and complete seals unless closures are self-sealing type. Makes 5 half pints.

RED RASPBERRY/APRICOT JAM

An interesting way to use the last of the year's dried apricots

1½ c. dried apricots
1 c. water
9 c. fresh red raspberries
6 c. sugar
½ tsp. salt

· Simmer apricots in water for 15 minutes. Add raspberries, sugar and salt. Cook rapidly for 15 minutes, or until of desired consistency.
· Ladle into sterilized hot jars to within ⅛" of jar top. Wipe jar rim; adjust lids. Process in boiling water bath 5 minutes. Remove from canner and complete seals unless closures are self-sealing type. Makes 5 pints.

RED RASPBERRY/PLUM JAM

Colorful medley of two red-ripe fruits

4 lbs. tart red plums, pitted (9 c.)
4 c. red raspberries (1 lb.)
Sugar

· Grind together plums and raspberries, using medium blade of food chopper. Measure pulp and for each cup of pulp, add 1 c. sugar. Combine in large saucepan. Bring to a boil and boil about 20 minutes, stirring frequently, until jellying point is reached.
· Ladle into sterilized hot jars to within ⅛" of jar top. Wipe jar rim; adjust lids. Process in boiling water bath 5 minutes. Remove from canner and complete seals unless closures are self-sealing type. Makes about 5 pints.

STRAWBERRY/RHUBARB JAM

Coat sponge cake slices with jam and sprinkle with confectioners sugar

1 qt. strawberries
3 c. diced rhubarb (½" dice)
6 c. sugar

· Mash strawberries. Combine with rhubarb in large saucepan. Add 4 c. sugar; bring to a rolling boil and boil 4 minutes, stirring constantly.
· Add remaining 2 c. sugar and boil 4 minutes more, stirring constantly.
· Ladle into sterilized hot jars to within ⅛" of jar top. Wipe jar rim; adjust lids. Process in boiling water bath 5 minutes. Remove from canner and complete seals unless closures are self-sealing type. Makes 2½ pints.

TRIPLE FRUIT JAM

Three fruits blend in this pretty red delicacy that wins praise

9 c. sugar
2 c. water
5 c. peeled, chopped apples (about 2 lbs.)
2 (8¼ oz.) cans crushed pineapple, drained
4 c. red raspberries, washed and drained

· Combine sugar and water in a large saucepan; boil until mixture spins a thread (230° F.).
· Add apples, pineapple and raspberries and boil until thick, about 20 to 30 minutes, stirring often.
· Ladle into sterilized hot jars to within ⅛" of jar top. Wipe jar rim; adjust lids. Process in boiling water bath 5 minutes. Remove from canner and complete seals unless closures are self-sealing type. Makes 5 pints.

PLUM/RASPBERRY JAM

A favorite in Eastern Kansas

3 lbs. prune plums
5 c. sugar
1 (10 oz.) pkg. frozen red
   raspberries

• Pit ripe, firm plums and put through food chopper using medium blade (should be 5 c. pulp).
• Combine plums and sugar in a 5- or 6-qt. pan. Add raspberries. Mix.
• Bring to a boil; reduce heat and simmer until thick, about 40 minutes. Stir occasionally.
• Ladle into sterilized hot jars to within 1/8 " of jar top. Wipe jar rim; adjust lids. Process in boiling water bath 5 minutes. Remove from canner and complete seals unless closures are self-sealing type. Makes 8 half pints.

LIME/PEAR HONEY

Rivals the gift of bees—the tasty medley of three fruit flavors

3 lbs. pears (about 12 medium)
1 (8¼ oz.) can crushed pineapple
Grated peel and juice of 1 lime
5 c. sugar

• Peel and core pears and put through food chopper.
• Mix all ingredients in a large saucepan and cook approximately 20 minutes; stir often.
• Ladle into sterilized hot jars to within 1/8 " of jar top. Wipe jar rim; adjust lids. Process in boiling water bath 5 minutes. Remove from canner and complete seals unless closures are self-sealing type. Makes 3 pints.

REBA'S PEAR HONEY

Honey-colored if cooked no longer than necessary for right consistency

4 c. peeled, cored, crushed pears
3 c. sugar
¼ tsp. salt
1 lemon, ground

• Combine all ingredients and cook in a heavy pan, stirring occasionally, for about 15 minutes or until of spreading consistency.
• Ladle into sterilized hot jars to within 1/8 " of jar top. Wipe jar rim; adjust lids. Process in boiling water bath 5 minutes. Remove from canner and complete seals unless closures are seal-sealing type. Makes 2½ pints.

VARIATION

ORANGE/PEAR HONEY: Use 1 orange, ground, instead of lemon.

PEACH HONEY

Try it on grapefruit or toast—makes winter breakfast taste like spring

1 large orange
12 large peaches, peeled and pitted
Sugar

• Quarter the orange; do not peel. Put orange pieces and peaches through food chopper. Measure the mixture; add 1 c. sugar for each cup fruit. Place in large saucepan and cook approximately 20 minutes, or until of desired consistency.
• Ladle into sterilized hot jars to within 1/8 " of jar top. Wipe jar rim; adjust lids. Process in boiling water bath 5 minutes. Remove from canner and complete seals unless closures are self-sealing type. Makes 5 pints.

Uncooked Jellies, Jams and Spreads

Uncooked, frozen jellies and jams have thousands of champions in country kitchens. They're made with thoroughly ripe fruit, either fresh or frozen. Because the fruit isn't cooked, it retains the maximum fresh, fruity taste and color.

These frozen sweets are easy to fix. You do not make pectin and acid tests, but merely add commercial pectin. And you skip the greatest hazard in jelly making—determining the jellying point. Here is the route to success:

Follow the directions in the recipes, measuring ingredients carefully. Use commercial pectin and do not attempt to substitute liquid and powdered pectins for each other unless the amount of each kind is specified. You will use more sugar than for cooked jellies and jams; you will have to add lemon juice or ascorbic acid for the less acid fruits.

Give the jams and jellies time to set before you store them in the freezer. It may take only a few minutes, but sometimes it takes 2 to 3 days. If they do not set within a few hours, put them in the refrigerator to finish setting.

Store uncooked jellies and jams in the freezer at 0° F. up to 6 months. If used within 3 weeks, store in refrigerator. Once you open an uncooked spread, keep it in the refrigerator and use continuously; it will spoil if held at room temperature.

If you notice a white moldlike for-

mation in the frozen products when you take them from the freezer to serve, don't be alarmed. It occasionally forms during storage of several months, is harmless and will melt quickly at room temperature.

TART CHERRY JELLY

Sparkling, cherry-red—serve with hot rolls when ham is on platter

4¾ c. sugar
  3 c. fresh cherry juice
  1 pkg. powdered fruit pectin
  ½ c. water

• Add sugar to 1¼ c. cherry juice; stir thoroughly.
• Slowly add pectin to water. Heat almost to boiling point, stirring constantly. Pour hot pectin into remaining 1¾ c. cherry juice. Stir until pectin is dissolved. Let pectin mixture stand 15 minutes, but stir it occasionally.
• Add the juice-sugar mixture; stir until all the sugar is dissolved.
• Pour into containers; cover with tight lids.
• Let stand at room temperature until set, overnight or at least 6 hours.
• Freeze or refrigerate. Makes 5 to 6 half pints.

Note: Frozen cherries may be used. Grind whole frozen cherries and thaw. Pour into jelly bag and allow juice to drain. Squeeze bag to obtain the maximum quantity of juice. If juice is from commercially frozen cherries, reduce sugar to 4¼ c.

ORANGE/LEMON JELLY

Fine flavor accent with eggs and egg dishes—wonderful with roast duck

1 pkg. powdered fruit pectin
2 c. lukewarm water
1 (6 oz.) can frozen orange juice concentrate
¼ c. strained fresh lemon juice
4½ c. sugar

• Slowly add pectin to lukewarm water in 2-qt. bowl, stirring constantly until completely dissolved. Let stand 45 minutes, stirring occasionally (stir, do not beat).
• Thaw orange juice concentrate and pour into 1-qt. bowl. Add lemon juice and 2½ c. sugar. Mix thoroughly (not all sugar will dissolve).
• Add the remaining 2 c. sugar slowly to the dissolved pectin. Stir until sugar dissolves.
• Add juice mixture to pectin mixture. Stir constantly until sugar dissolves.
• Pour into containers. Cover with tight lids. Let stand at room temperature overnight or until set. Freeze or refrigerate. Makes 5 to 6 half pints.

CONCORD GRAPE JELLY

Glows like amethysts—good with chicken, veal and cold roast pork

1 pkg. powdered fruit pectin
2 c. lukewarm water
1 (6 oz.) can frozen grape juice concentrate
3¼ c. sugar

• Slowly add pectin to lukewarm water in 2-qt. bowl, stirring constantly until completely dissolved. Let stand 45 minutes, stirring occasionally (stir, do not beat).

• Thaw grape juice concentrate and pour into 1-qt. bowl. Add 1½ c. sugar to juice. Mix thoroughly (not all sugar will dissolve).
• Add remaining sugar to dissolved pectin. Stir until sugar dissolves.
• Add the juice mixture to the pectin mixture. Stir until sugar dissolves.
• Pour into containers. Cover with tight lids. Let stand at room temperature about 24 hours or until set. Freeze or refrigerate. Makes 5 to 6 half pints.

STRAWBERRY JAM

Makes hot biscuits and vanilla ice cream taste equally special

2 (10 oz.) pkgs. frozen, sweetened strawberries
3 c. sugar
1 pkg. powdered fruit pectin
1 c. water

• Thaw frozen berries; purée in food mill or blender.
• Add sugar, mix thoroughly and let stand 20 minutes; stir occasionally.
• Combine powdered pectin and water; boil rapidly 1 minute, stirring.
• Remove from heat. Add the fruit to the pectin; stir for 2 minutes.
• Pour into containers; cover with tight lids. Let stand at room temperature 24 hours. If jam does not set, refrigerate until it does.
• Freeze or refrigerate. Makes 4 to 5 half pints.

Note: You can substitute ½ c. liquid pectin for 1 pkg. powdered pectin and water. Do not heat. Add strawberries and sugar to pectin; stir for 2 minutes.

VARIATIONS

CHERRY JAM: Put 2 (10 oz.) pkgs. frozen tart cherries through food chopper. Add 2 tsp. lemon juice. Substitute for strawberries.

PEACH JAM: Thaw and mash 2 (10 oz.) pkgs. frozen peaches. Stir in 1 tsp. powdered ascorbic acid or 3 tblsp. lemon juice. Substitute for strawberries.

RED RASPBERRY JAM: Use 3 (10 oz.) pkgs. of frozen raspberries and 4 c. sugar. Prepare like Strawberry Jam.

FRESH STRAWBERRY JAM

When spring gives you plump, juicy berries, use this recipe

**2 c. finely mashed or sieved fully
   ripe strawberries
4 c. sugar
1 pkg. powdered fruit pectin
1 c. water**

· Combine fruit and sugar. Let stand 20 minutes, stirring occasionally.
· Boil powdered pectin and water rapidly for 1 minute, stirring constantly. Remove from heat. Add berries and stir about 2 minutes.
· Pour into containers and cover. Let stand at room temperature 1 hour. Refrigerate until set.
· Store in freezer. Once opened, refrigerate. Makes 5 to 6 half pints.

Note: You can substitute ½ c. liquid pectin for 1 pkg. powdered pectin and water. Do not heat. Add strawberries and sugar to pectin; stir for 2 minutes.

VARIATIONS

APRICOT JAM: Add 1 tsp. powdered ascorbic acid or 3 tblsp. lemon juice to finely mashed and measured fresh fully ripe apricots. Substitute for strawberries.

BLACK RASPBERRY JAM: Substitute fully ripe black raspberries for strawberries.

CHERRY JAM: Pit fully ripe tart cherries and put through food chopper. Measure and substitute for strawberries.

PEACH JAM: Add 1 tsp. powdered ascorbic acid or 3 tblsp. lemon juice to fully ripe mashed peaches. Substitute peaches for strawberries.

PLUM JAM: Pit *mild* fully ripe plums and put through food chopper. Add 2 to 3 tsp. lemon juice, if desired. Measure and substitute for strawberries.

FRESH RED RASPBERRY JAM

Luscious when spread on warm-from-the-oven cakes and breads

**3 c. finely mashed or sieved fully
   ripe red raspberries
6 c. sugar
1 pkg. powdered fruit pectin
1 c. water**

· Combine berries and sugar. Let stand at room temperature about 20 minutes, stirring occasionally.
· Boil pectin and water rapidly for 1 minute, stirring. Remove from heat.
· Add fruit and stir about 2 minutes.
· Pour into containers; cover. Let

stand at room temperature 24 hours. If jam does not set, refrigerate until it does.

• Store in freezer. Makes about 8 to 9 half pints.

Note: You can substitute ½ c. liquid pectin for 1 pkg. powdered pectin and water. Do not heat. Add raspberries and sugar to pectin; stir for 2 minutes.

VARIATIONS

BLACKBERRY JAM: Substitute fully ripe blackberries for red raspberries, but reduce sugar from 6 c. to 5½ c.

CONCORD GRAPE JAM: Heat fully ripe stemmed grapes without adding water to break skins and soften grapes. Put through a food mill or colander to remove seeds. Measure pulp and substitute for raspberries.

TART PLUM JAM: Pit fully ripe tart plums and put through food chopper or blend in blender. Measure and substitute pulp for raspberries.

Uncooked Spreads

Spreads are less sweet than jams and they hold more of the delicate fresh fruit taste and aroma. You add more pectin but since it controls the consistency, you can use a little less of it if you prefer thinner spreads, more if you like thicker spreads. Make up a test batch to determine if you wish to use more or less. Add light corn syrup to reduce formation of sugar crystals during storage.

The best way to make the fruit purée is to use a food mill, an electric mixer with a 1-qt. bowl, or a blender. Or you can mash the fruit very fine.

To make a luscious topping for ice cream, thin the spread when you open it with light corn syrup to the consistency you like.

Store uncooked spreads up to 3 weeks in the refrigerator. Or store in the freezer for up to 1 year.

BLUEBERRY SPREAD

Make with wild berries, sweetened by sunshine, or "tame" berries

¼ c. powdered fruit pectin
2 tblsp. sugar
1 c. sieved blueberries
¾ c. sugar
2 tblsp. light corn syrup
2 tblsp. lemon juice

• Combine pectin and 2 tblsp. sugar in small mixer bowl; mix thoroughly. Add the finely sieved berries and mix at low speed for 7 minutes.

• Add remaining ¾ c. sugar, corn syrup and lemon juice. Mix 3 minutes longer.

• Pour into freezer containers and secure lids. Let stand at room temperature overnight or until jellied.

• Freeze. Makes 2 half pints.

CONCORD GRAPE SPREAD

Fresh-flavored—perfect teamed with pork roast, fried chicken or ham

¾ c. powdered fruit pectin
3 c. sugar
4 c. unsweetened grape juice
½ c. light corn syrup

• Mix pectin and ½ c. sugar in large mixer bowl. Add grape juice and mix at low speed for 7 minutes. Add the remaining sugar and the corn syrup. Mix 3 minutes longer.
• Pour into containers, secure lids. Let stand at room temperature for 24 hours.
• Store in freezer or refrigerator. Makes 6 half pints.

STRAWBERRY SPREAD

Brings the taste of May to Christmas —to all months on the calendar

1 c. powdered fruit pectin
3½ c. sugar
4 c. puréed or finely mashed strawberries, fresh or frozen without sugar

½ c. light corn syrup
2 to 3 tblsp. lemon juice

• Mix the pectin and ½ c. sugar in large mixer bowl. Add the puréed berries and mix at low speed for 7 minutes.
• Add remaining ingredients and mix 3 minutes longer.
• Pour into containers, fasten lids. Let stand at room temperature 24 hours.
• Store in freezer or refrigerator. Makes 6 half pints.

VARIATIONS

APRICOT SPREAD: Add 4 tsp. ascorbic acid powder to 4 c. puréed apricots; substitute for strawberries.

BERRY SPREADS: Substitute raspberries, loganberries, boysenberries or blackberries for the strawberries.

Note: Any fruit you can purée or chop very finely may be used to make these spreads. Peach, nectarine and pear spreads are delicious when fresh, but they are not good keepers.

Putting Up Preserves

Good preserves are made from whole small fruits, like strawberries and cherries, or uniform large pieces of larger fruits, like peaches and pears. When cooked, the fruit is tender and clear; it retains its shape, natural flavor and color. The consistency of the syrup varies from that of honey to a semijelly.

The problem is to keep the fruit plump and distributed through the syrup. The solutions are (1) to let soft fruits stand with the sugar added until the juice starts to flow, and then cook; and (2) to precook hard fruit until barely tender in a little water before adding the sugar.

The method used in making preserves differs somewhat with different fruits.

You need to stir some preserves gently and frequently for 5 minutes after removing them from the heat, skimming them at the same time. Then cover them and let stand several hours or better still, overnight. Next morning, heat them to boiling, fill into sterilized hot jars and process in the boiling water bath.

The proportion of sugar to use depends on how sweet the fruit is. For tart fruits, use 1 pound sugar to 1 pound of fruit; for sweeter fruits, ¾ pound sugar to 1 pound fruit. (One pound of sugar measures approximately 2 cups. The weight of fruits varies, but a rule of thumb is: 1 pound of prepared fruit measures about 3 cups. It's best to weigh for accuracy.)

It is desirable to cook small quantities at a time. And if you cook preserves rapidly once they come to a boil, they will retain their color, flavor and good texture. You need to stir them frequently as they cook, to prevent scorching.

APPLE PRESERVES

For those who like sweets—inexpensive if you grow your own apples

2 c. sugar
1 c. hot water
4 c. apples, peeled, cored and
    quartered

· Combine sugar and water in a saucepan. Bring to a boil, stirring until sugar dissolves. Cook until syrup forms a hard ball in cold water (250° F. on candy thermometer).
· Stir in apples and simmer until they are transparent, stirring constantly.
· Ladle into sterilized hot jars to within ⅛" of jar top. Wipe jar rim; adjust lids. Process in boiling water

bath 5 minutes. Remove from canner and complete seals unless closures are self-sealing type. Makes 3 pints.

CANTALOUPE PRESERVES

A farm cook captures sunshine in jars —ribbons at fairs with this recipe

2 lbs. firm, ripe cantaloupe
1¾ lbs. sugar (4 c.)
Juice of 1 lemon

· Peel cantaloupe and cut in thin slices 1" long. Mix sugar and cantaloupe; let stand overnight.
· Combine cantaloupe mixture and lemon juice in saucepan; cook until clear.
· Ladle into sterilized hot jars to within ⅛" of jar top. Wipe jar rim; adjust lids. Process in boiling water bath 5 minutes. Remove from canner and complete seals unless closures are self-sealing type. Makes about 2 pints.

CHERRY/PLUM PRESERVES

Beautiful burgundy-red preserves

10 red plums, pitted
 4 c. stemmed, halved and pitted
    Bing cherries
1 c. water
Sugar
½ tsp. salt
1 pkg. powdered fruit pectin

· Combine fruits, add water; bring to slow boil. Simmer 3 minutes. Remove from heat; measure.
· Add amount of sugar equal to cooked fruit; mix well. Add salt and pectin; stir until dissolved.
· Boil rapidly 2 minutes, stirring constantly. Remove from heat and skim.

• Ladle into sterilized hot jars to within ⅛″ of jar top. Wipe jar rim; adjust lids. Process in boiling water bath 5 minutes. Remove from canner and complete seals unless closures are self-sealing type. Makes 9 half pints.

FIG/LEMON PRESERVES

Ginger spices the fruit delightfully

6 qts. figs, peeled
6 qts. boiling water
8 c. sugar
3 qts. water
3 lemons, thinly sliced
4 tblsp. sliced preserved ginger

• Wash and stem figs. Pour boiling water over, let stand 15 minutes.
• Drain and rinse figs in clear, cold water. Drain again.
• Combine sugar and 3 qts. water; bring to a boil; add lemon slices (seeds discarded) and ginger; boil rapidly 10 minutes. Skim. Lemon slices may be removed.
• Add figs, a few at a time, to the syrup, so as not to stop the boiling. Cook rapidly until transparent.
• Lift figs out of syrup and place in a shallow pan. Boil syrup down until thick as honey. Pour over figs and let stand overnight.
• In the morning, bring to a boil. Ladle into sterilized hot jars to within ⅛″ of jar top. Wipe jar rim; adjust lids. Process in boiling water bath 5 minutes. Remove from canner and complete seals unless closures are self-sealing type. Makes about 5 pints.

SESAME FIG PRESERVES

Toasted sesame seeds add new touch

1½ lbs. peeled firm ripe figs
3½ c. sugar
2 c. water
4 lemon slices, seeds discarded
2 tblsp. toasted sesame seeds

• Prepare figs; weigh after peeling.
• Combine sugar, water and lemon slices in preserving kettle. Cook about 20 minutes. Drop figs carefully into syrup and cook until clear, about 30 minutes.
• Pack figs, lemon slices and sesame seeds into sterilized hot jars. Cook syrup to thickness desired and pour over fruit to within ⅛″ of jar top. Wipe jar rim; adjust lids. Process in boiling water bath 5 minutes. Remove from canner and complete seals unless closures are self-sealing type. Makes 4 half pints.

Note: Toast sesame seeds in moderate oven (350° F.) about 20 minutes.

GROUND CHERRY PRESERVES

Rather thin in consistency but very rewarding in taste

6 c. husked ground cherries
  (husk tomatoes)
1 c. water
8 c. sugar
¼ c. lemon juice (about 2 lemons)
1½ c. light corn syrup

• Put the prepared fruit in a large kettle. Add water. Bring to a boil and simmer for 10 minutes. Add sugar, lemon juice and syrup.
• Bring to a boil again and simmer 30 minutes. Remove from heat and let cool overnight.
• Next morning, heat to boiling, pour

into sterilized hot jars to within ⅛" of jar top. Wipe jar rim; adjust lids. Process in boiling water bath 5 minutes. Remove from canner and complete seals unless closures are self-sealing type. Makes 8 half pints.

Note: If thicker preserves are desired, the boiling mixture may again be cooled overnight, heated to boiling and canned the second morning.

KUMQUAT PRESERVES

Serve these sunny preserves to complement duck, other fowl or pork

1 qt. kumquats
2 c. sugar
3 c. water
1½ c. light corn syrup

• Wash kumquats and make ½" slit in one side of each kumquat. Simmer fruit gently in a little water about 10 minutes to tenderize skins. Drain.
• Combine sugar and water; bring to a boil and boil 5 minutes. Add kumquats and bring to a boil. Remove from heat, set aside overnight.
• Next morning, add ½ c. corn syrup to kumquats and bring to a boil. Remove from heat and let stand overnight the second time. Repeat the process two more times, adding ½ c. corn syrup each time, bringing to a boil, removing from heat and letting stand overnight. The last morning, heat to a boil, ladle into sterilized hot jars to within ⅛" of jar top. Wipe jar rims; adjust lids. Process in boiling water bath 5 minutes. Remove from canner and complete seals unless closures are self-sealing type. Makes 2 pints.

PEACH PRESERVES

Heat and blend ¼ c. butter and ½ c. preserves to dress up waffles

1 large orange
1½ qts. peeled, diced peaches
Juice of 2 lemons
1 c. chopped maraschino
   cherries
6 c. sugar

• Grate orange peel; dice orange fruit. Combine with peaches, lemon juice, maraschino cherries and sugar in large saucepan. Cook until of desired consistency.
• Ladle into sterilized hot jars to within ⅛" of jar top. Wipe jar rim; adjust lids. Process in boiling water bath 5 minutes. Remove from canner and complete seals unless closures are self-sealing type. Makes 3½ pints.

STRAWBERRY PRESERVES

An heirloom recipe for a great national favorite—colorful and luscious

4 c. hulled strawberries
3 c. sugar

• Add sugar to berries and let stand 10 minutes or until juices start to flow. (Some cooks like to cover them and leave in the refrigerator overnight.)
• Put berry-sugar mixture in a 4-qt. kettle and bring them to a boil, stirring constantly until sugar dissolves. Cook until berries are tender, about 3 minutes. Let stand overnight.
• Next morning bring preserves to a boil and boil 1 minute. Cover the preserves, remove from heat and let stand 2 minutes. Stir gently about 5 minutes, skimming if necessary.

• Ladle into sterilized hot jars to within ⅛" of jar top. Wipe jar rim; adjust lids. Process in boiling water bath 5 minutes. Remove from canner and complete seals unless closures are self-sealing type. Makes 3 half pints.

ROSY STRAWBERRY PRESERVES

Cook a batch daily while getting lunch during the strawberry season

1 qt. halved strawberries
4 c. sugar
1 tblsp. lemon juice

• Combine strawberries and sugar. Stir lightly and let stand 1 hour in flat bottomed shallow pan.
• Bring to a boil and cook 10 minutes stirring constantly.
• Add lemon juice; boil 3 minutes longer.
• Remove from heat, cover and let stand 24 hours.
• Bring to a boil and ladle into sterilized hot jars to within ⅛" of jar top. Adjust lids and process in boiling water bath 5 minutes. Remove from canner and complete seals unless closures are self-sealing type. Makes 2 pints.

RED RASPBERRY PRESERVES

A state fair champion. Spoon it on vanilla ice cream for a treat

4 c. whole red raspberries
4 c. sugar
Juice of 1 lemon

• Place raspberries in kettle with sugar and lemon. Bring slowly to a

boil over low heat, *shaking* all the while. *Do not stir.* Continue shaking and boil for 5 minutes.
• Remove from heat. Cover; let stand several hours or overnight. Heat and pour into sterilized hot jars to within ⅛" of jar top. Wipe jar rim; adjust lids. Process in boiling water bath 5 minutes. Remove from canner and complete seals unless closures are self-sealing type. Makes 7 pints.

RIPE TOMATO PRESERVES

An inexpensive sweet for hearty farm eaters if you grow your tomatoes

2 lbs. tomatoes
4 c. sugar
2 lemons, thinly sliced
½ tsp. salt
½ tsp. whole cloves
½ tsp. whole allspice
1 stick cinnamon
½ tsp. ground ginger

• Scald, skin and weigh small firm red or yellow tomatoes. (If larger than small hen egg, cut in halves or quarters.)
• Cover with sugar and let stand overnight in a cool place.
• Next morning, drain juice into large saucepan; add lemon (seeds discarded), salt and spices, tied in cheesecloth bag. Boil 10 minutes.
• Add tomatoes and cook, stirring frequently until they are clear and syrup fairly thick. Remove spices.
• Ladle into sterilized hot jars to within ⅛" of jar top. Wipe jar rim; adjust lids. Process in boiling water bath 5 minutes. Remove from canner and complete seals unless closures are self-sealing type. Makes 1½ pints.

VARIATIONS

PINEAPPLE/TOMATO PRESERVES: Add with tomatoes 1 c. pineapple tidbits. Makes 2 pints.

TOMATO CONSERVE: With tomatoes add 1 c. light raisins. Makes 2 pints.

Special Occasion Conserves

Conserves look like jams and they have the same consistency. But they are made with a combination of fruits. Often nuts and/or raisins are added. They are favored especially for meat and poultry accompaniments.

Note: To blanch almonds, cover with cold water and bring slowly to a rolling boil. Drain; slip off skins by pressing between forefinger and thumb.

APRICOT/ALMOND CONSERVE

Pots of gold on the shelf. Especially good with chicken or pork

3 oranges
Water
5 lbs. apricots (40 to 50)
¾ c. blanched almonds (about
      4 oz.)
10 c. sugar

· Peel oranges. Cover peel with cold water and bring to a boil. Boil 3 or 4 minutes. Pour off water. Remove white from peel and from oranges.
· Dice apricots and oranges, discarding seeds. Grind orange peel and nuts; mix with fruit. Add sugar.
· Cook rapidly for 45 minutes or until desired thickness. Pour into sterilized hot jars to within ⅛" of jar top. Wipe jar rim; adjust lids. Process in boiling water bath 5 minutes. Remove from canner and complete seals unless closures are self-sealing type. Makes 6½ pints.

SWEET APPLE CONSERVE

Delightful in filled cookies and equally fine on hot biscuits

4 c. sweet apples, peeled, cored
      and chopped (1½ lbs.)
2 c. sugar
2 c. raisins
Grated peel of 1 orange
Juice of 2 oranges
Grated peel and juice of 1 lemon
½ c. nuts (optional)

· Combine all ingredients in a large saucepan. Cook slowly until thick, about 45 minutes.
· Ladle into sterilized hot jars to within ⅛" of jar top. Wipe jar rim; adjust lids. Process in boiling water bath 5 minutes. Remove from canner and complete seals unless closures are self-sealing type. Makes 4 half pints.

CANTALOUPE/PEACH CONSERVE

Two summer friends, peaches and melon, unite to give winter pleasure

4 c. chopped peeled cantaloupe
4 c. chopped peeled peaches
6 c. sugar
¼ c. lemon juice
½ tsp. ground nutmeg
¼ tsp. salt
1 tsp. grated lemon peel
½ c. chopped walnuts (optional)

• Combine cantaloupe and peaches in large saucepan. Simmer 20 minutes, stirring until there is enough liquid to prevent fruit from sticking.
• Add sugar and lemon juice. Boil until thick. Add nutmeg, salt, lemon peel and nuts; boil 3 minutes.
• Ladle into sterilized hot jars to within ⅛" of jar top. Wipe jar rim; adjust lids. Process in boiling water bath 5 minutes. Remove from canner and complete seals unless closures are self-sealing type. Makes 4 to 5 half pints.

Note: Substitute orange peel for the lemon peel. Papaya may be used instead of cantaloupe.

CHERRY CONSERVE

Whole cherries add texture and flavor —try this favorite with ham

6 c. pitted, tart cherries
2 c. black raspberries
2⅔ c. sugar

• Combine all ingredients in saucepan and cook until thick, about 20 minutes, stirring often.
• Ladle into sterilized hot jars to within ⅛" of jar top. Wipe jar rim; adjust lids. Process in boiling water bath 5 minutes. Remove from canner and complete seals unless closures are self-sealing type. Makes 2 pints.

BING CHERRY CONSERVE

Thanks to the Idaho homemaker for her recipe for this gourmet special

1¾ lbs. Bing cherries, halved and pitted
1 qt. loganberries
2 slices pineapple, diced
12 apricots, cut fine
½ lemon, juice and grated peel
½ orange, juice only
Sugar equal to weight of combined fruits

• Place all fruits and juices in a large kettle. Bring to a boil and cook until cherries are tender. Add sugar. Boil for 10 minutes.
• Ladle into sterilized hot jars to within ⅛" of jar top. Wipe jar rim; adjust lids. Process in boiling water bath 5 minutes. Remove from canner and complete seals unless closures are self-sealing type. Makes 6 pints.

CRANBERRY/APPLE CONSERVE

Winter holiday special that's fine for Christmas giving and eating

3 c. cranberries
5 apples
1 orange
1 lemon
1 (1 lb. 4 oz.) can crushed pineapple (2½ c.)
1 pkg. powdered fruit pectin
5½ c. sugar
1 (2 oz.) pkg. almonds, slivered

• Discard soft cranberries. Quarter and core unpeeled apples. Put through food chopper, using medium blade.
• Extract orange and lemon juice; dis-

card seeds. Grind orange and lemon peels, using fine blade.
• Combine ground cranberries, apples, orange, lemon and pineapple in kettle. Bring to a boil.
• Add pectin and bring to a boil.
• Add sugar, stirring constantly; bring to a boil and boil hard 1 minute.
• Remove from heat; stir and skim for 5 minutes. Add almonds.
• Ladle into sterilized hot jars to within ⅛″ of jar top. Wipe jar rim; adjust lids. Process in boiling water bath 5 minutes. Remove from canner and complete seals unless closures are self-sealing type. Makes 9 half pints.

FIG/CITRUS CONSERVE

California hostesses keep a few jars handy for fast-fix guest meals

5   lbs. figs
Water
10 c. sugar
2   lemons, coarsely ground
4   oranges, coarsely ground
2   tsp. ground ginger

• Cover figs with a small quantity of water and boil until liquid is tan. Drain and reserve about 2 c. liquid. Mix with sugar, stir and heat to make a heavy syrup.
• Combine figs, sugar syrup, lemon and orange fruit pulp. Simmer gently about 1½ hours, stirring frequently. Skim foam from time to time.
• Add ginger, mixing well. Pour into sterilized hot jars to within ⅛″ of jar top. Wipe jar rim; adjust lids. Process in boiling water bath 5 minutes. Remove from canner and complete seals unless closures are self-sealing type. Makes about 5 pints.

Note: Four tblsp. finely chopped nuts may be added with the ginger.

GOOSEBERRY CONSERVE

The reader who shares this recipe gathers the berries in the woods

2   qts. gooseberries
3½ c. sugar
1   lb. candy orange slices
½   c. chopped walnuts

• Remove stems and tails from gooseberries. Wash carefully, put through a food chopper, place in large saucepan and bring to a boil. Add sugar, stirring constantly.
• Bring again to a boil. Add candy orange slices, cut in eighths. Cook, stirring, until mixture thickens to desired consistency. Remove from heat; skim; add walnuts.
• Ladle into sterilized hot jars to within ⅛″ of jar top. Wipe jar rim; adjust lids. Process in boiling water bath 5 minutes. Remove from canner and complete seals unless closures are self-sealing type. Makes 6 half pints.

GRAPE CONSERVE

Harmonious mingling of orange and grape flavors with rich raisin taste

8   c. stemmed Concord grapes
      (about 5 lbs.)
¼   c. water
2   oranges
1   c. seeded raisins
6   c. sugar
¼   tsp. salt
1   c. chopped walnuts

• Slip skins from washed grapes. Put skins in large kettle, pulp in a small one. Add water to skins and simmer gently 20 minutes. Cook pulp until soft enough to loosen seeds. Put pulp through food mill or sieve to remove seeds; mix with skins.

• Peel oranges, remove white membrane; cut orange in thin slices (discard seeds), cut peel in fine slivers. Stir orange slices and peel, raisins and sugar into grape mixture. Bring to a brisk boil and cook until conserve is thick (until jellying point is reached). Stir in salt and nuts.

• Bring to a boil and ladle into sterilized hot jars to within ⅛" of jar top. Wipe jar rim; adjust lids. Process in boiling water bath 5 minutes. Remove from canner and complete seals unless closures are self-sealing type. Makes 10 half pints.

PLUM/CRAB APPLE CONSERVE

Try this tasty tri-flavored special

2 c. diced crab apples
⅔ c. crushed pineapple
1 qt. pitted red plums
5½ c. sugar
⅓ c. slivered almonds

• Core, but do not peel, firm, ripe crab apples; cut in ¼" cubes. Combine crab apples, drained pineapple and plums in a large saucepan. Heat to boiling.

• Add sugar and cook, stirring constantly, until mixture is desired consistency. Remove from heat; skim; add almonds.

• Ladle into sterilized hot jars to within ⅛" of jar top. Wipe jar rim; adjust lids. Process in boiling water bath 5 minutes. Remove from canner and complete seals unless closures are self-sealing type. Makes 7 half pints.

PLUM AND PEACH CONSERVE

Fruits complement each other—the flavor contrasts make the difference

3 c. ground sweet plums (Italian prunes)
3 c. mashed peaches
6 c. sugar
2 medium oranges

• Combine plums, peaches and sugar in large saucepan. Add grated peel from 1 orange, juice from 2 oranges. Cook, stirring frequently, about 25 minutes, or to desired consistency.

• Ladle into sterilized hot jars to within ⅛" of jar top. Wipe jar rim; adjust lids. Process in boiling water bath 5 minutes. Remove from canner and complete seals unless closures are self-sealing type. Makes 4 pints.

RAISIN CONSERVE

Keep a few jars on hand to team with ham—a splendid combination

1 (15 oz.) pkg. seedless raisins (4 c.)
2½ c. water
½ c. lemon juice (4 lemons)
½ c. finely chopped walnuts
1 pkg. powdered fruit pectin
4 c. sugar

• Combine raisins, water and lemon juice. Cover and let stand overnight or at least 4 hours.

• Bring mixture to a boil and simmer, covered, 30 minutes, stirring often.

• Drain raisins, reserving juice. Put raisins through food chopper or chop them very fine.

• Measure 4 c. raisins and juice into a large kettle. Add nuts and pectin

Place over high heat and cook, stirring, until mixture comes to a hard boil. Stir in sugar. Bring to a full rolling boil; boil hard 1 minute.
· Remove from heat and skim off foam. Ladle into sterilized hot jars to within ⅛" of jar top. Wipe jar rim; adjust lids. Process in boiling water bath 5 minutes. Remove from canner and complete seals unless closures are self-sealing type. Makes about 7 half pints.

RHUBARB/ORANGE CONSERVE

Quickly made with strawberry-red rhubarb and candy . . . very colorful

5 c. red rhubarb
1 lb. candy orange slices
3 c. sugar

· To measure rhubarb, cut into ⅓" cubes. Cut candy slices into eighths.
· Mix rhubarb and sugar in a wide saucepan. Stir constantly over high heat. When the mixture begins to boil, add the orange pieces. Continue boiling and stirring until the mixture thickens to a jelly-like consistency.
· Ladle into sterilized hot jars to within ⅛" of jar top. Wipe jar rim; adjust lids. Process in boiling water bath 5 minutes. Remove from canner and complete seals unless closures are self-sealing type. Makes 6 half pints.

RASPBERRY/APPLE CONSERVE

More summer apples than you need for pies? Then try this good jam

9 c. sugar
2 c. water

6 c. Transparent or other early apples, diced
3 c. red raspberries

· Boil sugar and water until it spins a thread (230° F.). Add apples; cook 2 minutes.
· Add raspberries; cook 10 minutes.
· Ladle into sterilized hot jars to within ⅛" of jar top. Wipe jar rim; adjust lids. Process in boiling water bath 5 minutes. Remove from canner and complete seals unless closures are self-sealing type. Makes 3½ pints.

RHUBARB/APRICOT CONSERVE

If you have frozen rhubarb, you can make this spread on wintry days

1 (12 oz.) pkg. dried apricots
2 c. water
1 qt. diced rhubarb
5 c. sugar
⅓ c. slivered almonds

· Wash the dried apricots carefully. Cut with scissors into small pieces. Add water and soak about 2 hours.
· Combine rhubarb and apricots and cook 10 minutes, stirring constantly.
· Add sugar gradually, stirring to mix. Continue cooking until of jam-like consistency. (This will take about 20 minutes of careful stirring.) Add almonds.
· Ladle into sterilized hot jars to within ⅛" of jar top. Wipe jar rim; adjust lids. Process in boiling water bath 5 minutes. Remove from canner and complete seals unless closures are self-sealing type. Makes 6 pints.

FOUR-FRUIT CONSERVE

Buffet supper special. Adds color and a taste surprise that wins praise

1   qt. sweet cherries (2¾ c. stemmed and pitted)
1   qt. strawberries (3½ c.)
1   large orange, peeled and diced
2   c. drained, crushed pineapple
7   c. sugar
¼   c. chopped walnuts (optional)

• Combine all ingredients, except nuts, in large saucepan. Stir until sugar is dissolved.
• Cook over low heat, stirring frequently, until mixture sheets from spoon. Skim. Add nuts.
• Ladle into sterilized hot jars to within ⅛" of jar top. Wipe jar rim; adjust lids. Process in boiling water bath 5 minutes. Remove from canner and complete seals unless closures are self-sealing type. Makes 4 pints.

STRAWBERRY/CHERRY CONSERVE

Bright as strawberries plus cherries and twice as good as either alone

1   lb. Bing cherries (2½ c. pitted)
2   c. sliced strawberries
3   c. sugar
¼   c. lemon juice
¼   tsp. almond extract

• Combine cherries, strawberries and sugar and cook over low heat, shaking pan until sugar is dissolved. Increase heat and boil 8 minutes, stirring occasionally.
• Add lemon juice and almond extract

and boil 3 to 5 minutes or until jellying point is reached. Skim.
• Ladle into sterilized hot jars to within ⅛" of jar top. Wipe jar rim; adjust lids. Process in boiling water bath 5 minutes. Remove from canner and complete seals unless closures are self-sealing type. Makes 2 pints.

GREEN TOMATO CONSERVE

Loaded with that marvelous old-fashioned flavor—a treasured recipe

2   lemons
1½   c. water
16   green tomatoes
1   (2") stick cinnamon
½   tsp. whole cloves
1   tblsp. mixed pickling spices
2   c. diced peeled tart apples
3   c. sugar

• Peel lemons lengthwise. Cut peel into sliver-thin pieces. Cook peel in water in large saucepan for 30 minutes.
• Parboil tomatoes for 5 minutes. Chop coarsely.
• Tie spices in cheesecloth bag.
• Add apples, tomatoes, sugar and spices to cooked lemon peel. Bring to full boil and boil 20 minutes, stirring often. Remove spices. Reduce heat to simmer and cook 20 minutes, stirring often.
• Cut peeled lemons into very thin slices; discard seeds. Add to mixture and cook 20 minutes, stirring often.
• Ladle into sterilized hot jars to within ⅛" of jar top. Wipe jar rim; adjust lids. Process in boiling water bath 5 minutes. Remove from canner and complete seals unless closures are self-sealing type. Makes 7 pints.

YOUNGBERRY CONSERVE

Crunchy texture and fruity flavors unite to make this distinctive

3   large oranges
4   c. crushed youngberries (about 2 qts.)
7½   c. sugar
½   bottle liquid fruit pectin
1   c. finely chopped walnuts

• Grate oranges, removing only the orange part of peel. Squeeze 1 of the oranges to get ½ c. juice. Remove white peel from remaining 2 oranges and cut in pieces.

• Combine oranges, orange peel, juice, berries and sugar in a large saucepan. Place over high heat and bring to a full, rolling boil, stirring constantly. Boil hard 1 minute. Remove from heat and stir in liquid pectin. Skim, if necessary. Add nuts.

• Ladle into sterilized hot jars to within ⅛" of jar top. Wipe jar rim; adjust lids. Process in boiling water bath 5 minutes. Remove from canner and complete seals unless closures are self-sealing type. Makes 4 pints.

Old-fashioned Fruit Butters

Grandmother made fruit butters for two important reasons that are just as valid today: (1) they taste exceptionally good and (2) they make use of the sound parts of windfalls or culls. Among the favorite fruits for butters are apples, apricots, grapes, peaches, pears, plums, quinces, guavas and combinations of fruits.

HOW TO PREPARE FRUITS FOR BUTTERS

Cook fruits until soft, stirring constantly. Put cooked fruit through food mill or colander. For a superior, smooth butter, sieve the pulp to remove fibrous material.

APPLES: Peel and slice or quarter. Cook in an equal amount of water, cider, half water and half cider or Concord grape juice.

APRICOTS: Remove pits; peel if desired, crush fruit and cook in its own juice. (Grated orange peel makes a delightful seasoning.)

GRAPES: Crush and cook in own juice.

GUAVAS: Remove blossom and stem ends. Peel if skins are tough or blemished. Slice, put through food mill.

PEACHES: Scald and remove skins if desired. Pit, crush fruit and cook in its own juice.

PEARS: Remove stems, but do not core or peel. Quarter or slice. Cook in half as much water as fruit. Add 3 tblsp. lemon juice to each gallon fruit pulp.

PLUMS: Halve or quarter, remove pits and cook in their own juice.

QUINCES: Remove blossom ends, but do not core or peel. Cut in small pieces. Cook with half as much water as fruit.

FRUIT COMBINATIONS: Use equal parts of apples and quinces or apples and plums or apples and pears.

ADDITIONS TO FRUIT PULP

Sugar: Use white or brown sugar. Brown sugar darkens the light fruits; it gives a pronounced flavor to bland ones. The amount of sugar to add depends on personal tastes, but the general rule is half as much sugar as pulp.

Salt: Add ¼ to ½ tsp. for every gallon of fruit butter.

Spices: Usually ground spices are added, although some people prefer to omit them. About 1 tsp. ground cinnamon and ½ tsp. each ground ginger and ground allspice to 1 gal. of butter is a good proportion. Whole spices tied loosely in a cheesecloth bag may be substituted for ground spices in making light-colored fruit butters. Ginger is an especially tasty spice with pears. Also, adding 3 tblsp. lemon juice to 1 gallon of fruit pulp steps up the flavor.

COOKING FRUIT BUTTERS

1. Measure the pulp and sugar into large kettle; add the salt. Boil rapidly, stirring constantly to prevent scorching. As the butter becomes thick, lower heat to reduce spattering.

2. Add spices and lemon juice.
3. Continue cooking until butter is thick enough almost to flake off the spoon, or as Grandmother used to say: "Until it is thick enough to spread." Another test for consistency is to pour a tablespoon of the hot butter onto a chilled plate—if no rim of liquid forms around the edge of the butter, it is ready for canning.
4. Ladle into sterilized jars to within ⅛" of jar top. Wipe jar rim; adjust lids. Process in boiling water bath 5 minutes. Remove from canner and complete seals unless closures are self-sealing type.

APPLE/PEAR BUTTER: A FARM JOURNAL reader's favorite spread is fruit butter made with equal parts of pear and apple pulp, spiced and sweetened to taste. The fruit blend is delicious.

20-MINUTE APPLE BUTTER

Candy tints and spices—vinegar points up the apple flavor

3 qts. canned applesauce
10 c. sugar
½ c. vinegar
1 c. red cinnamon candies

• Combine all ingredients in saucepan and cook until candies dissolve and mixture is thick, 20 to 25 minutes.
• Ladle into sterilized hot jars to within ⅛" of jar top. Wipe jar rim adjust lids. Process in boiling wate bath 5 minutes. Remove from canne and complete seals unless closures ar self-sealing type. Makes 6 pints.

LEMON/APPLE BUTTER

*Brings a change to an old-time spread
—lemon takes the place of spices*

4 lbs. apples (about 12 medium)
2 c. sugar
Grated peel of 1 lemon

• Wash apples and cut in eighths.
• Cook in small amount of water until
tender, then put through food mill.
• Add sugar and grated lemon peel;
cook until thick.
• Ladle into sterilized hot jars to
within ⅛" of jar top. Wipe jar rim;
adjust lids. Process in boiling water
bath 5 minutes. Remove from canner
and complete seals unless closures are
self-sealing type. Makes 4 pints.

GRAPE/APPLE BUTTER

*Apples, grape juice and brown sugar
make a dark butter with rich taste*

2 lbs. apples
1½ c. grape juice
7½ c. brown sugar, firmly packed
½ tsp. butter
½ bottle liquid fruit pectin
½ tsp. ground cinnamon

• Wash and remove stems from
apples. Do not core. Dice finely into
large kettle.
• Add grape juice; simmer 15 minutes
or until tender.
• Sieve mixture. Measure 5 c., adding
more grape juice if needed, and return
to kettle. Add sugar and butter. Bring
to rolling boil; boil 1 minute. Remove
from heat.
• Add pectin and cinnamon. Stir frequently for 5 minutes.
• Ladle into sterilized hot jars to
within ⅛" of jar top. Wipe jar rim;
adjust lids. Process in boiling water

bath 5 minutes. Remove from canner
and complete seals unless closures are
self-sealing type. Makes 4 pints.

RHUBARB BUTTER

*Kind to the pocketbook, exquisite
rosy-red color, on the tart side*

6 c. cut-up rhubarb
½ c. water
2½ c. sugar
4 drops red food color (optional)

• Cut rhubarb in 1" lengths. Add
water. Purée in blender. This should
make 3 c. of pulp.
• Heat to a boil; mix in sugar. Cook,
stirring, until mixture is consistency of
fruit butter. Add food color.
• Ladle into sterilized hot jars to
within ⅛" of jar top. Wipe jar rim;
adjust lids. Process in boiling water
bath 5 minutes. Remove from canner
and complete seals unless closures are
self-sealing type. Makes 3 half pints.

ALMOND/CRAB APPLE
BUTTER

*A Minnesota State Fair winner—the
almond taste is subtle*

4 qts. crab apples
Water
3 c. sugar
½ tsp. almond extract

• Quarter crab apples. Cover with
cold water. Bring to a boil; simmer
until tender. Put through colander or
food mill. You should have 4 c. pulp.
• Put pulp in a shallow kettle; add
sugar. Boil rapidly, stirring constantly.
As butter becomes thick, reduce heat
to prevent spattering.

· Add almond extract. Continue cooking until butter is thick.

· Ladle into sterilized hot jars to within ⅛″ of jar top. Wipe jar rim; adjust lids. Process in boiling water bath 5 minutes. Remove from canner and complete seals unless closures are self-sealing type. Makes 8 half pints.

Note: Use red crab apples to get the best color.

VARIATION

SPICED CRAB APPLE BUTTER: Omit almond extract and add ¼ tsp. each, ground cinnamon and nutmeg.

TOMATO BUTTER

Many children walked fast to Grandmother's to spread this on bread

5 qts. ground ripe tomatoes
3 medium onions, ground
1 pt. vinegar
3 c. brown sugar, firmly packed
2 c. sugar
1 tsp. ground cinnamon
1 tsp. ground cloves
1 tsp. ground allspice
1 tblsp. salt

· Peel and grind tomatoes and onions.

· Combine vinegar, sugars, spices and salt in heavy saucepan; bring to a boil.

· Add the vegetables and simmer over low heat until thick, stirring frequently to prevent sticking.

· Ladle into sterilized hot jars to within ⅛″ of jar top. Wipe jar rim; adjust lids. Process in boiling water bath 5 minutes. Remove from canner and complete seals unless closures are self-sealing type. Makes about 5 pints.

Cheerful Marmalades

Speak of marmalades and almost everyone thinks oranges. Citrus fruits play an important role in this sweet, but many other fruits make delicious marmalades, too—cherries, apples, pineapple. And tomatoes. Marmalades look like jams, but they contain thin fruit slices or diced fruit distributed through clear, translucent jelly.

CRANBERRY/BANANA MARMALADE

Spring a breakfast surprise—gives buttered toast a festive taste

4 c. cranberries
1 c. water

6 medium bananas, mashed (about 2 c.)
7 c. sugar
1 bottle liquid fruit pectin

· Wash cranberries; add water. Simmer in covered kettle for 10 minutes.

· Add bananas and sugar; mix well. Bring to full rolling boil. Boil 1 minute, stirring constantly over entire bottom of kettle.

· Remove from heat. Stir in pectin; skim. Pour into sterilized hot jars to within ⅛″ of jar top. Wipe jar rim; adjust lids. Process in boiling water bath 5 minutes. Remove from canner and complete seals unless closures are self-sealing type. Makes 4½ pints.

Note: This marmalade may be made without pectin. If so, use only 3 c. sugar.

GROUND CHERRY MARMALADE

Lovely yellow color, crystal clear and delicate, yet rich in flavor

3   c. ground cherries (also called husk tomatoes)
2   c. cooked pears, drained and diced finely
¾   c. water
½   c. drained, crushed pineapple
¼   c. lemon juice (2 lemons)
7   c. sugar
½   bottle liquid fruit pectin

• Husk and wash the ground cherries. Combine with pears and water; simmer 25 minutes.
• Add pineapple, lemon juice and sugar; bring quickly to full rolling boil. Add pectin; boil rapidly 3 minutes.
• Remove from heat; skim.
• Ladle into sterilized hot jars to within ⅛" of jar top. Wipe jar rim; adjust lids. Process in boiling water bath 5 minutes. Remove from canner and complete seals unless closures are self-sealing type. Makes 7 half pints.

GOOSEBERRY/ORANGE MARMALADE

Gives a special flavor boost to lamb

2   large oranges
½   c. water
1½ qts. gooseberries
1   pkg. powdered fruit pectin
7   c. sugar

• With vegetable peeler, remove the orange part of the peel; cut into thin strips; place in saucepan with water and cook until tender.
• Remove remaining white peel from oranges and discard. Chop fruit pulp (discard seeds) and add to peel. Cook 10 minutes longer.
• Remove stems and tails from gooseberries; grind and combine with orange mixture. Measure 5½ c. into large saucepan. Add water, if necessary, to make up volume.
• Stir in pectin and bring to a boil; boil 1 minute.
• Stir in sugar and return to a full rolling boil; boil 1 minute.
• Ladle into sterilized hot jars to within ⅛" of jar top. Wipe jar rim; adjust lids. Process in boiling water bath 5 minutes. Remove from canner and complete seals unless closures are self-sealing type. Makes 4 pints.

LEMON MARMALADE

Team with toast or fruit bread and tea —your guests will like it

2   c. thinly sliced lemon
6   c. water
4½ c. sugar

• Combine lemon slices (seeds discarded) and water in large preserving kettle. Cook rapidly until tender, about 20 minutes; stir occasionally.
• Drain and measure liquid; add enough water to make 6 cups. Add with sugar to lemons, mixing well. For a more delicate flavor, divide mixture in half and cook separately.
• Boil rapidly, stirring frequently about 15 to 20 minutes, or until jellying point is reached. Skim.

• Ladle into sterilized hot jars to within ⅛″ of jar top. Wipe jar rim; adjust lids. Process in boiling water bath 5 minutes. Remove from canner and complete seals unless closures are self-sealing type. Makes 2 pints.

ORANGE MARMALADE

Spread a thin layer on your pumpkin pie and top with whipped cream

4 oranges
2 lemons
Water
Sugar

• Wash unpeeled fruit and slice very thin. Remove seeds and cores.
• Measure the sliced fruit and for each cup add 3 c. cold water; let stand 24 hours.
• Heat mixture to boiling; boil 15 minutes and again let stand 24 hours.
• On the third day, measure 3 c. of the mixture into a large saucepan; add 3 c. sugar and boil rapidly, stirring frequently, about 20 minutes or to the jellying point. Stir and skim.
• Ladle into sterilized hot jars to within ⅛″ of jar top. Wipe jar rim; adjust lids. Process in boiling water bath 5 minutes. Remove from canner and complete seals unless closures are self-sealing type.
• Repeat in 3-c. batches until all has been used. Cooking in small amounts gives a more delicate marmalade. Makes about 5 pints.

HONEYED MARMALADE TOPPING

Just right on hot pancakes and waffles —good on all hot breads

1 c. orange marmalade
½ c. honey

• Combine marmalade and honey in small saucepan. Bring to a boil over low heat, stirring constantly. Serve hot. Makes about 1⅓ cups.

PINEAPPLE/CARROT MARMALADE

Make this spun-gold spread in spring when fruit closet supplies are low

4 lbs. carrots, peeled
3 lemons
1 (1 lb. 4 oz.) can crushed pineapple (2 c.)
4 c. sugar
1 c. orange juice

• Put carrots and lemons (seeded but not peeled) through food chopper. Combine with pineapple, sugar and orange juice in a large saucepan. Cook until clear, stirring occasionally.
• Ladle into sterilized hot jars to within ⅛″ of jar top. Wipe jar rim; adjust lids. Process in boiling water bath 5 minutes. Remove from canner and complete seals unless closures are self-sealing type. Makes 4½ pints.

TOMATO/CITRUS MARMALADE

Beautiful color and taste and a compliment-catcher when served

4 qts. ripe tomatoes, peeled
Sugar
2 lemons
3 oranges
½ oz. stick cinnamon
¼ oz. whole cloves

• Cut tomatoes into small pieces. Drain off half the juice. Weigh tomatoes; add equal amount of sugar.

• Slice lemons and oranges very thin and cut slices into quarters; discard seeds. Tie spices loosely in cheesecloth bag. Combine tomato-sugar mixture, lemons, oranges and spices in a large saucepan. Cook rapidly, stirring frequently, until mixture reaches the jellying point. Remove spices.

• Ladle into sterilized hot jars to within ⅛″ of jar top. Wipe jar rim; adjust lids. Process in boiling water bath 5 minutes. Remove from canner and complete seals unless closures are self-sealing type. Makes 6 pints.

Note: For best results use a large shallow kettle and never cook more than the above amount at one time.

FRUIT/TOMATO MARMALADE

Let this brighten your next potluck or buffet supper—colorful and tasty

 6 **tart apples, peeled and cored**
12 **peaches, peeled and pitted**
 2 **large ripe tomatoes, peeled**
 4 **oranges, seeded**
Sugar

• Chop apples, peaches and tomatoes. Put oranges through food chopper. Combine fruits; measure.

• Bring to full boil; reduce heat, cook 10 minutes. For each cup of fruit, as measured before boiling, add ¾ c. sugar; stir well. Cook 45 minutes, or until thick, stirring often.

• Ladle into sterilized hot jars to within ⅛″ of jar top. Wipe jar rim; adjust lids. Process in boiling water bath 5 minutes. Remove from canner and complete seals unless closures are self-sealing type. Makes about 6 half pints.

The Gourmet Corner

When you visit department stores in metropolitan cities, you usually find a Gourmet Food Corner. And you also pass exclusive gourmet food shops in big urban centers. On the shelves of many of these luxury markets are rows of jellies, jams, marmalades, preserves and relishes. These specialties of good cooks, many of them from country kitchens, have high—even exorbitant—price marks. Yet they move from shelves to tables throughout the whole year.

We borrowed the idea of having a gourmet section in this chapter. These recipes rate high with our taste-testers. Their inclusion in this exclusive place signifies they are worth a good price, if you want to sell them, and that they'll win compliments from people who taste them.

BLUSHING PEACH JAM

Follow recipe with loving care for a beauty with that luscious Melba taste

2 c. crushed, peeled and pitted
   peaches (about 1½ lbs.)
¼ c. lemon juice
2 c. red raspberries
7 c. sugar
1 bottle liquid fruit pectin
Few drops almond extract

• To the crushed peaches add 2 tblsp. lemon juice. Let stand.
• Crush berries and add remaining 2 tblsp. lemon juice.
• Combine peaches and raspberries with sugar in heavy kettle; mix well, and bring to a full rolling boil, stirring constantly. Boil 1 minute, remove from heat and add pectin.
• Stir and skim. Add extract.
• Ladle into sterilized hot jars to within ⅛" of jar top. Wipe jar rim; adjust lids. Process in boiling water bath 5 minutes. Remove from canner and complete seals unless closures are self-sealing type. Makes 4 half pints.

THREE-BERRY JAM

An extra-delicious spread that captures the taste of summer

3 c. red raspberries
3 c. youngberries
3 c. loganberries
1 pkg. powdered fruit pectin
7 c. sugar

• Crush berries, one layer at a time; measure 5 c. into large saucepan.
• Stir in pectin. Place over high heat and bring to a hard boil, stirring.
• Add sugar and stir until mixture comes to a full rolling boil; boil hard 1 minute. Skim off foam with spoon.
• Ladle into sterilized hot jars to within ⅛" of jar top. Wipe jar rim; adjust lids. Process in boiling water

bath 5 minutes. Remove from canner and complete seals unless closures are self-sealing type. Makes 4 pints.

BLACKBERRY/LEMON JAM

The lemon masks some of the strong berry taste. Adds a tang to meals

2 lemons, seeded and coarsely
   ground or chopped
1½ c. water
6 c. blackberries
7 c. sugar

• Combine lemons and water and cook for 20 minutes.
• Add berries and sugar and continue cooking 20 minutes, or until thickened.
• Ladle into sterilized hot jars to within ⅛" of jar top. Wipe jar rim; adjust lids. Process in boiling water bath 5 minutes. Remove from canner and complete seals unless closures are self-sealing type. Makes 4 pints.

AMBROSIAL JAM

For special guests on festive occasions —lovely color, mingled flavors

8 peaches, peeled and pitted
3 large oranges, seeded
Pulp of 1 peeled medium cantaloupe
1 lemon
1 (8¼ oz.) can crushed pineapple
   (1 c.)
Sugar

• Chop all ingredients fine, putting oranges through food chopper, and combine. Add ¾ c. sugar for every cup of fruit. Let stand overnight.

• Next morning, gently cook mixture for 1 hour, stirring frequently.
• Ladle into sterilized hot jars to within ⅛″ of jar top. Wipe jar rim; adjust lids. Process in boiling water bath 5 minutes. Remove from canner and complete seals unless closures are self-sealing type. Makes about 8 pints.

Note: You can add 6 maraschino cherries, drained and sliced thin, just before removing jam from heat.

RUBY PRESERVE

A gem from a Montana kitchen

Sugar
1 qt. strawberries
1 qt. red raspberries
1½ lbs. pitted stemmed cherries
¼ c. lemon juice

• Add sugar equal in weight to all fruits. Mix. Boil for 25 minutes.
• Add lemon juice and boil 2 minutes.
• Ladle into sterilized hot jars to within ⅛″ of jar top. Wipe jar rim; adjust lids. Process in boiling water bath 5 minutes. Remove from canner and complete seals unless closures are self-sealing type. Makes 6½ pints.

AUTUMN CHERRY CONSERVE

Five fruits blend their flavors with ground cherries in this golden treat

¾ c. ripe ground cherries (also called husk tomatoes)
1⅔ c. crushed pineapple
Juice of ½ lemon
Grated peel and juice of 1 orange
5 medium apples, finely diced

1 c. cranberries
1 pkg. powdered fruit pectin
4½ c. sugar

• Hull, wash and prick ground cherries. Combine with pineapple, lemon juice, grated orange peel and juice and apples. Add cranberries.
• Bring to a boil, add pectin and stir while bringing to a boil.
• Add sugar, stirring constantly. Bring to a full rolling boil over high heat; boil hard 1 minute, stirring constantly.
• Remove from heat; skim.
• Ladle into sterilized hot jars to within ⅛″ of jar top. Wipe jar rim; adjust lids. Process in boiling water bath 5 minutes. Remove from canner and complete seals unless closures are self-sealing type. Makes 5 to 6 half pints.

CURRANTS LUSCIOUS

Luscious to look at—luscious to eat

1 qt. red currants
Sugar
1 qt. loganberries
1 pt. strawberries
1 lb. black cherries, pitted (about 2¾ c.)
1 pt. red raspberries

• Wash currants (not necessary to remove stems); mash slightly to start juice; cook slowly until currants look white. Drain in jelly bag. Add 1 c. sugar to each cup of juice.
• Weigh remaining fruits and add 2 c. sugar for each pound of fruits. Pour sweetened currant juice over fruits and sugar and let stand overnight.
• Boil rapidly for 15 minutes. Ladle into sterilized hot jars to within ⅛″ of jar top. Wipe jar rim; adjust lids.

Process in boiling water bath 5 minutes. Remove from canner and complete seals unless closures are self-sealing type. Makes 6½ pints.

AMBER MARMALADE

*"The best of all citrus marmalades,"
says the contributor of this recipe*

1 grapefruit
1 orange
1 lemon
Water
Sugar

· Wash unpeeled fruit and slice very thin. Remove seeds and cores.
· Measure the sliced fruit and for each cup add 3 c. cold water; let stand 24 hours.
· Heat mixture to boiling; boil 15 minutes and again let stand 24 hours.
· On the third day, measure 3 c. of the mixture into a large saucepan, add 3 c. sugar and boil rapidly, stirring frequently about 20 minutes or until the jellying point is reached. Stir and skim.
· Ladle into sterilized hot jars to within ⅛" of jar top. Wipe jar rim; adjust lids. Process in boiling water bath 5 minutes. Remove from canner and complete seals unless closures are self-sealing type.
· Repeat in 3-c. batches until all has been used. Makes about 5 pints.

You never will taste more delicious strawberry preserves than those you make with this recipe. The excellent canner, famed in her county for these preserves, specifies that you must follow the recipe to the letter. The *timing is especially important.*

You will notice the first cup of berries cooks 16 minutes, the last, 4.

Red raspberries also make a superior preserve. Fragile as they are, some of the berries remain whole if you follow the same directions.

CARDINAL STRAWBERRY PRESERVES

Juicy-ripe berries keep their plumpness in the glossy syrup

1 c. crushed strawberries
1 qt. whole strawberries
4 c. sugar

· Place crushed berries in an 8-qt. heavy kettle. Divide whole berries into three equal parts.
· Add 1 c. sugar to crushed berries. Cook, stirring constantly until sugar dissolves and mixture comes to a boil. When bubbles start to appear around side of kettle, start counting time and boil exactly 4 minutes.
· Add one third of whole berries and 1 c. sugar, stirring until sugar dissolves. When boiling starts, count time and boil exactly 4 minutes.
· Add the second third of the whole berries and 1 c. sugar, stirring until the sugar dissolves; boil 4 minutes.
· Add the last third of the berries and 1 c. sugar and boil exactly 4 minutes. Stir gently to prevent scorching. Use care not to break berries.
· Remove from heat, skim and pour preserves into a flat glass utensil (like a large cake pan) and let stand 10 to 12 hours, giving fruit a chance to plump.
· Ladle into sterilized hot jars to within ⅛" of jar top. Wipe jar rim; adjust lids. Process in boiling water

bath 5 minutes. Remove from canner and complete seals unless closures are self-sealing type. Makes about 4 half pints.

VARIATION

GOURMET RED RASPBERRY PRE-SERVES: Substitute red raspberries for strawberries.

Wild Fruit Specialties

Summer in the country finds good cooks and children coming home from the woods, mountains and pastures with buckets and baskets full of wild fruits. What marvelous fruity spreads are put up the next day! Here is a sampling of these treats.

A Yakima County, Washington, home economist, who tested some of the jam, jelly and other fruit recipes in this cookbook, wrote: "Every summer I make a trip to the west side of the Cascade Mountains to pick wild blackberries. Wild Blackberry Sundae Topping is a specialty at our house. Members of my club would be disappointed if I did not serve it on vanilla ice cream when they meet with me. It's so easy to make. The seeds in the berries are so small that you do not have to strain them out."

WILD BLACKBERRY SUNDAE TOPPING

One of the best-tasting sauces on ice cream that you'll ever encounter

1 qt. wild blackberries
1 c. sugar

• Cook blackberries 5 minutes. Mash while cooking. Add sugar and cook until slightly thickened.

• Ladle into sterilized hot jars to within ⅛" of jar top. Wipe jar rim;

adjust lids. Process in boiling water bath 5 minutes. Remove from canner and complete seals unless closures are self-sealing type. Makes 3½ cups.

BLACKBERRY/ HUCKLEBERRY JAM

Cook some of the jam less—it's thinner, wonderful on vanilla ice cream

6 c. wild blackberries
¼ c. water
1 c. huckleberries
7 c. sugar
½ bottle liquid fruit pectin

• Wash and pick over blackberries. Crush. Combine with water in saucepan. Bring to a boil and simmer, covered, 5 minutes.

• Force mixture through coarse sieve or food mill.

• Wash and sort over huckleberries, leaving in green ones. Add to blackberry pulp; measure 4 cups (add water if necessary to make full amount).

• Combine fruit and pulp with sugar in very large saucepan, mixing well. Heat to full rolling boil; boil hard 1 minute, stirring constantly. Remove from heat; stir in pectin; skim.

• Ladle into sterilized hot jars to within ⅛" of jar top. Wipe jar rim;

adjust lids. Process in boiling water bath 5 minutes. Remove from canner and complete seals unless closures are self-sealing type. Makes 10 half pints.

VARIATION

BLACKBERRY/BLUEBERRY JAM: Substitute blueberries for huckleberries.

WILD BLACKBERRY/PLUM BUTTER

Taste and you'll know why it's good to have wild berries grow nearby

1	qt. wild blackberries
2	lbs. tart plums
½	c. water
7½	c. sugar
½	bottle liquid fruit pectin

· Wash and pick over berries. Pit plums but do not peel. Cut in small pieces. Combine fruit and water in saucepan. Bring to a boil and simmer, covered, 5 minutes.

· Force mixture through coarse sieve or food mill to remove seeds. Measure 4½ c. pulp into a very large saucepan. If necessary, add water to make full amount.

· Add sugar; mix well. Heat to full rolling boil; boil hard 1 minute, stirring constantly. Remove from heat; stir in pectin; skim.

· Ladle into sterilized hot jars to within ⅛″ of jar top. Wipe jar rim; adjust lids. Process in boiling water bath 5 minutes. Remove from canner and complete seals unless closures are self-sealing type. Makes 8 half pints.

Chokecherry jelly and butter have a bittersweet flavor that many country people relish. But some families prefer to let the birds have the bright red cherries! One farm cook explains it this way: "If you like game, you usually enjoy wild cherry spreads. We think they are a perfect accent for wild duck dinners."

Use fully ripe chokecherries to prepare the juice for jelly making. Stem about 3½ lbs. cherries and place in a large kettle with 3 c. water. Cover and cook 15 minutes. Place in jelly bag and squeeze out the juice.

CHOKECHERRY JELLY

Wild cherry jelly—serve it proudly, for few hostesses can

3	c. chokecherry juice
6½	c. sugar
1	bottle liquid fruit pectin
¼	tsp. almond extract (optional)

· Pour juice into large kettle. Add sugar and stir to mix.

· Place over high heat and bring to a boil, stirring constantly. Stir in pectin, bring to a full rolling boil and boil hard 1 minute, stirring constantly.

· Remove from heat; skim. Add extract.

· Pour into sterilized hot jars to within ⅛″ of jar top. Wipe jar rim; adjust lids. Process in boiling water bath 5 minutes. Remove from canner and complete seals unless closures are self-sealing type. Makes about 9 half pints.

· *Or* pour into sterilized jelly glasses to within ½″ of jar top. Cover immediately with ⅛″ melted paraffin.

Note: Almond extract gives a stronger cherry taste.

CHOKECHERRY SYRUP: Sometimes a Colorado farm homemaker omits pectin when making chokecherry jelly. "Of course, it doesn't jell," she says, "but you have a wonderful table syrup to serve on pancakes."

CHOKECHERRY/APPLE BUTTER

A favorite with grownups who had it on their tables in childhood

4 c. apple pulp
2 c. chokecherry pulp
5 c. sugar
½ tsp. almond extract

• Prepare pulp of both fruits first by putting cooked fruit (unsweetened) through a sieve or food mill.
• Heat to a boil, stirring carefully.
• Add sugar. Stir constantly until it just begins to thicken.
• Add extract and blend.
• Ladle into sterilized hot jars to within ⅛" of jar top. Wipe jar rim; adjust lids. Process in boiling water bath 5 minutes. Remove from canner and complete seals unless closures are self-sealing type. Makes 8 half pints.

ELDERBERRY JELLY

Something different and delicious to serve with game. Grandpa's delight

3½ c. elderberry juice (about 3½ lbs. ripe berries)
Apple juice (optional)
½ c. fresh lemon juice, strained
7½ c. sugar
1 pkg. powdered fruit pectin

• Prepare elderberries by removing large stems. Place in large kettle; crush. Cover and simmer about 15 minutes. Strain through jelly bag.
• Measure juice. If you do not have quite enough, add apple juice. Add lemon juice and pour into kettle.
• Heat, adding sugar, and bring to a boil, stirring constantly.
• Add pectin. Bring to a full rolling boil and boil hard 1 minute. Remove from heat, skim off foam.
• Pour into sterilized hot jars to within ⅛" of jar top. Wipe jar rim; adjust lids. Process in boiling water bath 5 minutes. Remove from canner and complete seals unless closures are self-sealing type. Makes about 5 half pints.
• *Or* pour into sterilized jelly glasses to within ½" of jar top. Cover immediately with ⅛" melted paraffin.

One of our Massachusetts readers gave us her Beach Plum Jelly recipe with the following comment: "Our family thinks this red jelly has few equals. We gather the tiny fruit from bushes that grow in the sand along the ocean. It has such interesting individual traits. Although the bushes bloom every spring, they bear fruit only once in three years."

"When ripe, the fruit has a tough skin like wild grapes, but its shape and pulp is more like that of a miniature plum. The refreshing taste is suggestive of wild cherries—a touch of bitterness that makes the jelly wonderful to serve with chicken and meats. And like many women near the ocean, I sell homemade Beach Plum Jelly at my roadside stand. Tourists who come to New England to see our colored foliage in autumn are good customers."

BEACH PLUM JELLY

Sparkling red jelly that tastes more nearly like wild cherries than plums

4 c. beach plum juice
4 c. sugar

• Use red beach plums (not ripe); cover them with water and bring to a boil. Drain and discard water. Pour on more hot water, not quite enough to cover. Cook until plums are soft.
• Drip juice through jelly bag. Measure and add an equal amount of sugar. Boil over high heat until jelly point is reached. Remove from heat; skim.
• Pour into sterilized hot jars to within ⅛" of jar top. Wipe jar rim; adjust lids. Process in boiling water bath 5 minutes. Remove from canner and complete seals unless closures are self-sealing type. Makes 3 to 4 half pints.
• *Or* pour into sterilized jelly glasses to within ½" of jar top. Cover immediately with ⅛" melted paraffin.

PLUM BUTTER

It's "plum good" and makes a tasty, bright topping for vanilla ice cream

3¾ qts. pitted wild plums
Water
2¾ c. sugar

• Cover plums with water and boil until skins are tender. Put through a sieve. Makes 3⅓ c. pulp.
• Bring to a boil; add sugar, stirring as you do. Boil until jelly-like consistency, stirring constantly.
• Ladle into sterilized hot jars to within ⅛" of jar top. Wipe jar rim; adjust lids. Process in boiling water bath 5 minutes. Remove from canner and complete seals unless closures are self-sealing type. Makes 4 half pints.

Note: "Tame" plums may be used instead of the wild fruit.

VARIATION

SPICY PLUM BUTTER: For a spicy butter, add ground cinnamon, ½ tsp. to 3 c. pulp.

PLUM/PINEAPPLE CONSERVE

Serve when your club meets—it will be the talk of the afternoon

5 c. pitted wild plums
Grated peel of 1 orange
1 tsp. powdered ascorbic acid
1½ c. drained crushed pineapple
1 pkg. powdered fruit pectin
5 c. sugar
⅓ c. shredded almonds (optional)

• To plums, add orange peel, ascorbic acid and pineapple. Heat to boiling and add pectin, stirring constantly. Heat to a boil; add sugar, stirring.
• Bring to a full rolling boil, stirring constantly. Boil hard 1 minute.
• Remove from heat; skim. Stir in nuts.
• Ladle into sterilized hot jars to within ⅛" of jar top. Wipe jar rim; adjust lids. Process in boiling water bath 5 minutes. Remove from canner and complete seals unless closures are self-sealing type. Makes 8 half pints.

Note: "Tame" plums may be used if wild ones are not available.

SPICED MULBERRY JAM

Cinnamon and lemon blend temptingly with the wild mulberry taste

1 qt. prepared mulberries
3 c. sugar
¼ c. lemon juice
½ tsp. ground cinnamon

• Stem mulberries (mixture of ripe and green) and cover with cold salted water. Use ¼ c. salt to 1 qt. water. Let stand 5 minutes. Drain. Rinse in cold water three times.

• Crush berries. Add sugar, lemon juice and cinnamon; cook slowly, stirring, until the sugar dissolves. Boil rapidly, stirring constantly to prevent scorching, until jellying point is reached. Remove from heat; skim.

• Ladle into sterilized hot jars to within ⅛" of jar top. Wipe jar rim; adjust lids. Process in boiling water bath 5 minutes. Remove from canner and complete seals unless closures are self-sealing type. Makes about 3 half pints.

OLD-FASHIONED MULBERRY JAM: Use cider vinegar for the lemon juice.

Minnesota Farm Woman's Wild Grape Butter

FIRST DAY

• Choose one beautiful, blue October day after the first light frost.
• Add a dear friend and go to your favorite haunt—be it the woods or railroad right-of-way where the wild grapes flourish.
• Leisurely pick your baskets full, stopping to enjoy the glory of autumn's coloring.
• Catch a glimpse and hear the honking of the wild geese as they fly in formation overhead.
• Listen to the chickadee as he flits near you, and to the wild ducks as they try to flee the hunters.
• As you work, remember the days when you and your brothers and sister used to pick grapes along fence rows on the way to Grandpa's.
• Remember, too, the days when you took your own little children to the woods and gave them nature lessons (indelible to them now), as you all

sauntered along, crunching leaves beneath your feet and with a song in your heart.
• Smell the tinge of horsemint and faint suggestion of smoke in the air.
• Think of the future of your forests.
• Pick a colorful bunch of bittersweet to send to that faraway sister who gave you some of your first lessons in art appreciation.
• Return home before the corn pickers do and set a colorful table, centered with autumn leaves and fresh vegetables right from your garden.
• Remember Psalm 19 as you work, and pray for continued peace.

SECOND DAY

• In the morning, cut the heavy stem ends from grape bunches and pull out the thistle or milkweed down. Wash grapes. Then follow this recipe.

WILD GRAPE BUTTER

Well worth the effort of picking, an ideal escort for wild game

6 qts. stemmed wild grapes
Water
4 qts. apples
4 c. sugar

• Cover grapes with water, bring to a boil and simmer 20 minutes. Drain. Save juice for Wild Grape Jelly (recipe follows).

• Place grape pulp in a loosely woven cheesecloth bag; put the bag in the kettle along with apples which have been quartered but not peeled. (Grape pulp, cooking with apples, adds flavor; the bag makes it easy to remove seeds.) Cover fruit with water; bring to a boil; simmer 20 minutes. Drain. Save this juice, too, for Wild Grape Jelly.

• Discard the bag with grape seeds in it. Put apple pulp through a sieve; it should measure 5 c.

• Place in kettle, add the sugar and heat to boiling, stirring constantly. Cook to desired consistency.

• Ladle into sterilized hot jars to within ⅛" of jar top. Wipe jar rim; adjust lids. Process in boiling water bath 5 minutes. Remove from canner and complete seals unless closures are self-sealing type. Makes 4 half pints.

WILD GRAPE JELLY

Make this with grape and grape-apple juices saved from Wild Grape Butter

Juice of 1 lemon
6   c. wild grape juice
1   pkg. powdered fruit pectin
7½ c. sugar

• Add the strained lemon juice to the grape juice; heat to boiling in a large kettle.

• Add the pectin and again bring to a boil. Stir in the sugar.

• Bring to a rolling boil, boil hard for 1 minute, stirring constantly. Remove from heat; skim.

• Pour into sterilized hot jars to within ⅛" of jar top. Wipe jar rim; adjust lids. Process in boiling water bath 5 minutes. Remove from canner and complete seals unless closures are self-sealing type. Makes 6 half pints.

• *Or* pour into sterilized jelly glasses to within ½" of jar top. Cover immediately with ⅛" melted paraffin.

CHAPTER 16

Curing Meat, Poultry, Fish

Tips on Curing Meats · Sugar-Cured Ham and Bacon · Home-Cured
Corned Beef · Smoked Chicken, Turkey, Fish · Recipes for Baked
Ham, "Boiled" Ham, Glazes for Ham, Creamed Country Ham and
Slick Tricks with Ham · Recipes for Making Sausage

When it's time to decorate the living room with holly and bring out glitter
to hang on the Christmas tree, some farmers hang sugar-cured hams and
bacon in the smokehouse. These early birds know that meat cured in No-
vember, December or January will be aging before spring arrives, thus
minimizing spoilage hazards and insect infestation which warm weather
brings.

These days, many other farmers have their meats custom-cured commercially. Still others freeze meats to bring out, thaw and cure when
needed. But thousands of families still fix their hams and bacons and many
of them do a good business selling country-style meats.

Ask the expert in home curing what he does to get good results and
you'll find he follows rules. One Kentucky farmer who cures excellent
meat, explains his philosophy like this: "Curing is a race between salt and
bacteria reaching the center of the meat. I try to help salt get there first."

This chapter also contains directions for making country-style sausage
and corned beef, and for smoking poultry and fish.

People have different ideas on aging ham, depending largely on where
they live. But just about everyone agrees that cured meats deserve priority
in real country meals.

Country-cured hams can be baked and glazed, or "boiled" and glazed.
We share ideas from country cooks for tasty and attractive glazes, plus
their best suggestions for making good use of leftover ham.

Sugar-Cured and Hickory-Smoked Meats

When a blizzard strikes in early morning and there are outdoor chores to do, about the happiest thing a farmer can face is a good country breakfast. With pan-fried ham, sizzling sausage cakes or ribbons of hickory-flavored bacon on the platter, the day gets off to a good start, regardless of the weather.

CURING MEAT

The aim in curing meat is a triple one —to preserve it, to color it and to give it the flavor we like—the characteristic ham, bacon or corned beef taste.

The main preserving agent in cured meat is salt. It works by drawing moisture from the meat and at the same time entering meat cells by diffusion. When it penetrates meat tissues, salt has the power to limit bacterial growth. Too much salt, while effective against bacteria, makes meat dry, hard and over-salty.

In the dry-curing process, salt is mixed with sugar (which adds flavor and improves meat texture) and with another preservative (nitrate, nitrite, or both—see information following). This mixture is rubbed on all surfaces of fresh ham or bacon.

Just as the Kentucky farmer says, curing is actually a race between the growth of spoilage bacteria and the penetration of the preserving salt into the center of the meat. Keeping meat cold during the curing process slows bacterial growth and gives salt the chance to win the race—to get into the meat in sufficient concentration to preserve it. The optimum curing temperature is 38° F. Higher temperatures may cause spoilage; lower temperatures may decrease the rate of penetration of the salt into the meat.

WHAT ABOUT NITRATE?

Country people have used saltpeter (potassium nitrate or sodium nitrate) for years to protect the color and keeping qualities of home-cured hams, bacon and corned beef. Its use actually dates back to Biblical times, when desert salts containing nitrate were used to cure meats.

If bacterial action is effective in the curing process, nitrate converts to nitrite. It is the latter compound which gives cooked, cured meat its typical red color and flavor. Nitrite in meat also insures against the development of botulinal toxin (botulism).

Commercial curing mixtures, available at farmer's co-ops, feed stores and some supermarkets, are a blend of salt, sugar, saltpeter and sodium nitrite, along with spices for flavoring. Not only do they save the bother of mixing ingredients, they also offer a known amount of the nitrite preservative.

If you choose a commercial curing mixture, follow package directions. In its booklet, "A Complete Guide to Home Meat Curing," the Morton Salt Company describes a combination cure. By this method, you use a meat pump to place a curing pickle inside large pieces of meat, next to the bone; then a dry mixture is rubbed on the outside of the meat. (For information, address Morton Salt Company at Box 355, Argo, IL 60501.)

The use of nitrite in commercial curing and preblended curing mixtures is being studied. Prodded by cancer researchers, meat scientists are trying to determine how little of the preservative is needed to protect against the botulinum bacteria. As a result of these studies, there may also be some alteration in the amount of nitrate permitted in the Sugar Cure Mixture for the country-cured hams and bacon. When further information is available, it will be announced to the public. Your County Extension Office should have up-to-date information on this. But until such time as a new recommendation is made (or an effective substitute preservative is found), it is necessary to continue to add nitrate to the Sugar Cure Mixture, as called for in directions that follow.

Country-cured Ham and Bacon

Directions that follow for curing and smoking country-style hams and bacon are a guide. It also takes judgment, knowledge—and watchfulness —to get good results. For example: if temperatures fall below freezing for several days during the curing, you should add the same number of days to the curing time. Too low a temperature greatly reduces the penetration of salt into the meat.

The weight and thickness of the pieces dictates the length of the cure. Meat should cure in proportion to the weight or thickness of the individual pieces—2 days per pound of meat, or 20 days for a 10-lb. ham, or 7 days per inch thickness. How you handle it after curing depends on whether or not you plan to smoke it.

Usually, cured meat is smoked before it is put into storage. Most people prefer the flavor that develops in smoking; also, the process helps dry the meat and furnishes additional protection against bacterial growth and fat oxidation (rancidity).

Here is a recipe for the Sugar-Cure Mixture and directions for curing pork recommended by North Carolina State University:

SUGAR-CURE MIXTURE

(*for 100 lbs. meat*)

8 lbs. salt
3 lbs. white or light brown granulated sugar
3 oz. saltpeter (potassium nitrate)

· Use medium granulated or flake salt (often called bag or sack salt)—do not use iodized salt.
· Mix the ingredients, using care to distribute saltpeter evenly through the salt. This is easier to accomplish in two steps: Mix the required amount of saltpeter with a handful of salt; then mix the handful into the rest of the curing mixture. Use only fresh ingredients for the mix; never use the old mix from previous curing.

Preparing Meat: Chill meat quickly and keep it cold. If hogs are to be killed on the farm, check the weather forecast; kill when your weatherman says: "Light frost tonight." A lean, meat-type hog makes the best cured meat.

Do not cut meat or apply cure until meat is *thoroughly* chilled.

Dry-Curing Hams and Shoulders: Salt does not penetrate until it dissolves in the moisture of the meat. Since the salt in the cure draws out the most moisture the first few days of curing, it is desirable to apply the Sugar-Cure Mixture to hams three times instead of rubbing it on once.

Weigh trimmed meat. Allow 1¼ oz. of Sugar-Cure Mixture per pound of ham; i.e., for a 15-lb. ham, use 19 oz. mix. This is the total amount for the three applications.

As soon as meat is chilled, rub on approximately one-third (or slightly more) of the Sugar-Cure Mixture. You need not rub it a lot; just be sure the salt covers all the meat surfaces. Use care not to break the outside membranes, or the hams may become hard and dry during aging. Pack some of the mix into the shank end along the bone. Treat shoulder cuts like hams.

For the most even cure, stack hams (no more than 3 deep) on a shelf. The ideal temperature for shelf curing is from 36° to 40° F. There is some danger of spoilage if temperatures go above 50° F. for any length of time.

On the third day, rub on half of the remaining Sugar-Cure Mixture. On the tenth day, rub on the rest of it. Use care not to knock salt off ham.

Leave a 15-lb. ham in the cure 2 days for each pound of meat, or 30 days. A 20-lb. ham should stay in cure 1¾ days per pound, or 35 days, and a 25-lb. ham, 1½ days per pound or 38–40 days. (Allow an extra day in cure for each day temperature is below freezing.)

Keep track of curing time. If you cut it short, the meat may spoil. When it is too long, the meat loses quality.

Dry-Curing Bacon: One application of the Sugar-Cure Mixture is enough for bacon. Weigh pork sides; allow ¾ to 1 oz. Sugar-Cure Mixture per pound of bacon. Cover bacon with the mixture, patting more of it on the thicker pieces of meat. Stack one upon another on the shelf, with skin side down.

Cure bacon 1½ days for each pound of meat, or 7 days per 1″ of thickness. Keep bacon cold all during the cure, 36 to 40° F.

AFTER CURING

Remove cured meat from the dry pack, smaller pieces first, as curing time is up. If you plan to smoke the meat, hold the small pieces under refrigeration or in a *cold* place (36 to 40° F.) until heavier cuts are cured.

If you do not desire a smoky flavor, brush off excess salt and hang meat in a cold place at least a week, or until it loses about 5 percent of its cured weight. Put in refrigerator. Plan to use unsmoked meat more promptly.

Hams that are to be smoked should be hung under refrigeration (40 to 45° F.) for 20 to 30 days, until the salt content has equalized or spread evenly all through the meat to the bone. This is a very essential procedure to prevent spoilage.

Hams should not be smoked or exposed to high temperatures until the salt content has equalized.

To Smoke Cured Meat: Scrub it first in warm water to remove excess salt and grease. This prevents salt streaking and allows the smoke to color the meat evenly. If meat has been in the cure a bit too long, soak it for an hour or two. Soak bacon 30 minutes.

Hang meat to dry 2 hours or longer. (Do not wash ham and hang it in a cooler; if hams are hung wet in a cooler, bacteria will begin to grow on the surface.) To hang, put a strong cord through shanks of hams and shoulders and tie. Reinforce flank end of bacon with a hardwood skewer to hold it square; fasten two loops of cord under the skewer just off center, so it will hang straight.

Hang the dry meat in the smokehouse so that no two pieces touch. Build under it a fire of green hickory wood—first choice—or other hardwood such as maple, oak, pecan or dried applewood. Or use hardwood sawdust. Do not use resinous woods such as pine and other evergreens.

The ideal smokehouse temperature is between 70 and 90° F. for hams to be smoked. Open the ventilator to let the moisture escape. On the second day, close the ventilators and smoke meat until it has the color you like best; light or dark mahogany or chestnut brown. Usually two days of smoking is enough. Remember that a thin haze of smoke is as effective as a dense cloud. Use care not to overheat the meat.

Testing Smoked Meat: Use a "ham trier" or any 10″ skewer or pointed wire. From both loin and hock ends, run it to the center of the ham along the bone. When you pull it out, smell it immediately. If the skewer has a sour, unpleasant odor, the meat should be cut and examined for spoilage; if it's putrefied (disagreeable odor), destroy it. Check shoulders from the shank end, at the shoulder point and under the blade bone. Sweet, pleasant-smelling meat can be wrapped for storage.

Storing Cured Meats

Cool the hams, shoulders and bacon upon removal from the smokehouse. Rub with black pepper. If desired, mix black with a little red pepper.

Pack meat pieces individually in cotton bags to protect them from insects, particularly skipper flies and beetles. Put 6 to 7″ crumpled newspaper in the bottom of a clean, tightly woven cotton bag and add the meat. Stuff more paper loosely around the meat to hold it away from the bag.

Do not pack the bag tightly, since moisture must be released from the ham to prevent spoilage. Tie the cloth bag securely to keep out insects, making a loop to hang meat. Do not hang by strings attached to meat—insects may use it as an entrance into meat.

Aging: Hang the hams to age in a dry, well-ventilated room for at least 6 months. Shoulders should be used before 6 months; they tend to crack open and are then subject to mold. A good aging temperature is from 70 to 80° F. Below 45° F., little or no aging flavor develops. Many farmers keep their hams a year or longer. When a ham is a year old, it becomes an "old country ham." Hams over a year old tend to become dry and hard and the fat may become rancid. Their flavor is more pro-nounced. Only heavy hams (over 25 lbs. at start of curing process) should be aged more than a year, but this practice is not very feasible.

To stop aging, you can put the meat in cold storage, or cut, wrap and freeze it. For best flavor, do not keep ham, if frozen, longer than 3 months. The aged flavor is lost thereafter.

Baked and Boiled Country Ham

A cured country ham, whether it's smoked or not, is still raw pork, and it must be completely cooked. (By con-trast, some commercially processed hams are partly or wholly cooked.)

To prepare a country ham for cook-ing: Wash and scrub with a stiff brush. Trim off dry, hard, dark edges. Cover with cold water and soak for 12 to 14 hours in a cool room. You can skip the soaking with well-cured hams of good quality, but most home-makers prefer to soak them.

BAKED COUNTRY-STYLE HAM

1. Place the prepared ham, skin side up, on rack in open pan.
2. Bake uncovered (without adding water) in a slow oven (300° F.) until tender or until meat thermom-eter registers 170° F. Insert it in meat so bulb touches no bone or fat. Allow 25 to 30 minutes per pound for whole hams, 45 to 50 minutes per pound for butts.
3. Serve the ham, or remove the skin and glaze ham by directions that follow.

"BOILED" COUNTRY-STYLE HAM

1. Place prepared ham on rack in kettle and cover with boiling water. Add 1 tblsp. molasses or brown sugar to each quart water, if desired.
2. Simmer, *but do not boil,* until ham is tender and thermometer regis-ters 160° F. Let ham cool in broth. It will finish cooking as it cools. Al-though the ham is cooked in simmer-ing (not boiling) water, it is called "boiled ham."
3. When ham is cool, remove skin, chill and serve or add a glaze and bake 30 minutes.

GLAZING COUNTRY-STYLE HAM

1. Use sharp knife to remove skin from baked or "boiled" ham. Score in 1 or 2" squares down to meat surface.
2. Cover with glaze (see direc-tions).
3. Stick long-stemmed clove into each square.
4. Bake in a moderate to hot oven

(375 to 400° F.) 15 to 30 minutes or until surface is browned and glazed.

GLAZES FOR COUNTRY-STYLE HAM

HONEY/CHERRY: Baste ham while it bakes with strained honey, to which you have added chopped maraschino cherries or the cherry juice.

MUSTARD: Coat ham with thin layer of prepared mustard; sprinkle generously with brown sugar.

PEACH: Coat ham with brown sugar and baste with peach pickle juice (or spiced crab apple juice).

CIDER: Coat ham with brown sugar and baste with sweet cider.

PINEAPPLE: Mix 1 c. drained crushed pineapple and 1 c. brown sugar, firmly packed. Spread over ham. Baste often with syrup from canned pineapple.

BROWN SUGAR SYRUP: Baste ham while it bakes with syrup made by boiling 1 c. brown sugar, firmly packed, with ½ c. water 5 minutes.

MAPLE SYRUP: Baste ham with 2 c. maple syrup.

JELLY: Spread over ham 1 c. currant or apple jelly, cooked with 1 tblsp. cornstarch until clear.

ORANGE: Spread 1 c. orange marmalade over ham. Or spread over ham 1 c. brown sugar, firmly packed, and juice and grated peel of 1 orange, mixed. Garnish with orange slices.

CHERRY: Spread 1 c. sweet cherry preserves over ham.

CREAMED COUNTRY-STYLE HAM

For a farm feast, serve on hot corn sticks, biscuits, waffles or toast

1   tblsp. ham fat or butter
1   tblsp. chopped onion
2   c. chopped cooked lean ham
¼   c. flour
2½   c. milk

· Place fat in heavy frying pan, add onion and cook until onion is tender, not browned. Add ham, stir and heat.
  Add flour, stir and cook about 1 minute. Add milk, ½ c. at a time, stirring constantly. Cook 2 to 5 minutes until mixture is thick as heavy cream. Makes 4 servings.

Note: You can use scraps of cooked ham. The hock makes excellent creamed ham.

Slick Tricks with Ham

Almost all good country cooks have pet ways with cooked ham. Here are a few favorites from FARM JOURNAL readers. You'll want to try them.

WAFFLES: Sprinkle 2 tblsp. finely chopped ham on each waffle just before baking.

BISCUITS: When baking biscuits (2 c. flour) add ⅔ c. finely ground ham to the flour-fat mixture.

SALAD: Moisten 4 c. diced ham, ¾ c. chopped salted peanuts and 1½ c. diced celery, mixed, with salad dressing. Serve on crisp lettuce.

DEVILED EGGS: Add ½ c. finely ground ham to the mashed yolks of 6 hard-cooked eggs. Season with salad dressing and mustard pickle relish and heap in egg whites. Garnish with minced parsley or dust with paparika.

HAM IN POPOVERS: Fill hot popovers (recipe in Chapter 5) with Creamed Country-Style Ham (recipe above). A wonderful luncheon dish.

Home-cured Corned Beef

Combine home-cured corned beef with potatoes for the world's best hash. Top it with poached eggs, add a green vegetable and a fruit salad. Dinner is ready!

Fat animals make the best corned beef. Use the fatter cuts like the flank, plate, chuck and rump. Remove all the bones so meat can be packed in even layers.

Cool meat and start the corning as soon as possible. Meat that is not really fresh may sour during the curing. Do not corn meat that is frozen. Cut meat in pieces about 6″ square, uniformly thick.

1. Weigh meat. For each 100 lbs. allow 8 lbs. granulated salt (not iodized table salt).

2. Sprinkle a layer of salt ¼″ deep in bottom of clean scalded stone jar or wooden barrel.

3. Pack meat pieces as closely as possible, in a layer 5 to 6″ thick.

4. Alternate layers of salt and meat, covering the top layer of meat with considerable salt.

5. Let the salted meat stand overnight in a cool place (under 45° F.).

6. In the morning, for 100 lbs. meat, dissolve 4 lbs. sugar, 2 oz. baking soda and 4 oz. saltpeter in 1 gal. of lukewarm water. Add 3 gals. cold water. Pour over salted meat.

7. Weight the meat down to keep it under the brine. Use a loose board cover with a weight on it. If any meat is not covered with brine, both the meat and the brine will spoil quickly.

8. Keep the meat in a cool place (under 45° F.), for the sugar in the brine has a tendency to ferment.

9. Keep the meat in the brine for 28 to 40 days to get a good cure.

10. Watch the brine closely for signs of spoilage. If it starts to appear ropy or stringy, remove the meat and wash vigorously with a stiff brush and warm water. Scald container. Repack meat with a light coating of salt, using 6 lbs. salt instead of 8, for 100 lbs. meat. Let stand 24 hours in a cool place. Then cover with fresh sugar so-

lution (see Step 6). The dry salt treatment before adding liquid will help reduce growth of the bacteria that has developed.

11. You can keep the beef in the brine, under refrigeration, until used. Or remove it, wash, drain and smoke or can it. (See Chapter 12, Canning Meat, Poultry and Game.)

Smoked Chicken
THE WAY VIRGINIANS LIKE IT

These directions for smoking chickens were developed by the Department of Food Science and Technology, Virginia Polytechnic Institute and State University:

1. Eviscerate chickens, wash thoroughly and cool to 38° F. Use your thermometer.

2. Dissolve 5 lbs. granulated salt and 5 lbs. sugar in 7 gals. water that has been boiled and cooled. Place in clean, scalded stone jar. Add approximately 25 lbs. chicken.

3. Let chickens remain in cure 1¼ days per pound, each bird. Thus, a 4-lb. chicken should remain in the cure 5 days. Cure the chickens in a cool place (38° F.).

4. Remove birds from cure, wash in cool water (running water, if available) 30 minutes. Let birds drip dry.

5. Place chickens in stockinet and hang breast side down in smokehouse, 5 to 6' above source of smoke. Smoke 4 to 5 hours or until chickens reach the desired degree of brown color.

6. Roast the birds like fresh chicken; or cool, package and freeze for later use. Serve hot or cold.

Smoked Turkey
THE WAY CALIFORNIA RANCHERS FIX IT

Turkeys, smoked by the following directions, have a decided smokehouse flavor. They generally are enjoyed more for snacks and appetizers than for meat in a meal.

1. Cure the prepared turkeys in brine for 2 to 3 weeks before smoking. Remove feathers, viscera, crop, the entire neck and all superfluous tissues from inside pockets and cavities. Wash thoroughly. Remove the leg tendons to permit better penetration of the brine.

2. To make brine for 100 lbs. of turkeys, dressed weight, add to 18 gals. water: 25 lbs. salt, 12½ lbs. white or brown sugar, 12½ oz. saltpeter. (Instead of 12½ oz. saltpeter, you can use 5 oz. sodium nitrite.)

3. Place turkeys in a clean wood barrel (do not use metal container) and cover them with brine. Weight turkeys down to keep them from floating. Keep the temperature of the brine as near *38° F. as possible.* Leave them in the brine 1¼ days per

pound, each bird, but not less than 2 weeks nor more than 3. Thus, a 12-lb. turkey would remain in brine 15 days.

4. After the fourth day of curing, remove turkeys and thoroughly mix the solution. Repack the birds, interchanging the top and bottom turkeys. Repeat this process every 7 days during the curing.

5. Remove turkeys from brine and wash them thoroughly. Hang to drain in a cool place for 24 hours. Enlarge the openings at both ends slightly, propping them open for ventilation. It is important that cavities dry out.

6. To smoke turkeys, hang by drumsticks. Keep openings propped open. Let birds hang without smoke for 6 hours in a warm place, 110° F. Then smoke them at a temperatur of 110° F. for 20 hours or at 140° F for 16 hours. The higher temperatur gives more attractive color, but slightl increases shrinkage.

To store turkeys, hang them in cool, dry place, with a temperature n higher than 68° F. They will kee from 1 to 2 weeks. If mold or yeas develops in the turkey cavity, swab out with cotton saturated with 95 per cent pure grain alcohol.

To cook turkeys, soak them in col water 2 hours or longer, changing th water once. Roast them as you woul fresh turkeys and serve hot or cold.

Wisconsin Smoked Fish

Many farm families have ardent fishermen who proudly bring home their catch. While most of the fish usually are enjoyed at once, or frozen, many people like the variety smoked fish adds to meals. Country cooks find their smoky flavor seasons many dishes delightfully. They add bits of the fish to scalloped potatoes and eggs, to scrambled eggs and potato salad, for instance. And on crackers, it makes a snack most men relish.

Equipment for Smoking Fish: If a smokehouse is not available, use a barrel with head removed. Remove about 1 foot from the lower end of one stave to permit stoking the fire in the pit below. If you do not have a barrel, use a cardboard carton about 30″ square and 48″ high. Remove one end from the carton; this makes the bottom of the improvised smoke house. Unfasten the flaps on the op posite end so they can be folded bac and used as a cover. You ca strengthen them, if you find it neces sary, by tacking ¾″ strips of woo vertically on the outside of eac corner, and horizontally on the sides top and bottom, to make it sturdy. O one side of the carton slash down 12 in two places 10″ apart. By foldin this piece up, you have an adjustabl flap "door."

Prevent Fire. Control the ventilatio and keep the fire covered to mak smoke, not flames. Then the carto will not catch on fire. Keep door an top flaps closed during smoking, le ting just enough air enter to keep fir smoldering.

Suspend Rods in Barrel or Carton. Space iron or wooden rods across top of the "smokehouse" far enough apart that the fish do not touch when hung. Or use a rack of coarse wire instead of rods. If you place a tray of wire mesh below fish when hung, it will catch any that may fall off during smoking.

Make Wire Hooks. You can use 8 or 10 gauge steel wire about 14" long. Make a hook in the center of the wire large enough to slide over the rods. Bend the ends so fish will not slip off during smoking. Or use wire coat hangers; cut off the bottom cross wires, bend ends to form hooks.

Prepare Fish. Scale and dress fish, remove heads but leave collar-bones attached. Wash thoroughly. Split fish to the back skin, but not through it. Cut large fish into steaks.

Place Fish in Brine. If fish are caught and cleaned in the evening for smoking the next day, let them stand in a brine overnight. As they are dressed, place them in a stone jar or an enameled container, in brine made of 1½ c. salt to 1 gal. water. When all the fish are dressed, remove them from the brine; drain. Lay them flat, flesh side up, in an enameled pan or crock (do not use metal) and salt each piece, using a salt shaker. Cover with waxed paper and store in the refrigerator or a very cold place overnight. In the morning, drain and rinse fish and let excess moisture drip off for 15 minutes before placing over fire.

If you catch and smoke the fish the same day, prepare them and place in a brine made of 4 c. salt to 1 gal. water. Leave in brine ½ to 1 hour, depending on size and thickness of fish. Remove from brine, rinse in cold, fresh water. Drain about 10 minutes and let hang in a breezy place about an hour. They are ready for smoking.

Fire for Smoking. Use oak, hickory, maple, alder, beech, apple, white birch or ash to make fire. Or use corncobs. Do not use wood containing resin. Cut wood pieces about 8" long and 1" in diameter. Sawdust burns slowly and makes a good smudge. Too much smoke overdoes the smoke flavor. Kindle the fire in pit under barrel or carton, using care to have it smoldering with smoke; do not let it flame. Stoke fire every half hour or as needed.

String Fish on Hooks. Thread them on wire just below the collarbone. This holds them flat and increases the capacity of the barrel or carton. It also makes for even smoking. Hang the hooks on the rods and space them evenly so that the fish do not touch.

Use Hot Smoke or Cold Smoke. Take your choice of methods.

Hot Smoke: Keep the temperature at the fish level at 100° F. for the first 4 or 5 hours of smoking. Hold your hand at the fish level. It should feel barely warm so the fish will dry slowly. Increase to about 180 to 200° F. for about an hour to cook fish. They are done when the backbone separates from the meat. Remove and cool—ready to eat.

Cold Smoke: Follow directions for Hot Smoke but do not use the high temperature at the end. These fish must be cooked before eating.

Store Properly: Keep fish in refrigerator or a cold place until used. Or wrap them and freeze.

Note: Some farmers use a barrel, set on the side of a hill, for the smokehouse. A stove pipe from a fire below carries the smoke. They remove both ends of the barrel and insert a wire shelf near the top; they cover the barrel top with burlap.

Note: To can smoked fish, process in pressure cooker at 10 pounds pressure (240° F.) 1 hour and 40 minutes.

Sizzling Country Sausage

If your family approves of the sausage you make, treat your recipe like the treasure it is. While homemakers often disagree on sausage seasonings, almost everyone believes nothing tastes better on a cold morning than sausage with fried apples, scrambled eggs, pancakes or golden gravy and hot biscuits. Seasonings for this country special vary from one farm to another and in different areas. Southern cooks almost always add a generous dash of red pepper, for instance.

Choose the seasonings your family prefers or buy the ready-to-use blends on the market. But do grind *1 part fat to 3 parts lean pork to make sausage.* More fat makes it shrink when cooked and too rich—less gives hard patties that do not brown readily.

Combine the ingredients properly: either cut the pork in small pieces or put it through the chopper, using a coarse plate. Then spread it on a table or other flat surface and evenly sprinkle on the seasonings. Grind seasoned meat with a ⅛ or ³⁄₁₆″ plate.

When you feel adventurous and are not pleased with the way your sausage tastes, try different seasonings. You may want to make up a small batch first and test it on the family. Here are some recipes to try from farm kitchens, favorites of excellent country cooks.

Note: See the chapters on Freezing: Meat, Poultry, Fish and Game; Canning: Meat, Poultry and Game for directions for freezing and canning sausage.

PENNSYLVANIA DUTCH SAUSAGE

A big family's favorite sausage recipe —note it contains brown sugar

20 lbs. fresh pork trimmings, cut into cubes
¾ c. salt
6 tblsp. black pepper
1 c. brown sugar, firmly packed

• Mix meat with seasonings before grinding; grind meat twice, using ⅛″ hole plate.
• Attach stuffer spout; force enough sausage into spout to fill it and prevent air pockets; slip casing over spout; feed on as much casing as spout will hold.
• Stuff casings to desired length; then

tie ends together with strong twine. Hang sausage to keep round shape. • Smoke or freeze. Makes 20 pounds.

Note: If you like stuffed smoked sausage but don't have casings, you may be able to order them at the meat market or locker plant. Or, make cloth casings from cheesecloth, stitching about 12 by 1½.

GEORGIA SAUSAGE: To 25 lbs. ground pork add 1 c. salt, 3 tblsp. black pepper, ½ tblsp. ground red pepper and 3 tblsp. rubbed sage.

NORTH DAKOTA SAUSAGE: To 25 lbs. ground pork add 1 c. salt, 2 tblsp. rubbed sage, 2 tblsp. black pepper, 1 tsp. sweet marjoram leaves, 1 tblsp. sugar, 1 tsp. ground mace, 1 tsp. red pepper and 1 c. flour.

HALF-AND-HALF SAUSAGE: Make North Dakota Sausage with equal parts of beef and pork. (Note that flour is sprinkled on with the seasonings.) Add a little cold water to make neat patties easily.

IOWA SAUSAGE: With 4 lbs. pork trimmings use 5 tsp. salt, 4 tsp. rubbed sage, 2 tsp. black pepper, ½ tsp. ground cloves or 1 tsp. ground nutmeg and 1 tsp. sugar. The country cook who champions this blend of seasonings warns that you may wish to omit the cloves or nutmeg.

BEEF PORK SAUSAGE: A Michigan country cook uses 1 lb. lean beef to every 3 lbs. pork for her sausage famous in her neighborhood. She especially recommends adding beef when the pork is fat.

NO-CRUMBLE SAUSAGE: The secret of this, an Illinois farm woman says, is to add a little cold water to the sausage. She kneads it until the mixture becomes sticky and doughlike. Then she packs the sausage in a pan and chills it to slice for cooking. Slices hold their shape—do not crumble.

Index

WHEN YOU CAN FRUITS
HOW MUCH WILL IT MAKE?

Fresh	Canned Quarts	Fresh	Canned Quarts
APPLES		**NECTARINES**	
48 lbs.	16 to 20	18 lb. flat box	6 to 9
2 to 3 lbs.	1	2 to 3 lbs.	1
2½ to 3½ lbs.	1 (sauce)		
		PEACHES	
		1 bushel	18 to 24
APRICOTS		2 to 3 lbs.	1
24 lb. lug	10 to 12		
2 to 2½ lbs.	1		
		PEARS	
		46 lbs. (box)	16 to 2?
BERRIES (EXCEPT STRAWBERRIES)—		2 to 2½ lbs.	1
4 to 8 (½-pint) baskets	1		
BLACKBERRIES		**PINEAPPLE**	
32 qts.	24	2 lbs.	1
BLUEBERRIES			
32 qts.	16	**PLUMS**	
		28 lb. lug	12 to 1
		2 to 2½ lbs.	1
CHERRIES, UNPITTED			
3 to 4 pts. or 2 to 3 lbs.	1		
Bushel (with stems)	22 to 32	**RHUBARB**	
		1 to 2 lbs.	1
FIGS			
5 to 6 lb. box	2 to 3	**TOMATOES**	
2 to 2½ lbs.	1	Bushel	20
		2½ to 3½ lbs.	1
GRAPEFRUIT		Bushel	12 to ?
4 to 6 grapefruit	1		(juic?
Bushel	12		